KEY CONCEPTS IN INNOVATION

Palgrave Key Concepts

Palgrave Key Concepts provide an accessible and comprehensive range of subject glossaries at undergraduate level. They are the ideal companion to a standard textbook, making them invaluable reading to students throughout their course of study, and especially useful as a revision aid.

Key Concepts in Accounting and Finance
Key Concepts in Business Practice
Key Concepts in Criminal Justice and Criminology
Key Concepts in Cultural Studies
Key Concepts in Drama and Performance (*second edition*)
Key Concepts in e-Commerce
Key Concepts in Human Resource Management
Key Concepts in Information and Communication Technology
Key Concepts in International Business
Key Concepts in Innovation
Key Concepts in Language and Linguistics (*second edition*)
Key Concepts in Law (*second edition*)
Key Concepts in Leisure
Key Concepts in Management
Key Concepts in Marketing
Key Concepts in Operations Management
Key Concepts in Philosophy
Key Concepts in Politics
Key Concepts in Public Relations
Key Concepts in Psychology
Key Concepts in Social Research Methods
Key Concepts in Sociology
Key Concepts in Strategic Management
Key Concepts in Tourism

Palgrave Key Concepts: Literature

General Editors: John Peck and Martin Coyle

Key Concepts in Contemporary Literature
Key Concepts in Creative Writing
Key Concepts in Crime Fiction
Key Concepts in Medieval Literature
Key Concepts in Modernist Literature
Key Concepts in Postcolonial Literature
Key Concepts in Renaissance Literature
Key Concepts in Romantic Literature
Key Concepts in Victorian Literature
Literary Terms and Criticism (*third edition*)

Further titles are in preparation
www.palgravekeyconcepts.com

Palgrave Key Concepts
Series Standing Order
ISBN 1–4039–3210–7
(*outside North America only*)

You can receive future titles in this series as they are published by placing a standing order. Please contact your bookseller or, in case of difficulty, write to us at the address below with your name and address, the title of the series and the ISBN quoted above.

Customer Services Department, Macmillan Distribution Ltd, Houndmills, Basingstoke, Hampshire RG21 6XS, England

Key Concepts in Innovation

Hamsa Thota
President, Innovation Business Development Inc.

Zunaira Munir
Managing Director, Strategize Blue, San Diego

First published 2011 by
PALGRAVE MACMILLAN

Palgrave Macmillan in the UK is an imprint of Macmillan Publishers Limited,
registered in England, company number 785998, of Houndmills, Basingstoke,
Hampshire RG21 6XS.

Palgrave Macmillan in the US is a division of St Martin's Press LLC,
175 Fifth Avenue, New York, NY 10010.

Palgrave Macmillan is the global academic imprint of the above companies
and has companies and representatives throughout the world.

Palgrave® and Macmillan® are registered trademarks in the United States,
the United Kingdom, Europe and other countries

ISBN 978-0-230-24462-7

This book is printed on paper suitable for recycling and made from fully
managed and sustained forest sources. Logging, pulping and manufacturing
processes are expected to conform to the environmental regulations of the
country of origin.

A catalogue record for this book is available from the British Library.

A catalog record for this book is available from the Library of Congress.

10 9 8 7 6 5 4 3 2 1
20 19 18 17 16 15 14 13 12 11

Printed and bound in Great Britain by
CPI Antony Rowe, Chippenham and Eastbourne

To our families
For unwavering support and love during the writing of this book

'A fundamental reference book for both novice and experienced professionals in the field of innovation. It should be on everyone's book shelf.'
– **Professor Peter Koen**, *Stevens Institute of Technology, USA*

'an accessible source of clear definitions supporting the work of global innovation teams and valuable to those in the field of product innovation management. It develops a foundation to support the use of each concept and stimulates each reader's thoughts about new approaches to their own challenges.'
– **Emeritus Professor Thomas P. Hustad**, *Indiana University, USA*

'the book is suitable for students, practitioners and even the policy makers, as a valuable desk reference for innovation management.'
– **Professor Xiaobo WU**, *Zhejiang University, China*

'Thota and Munir delve beneath the buzzword and deconstruct innovation into its operational parts. It creates a toolkit as well as a common terminology. Key Concepts in Innovation should be read by entrepreneurs seeking to embed innovation across their products, processes and business strategies.'
– **Debra van Opstal**, *Center for the Study of the Presidency and Congress, USA*

'a great reference book for all interested in formulating change strategies to address the multiple challenges faced by individuals and organizations alike and succeed in the 21st century. All students will benefit immensely by becoming familiar with the inter-disciplinary tools and concepts discussed in this book.'
– **Vivek Singhal**, *Founder of Strategic Business Management and Senior Fellow of Global Knowledge Initiative, USA*

'a comprehensive and careful listing of all the pertinent concepts related to bringing out a winning product. I strongly recommend this book to anyone managing products.'
– **Vivek Kumar**, *CEO, AlumClub, India*

'organized in a way that brings clarity to this complex subject, this book belongs on every innovator's desk.'
– **Mark Adkins**, *VP Marketing, Product Development and Management Association, USA*

'a one-stop reference for key concepts in innovation, addressing students, practitioners and propagators of innovation. This book will provide one language that the innovation professionals across any industry, culture or geography will speak!'
– **Gupta Stueti**, *Innovation Professional, India*

Contents

Acknowledgements

Compiling a comprehensive list of key terms and concepts related to innovation from diverse fields (especially from business and engineering) was a daunting task. We are grateful for the help of our colleagues from the Product Development and Management Association, the New Product Institute, family members and friends who guided us in the entry selection process.

We acknowledge the invaluable help received from Claire-Juliette Beale, Raimund Broechler, George Castellion, Kesavan Chandrasekaran, Brian Christian, Serhat Cicekoglu, Mary Drotar, Donavan Hardenbrook, Shuhua Hu, Tom Hustad, Rego Lyndon, Sufian Munir, Karthik Ramachandran, Abu Saleh Md, Apurva Shah, Shrihas Shah, Vivek Singhal, Tricia Sutton, Devi Thota, Sindhu Thota, Giri Vayalapalli, Xiangdong Ye and Sohail-uz-Zaman.

We also appreciate the encouragement for this project from Harvey Cheng, Giulio Gianni, Peter Harland, Shan He, Nasheeta Mohsin, YoungWon Park, Raihan Rasool, Sadia Aziz, Aruna Shekar, Shimon Shmueli, Jaeyong Song, Jagjit Singh Srai and Yufeng Zhang. Additionally, we would like to acknowledge Brendan Murphy for his support throughout the project.

We wish to give special thanks to Suzannah Burywood and Jennifer Schmidt, our editors, for their insightful comments and editorial feedback, manuscript reviewers for detailed analysis and constructive suggestions, as well as Mark Cooper and the Palgrave Macmillan publishing team for their commitment and enthusiastic support. Finally, we would like to express our gratitude to our families for their love, support and understanding and enduring late evenings of work on this manuscript.

It is quite possible that an informed reader might find some key entries omitted and the inclusion of others with marginal significance. The selection process was difficult, and we ask for our readers' understanding. Trade-offs were necessary to keep the size of Key Concepts in Innovation in line with other books in the Key Concepts series. We hope that this book contributes to the continued success of our readers.

Introduction

Innovation is a rapidly emerging field of study in business and engineering. In the early days, the scope of innovation largely fell under the realm of technology management or new product development, but during the past decade the scope of innovation has continued to grow. Discussions about innovation are common and innovation research is now an established area of academic research. Its key terms and concepts are often mentioned in business and the popular press as well as in political discourse. However, there is a lack of common understanding about the key terms and concepts and the context in which they are used in various fields. There has been a strong unmet need for key terms and concepts to be selected and explained clearly so that their applications in one area of study can be correctly interpreted by students in other fields.

The purpose of *Key Concepts in Innovation* is to create a comprehensive list of key concepts that bring together both engineering and business approaches to innovation. In business schools, innovation is taught by management and marketing faculty. In engineering schools, it is taught by engineering faculty, who focus more on the design and management of technology. This book clarifies and codifies the innovation terms used in these fields and is a valuable addition to textbooks used for teaching innovation. It provides an interdisciplinary framework necessary for undergraduate students enrolled in courses related to innovation in business and engineering. It also serves as a practical guide for entrepreneurs and other practitioners who want to understand and implement innovation principles to achieve firm growth. Historically, the practices of innovation led the development of theory.

To meet our readers' needs, we have positioned product/services development at the centre and explored its linkages to various specialised knowledge areas in a two-step process (see Figure 1, The Thota–Munir Innovation Integration Framework). In the first step we collected and processed key terms related to innovation from different specialised knowledge areas (product and services development, technology/engineering, creativity, entrepreneurship, business, policy, design, people and culture). In the second step we looked for specific categories such as innovation tools and methods; theories; innovation systems and standards; generally accepted terminology; and implementation of theory into practice as screens for the inclusion of terms and concepts in the book. We made an effort to clarify conflicting information, where appropriate, and develop a common understanding of how concepts are applied across multiple knowledge areas. Our output, which is based on the literature cited in this book, is listed in Appendix 1: The Thota–Munir Innovation Integration System. To our knowledge, this is the first systematic attempt to integrate the key concepts in innovation.

Practical examples of key concepts referenced in the book are listed in Appendix 2: Innovation in Practice, to demonstrate the conversion of theory into practice by

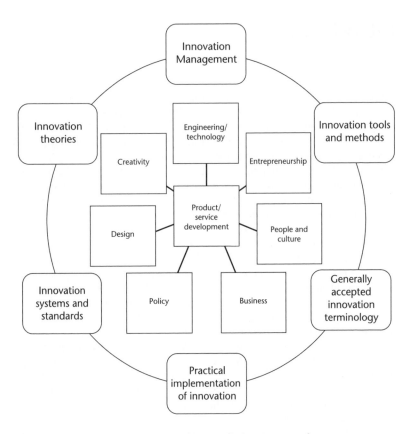

Figure 1 The Thota–Munir Innovation Integration Framework

leading organisations. We are confident that our readers who are innovation practitioners will draw inspiration from such examples, and become motivated to
apply relevant concepts to the benefit of their own organisations.

The structure of *Key Concepts in Innovation* is straightforward. The key concepts
are listed alphabetically in order to ensure that the reader can quickly find the
term or entry of immediate interest. The main concepts are in bold type (the
heading of the entry). There may be a reference within the text to another key
concept identified by a word or phrase that is in bold type so that the reader can
find it easily in the alphabetical listing. Cross-references are given in italics.
Because they are embedded in the text, the words in italics may not exactly match
the cross-reference listed in the index. However, the index will always take the
reader to the correct place in the alphabetical entries.

Most entries feature book or journal references. Some entries also have additional references for further learning. Others include website references; in most

cases long-established websites have been selected. Our aim is that this book both serves as an enduring reference for the practitioner and enriches the learning experience of students beginning their career in the inspiring and exciting field of innovation.

Absorptive capacity

Absorptive capacity refers to the capacity of an organisation to absorb new knowledge. It is a complex organisational construct. It incorporates the organisation's existing knowledge base, values and goals; technological infrastructure; leadership and knowledge sharing; and collaboration with other organisations to absorb or synthesise new knowledge. Absorptive capacity requires learning capability. It builds problem-solving skills within organisations and becomes a crucial resource that is hard for competitors to imitate. It is dependent on the level of knowledge in the corporation.

Further reading
Zahra, A.S. and George, G. (2002) 'Absorptive Capacity: A Review, Reconceptualization and Extension', *Academy of Management Review*, 27(2),185–203.

Accidental innovation

Many innovations are the by-product of accidents. Accidental innovations in new products, designs and processes occur when unexpected insights are obtained either within or outside of an organisation.

History has shown that accidents are crucial to innovation. For example, in 1928 Alexander Fleming accidentally discovered Penicillin mould and its antibacterial agent, which he called *penicillin*. This discovery revolutionised the way bacterial infections are treated with antibiotics and made medical history. While the discovery of penicillin may appear to be a chance occurrence, Fleming was highly attuned to his environment, which enabled him to recognise that the bacterium *Staphylococcus aureus* was killed by the mould growing inside the Petri dish. Over the centuries, many innovators like Fleming have made countless discoveries while looking for something else or nothing at all. Accidental innovations lead to new scientific and technological breakthroughs such as anaesthesia in medicine; photography in the chemical industry; and microwave ovens in appliances manufacturing.

Robert Austin and Devin Lee (2006) suggested that, to turn accidents into innovations, companies need to recognise serendipitous opportunities. Companies need to hire creative people and structure work to harness their creative energies. The invention of Post-It™ notes at 3M is a classic example of the realisation of innovation by harnessing employees' creative energies.

Further reading
Austin, Robert D. and Lee, Devin (2006) 'Accident, Intention and Expectation in the Innovation Process', working paper, Cambridge, MA, Harvard Business School.
http://docentes.fe.unl.pt/~satpeg/PapersInova/Accident%2006-4-5.pdf (accessed on March 5, 2010).

Acquisition strategy

Acquisition is an element of **product strategy**. This is a 'buy vs make' decision for innovation. For example, a new product platform can be brought to market faster through a strategic acquisition. The value of an acquisition can be assessed by comparing the strategy for internal development with that for the external acquisition. Acquisitions may be paid for in cash or by an exchange of stock.

There are several types of acquisition strategy for innovation:

- *Product platform* acquisition to expand into a new market. In a six-year period in the 1990s, Cisco acquired 42 smaller firms with new technologies in various stages of development so that it could enter into new markets faster. In implementing this strategy Cisco acted more as an integrator and project partner, acquiring or teaming up with smaller, more innovative firms for key technologies.
- Technology acquisitions to complement the existing product portfolio. In 2010 Kapsch CarrierCom AG acquired parts of Nortel's Carrier Networks Division, to transform itself from a regional supplier to a Global System for Mobile Communications (GSM) player.
- Technology acquisitions to develop really new products. Nortel partnered with Microsoft for unified communications and with IBM for web services to expand its presence in the enterprise IT market space. However, partnering in trendy R&D projects also exposed Nortel to the risk of market rejection. The enterprise IT market did not readily accept it as a Web 2.0 player.
- Acquisition to expand into a new market. BASF, a global chemical company, acquired BCD Rohstoffe für Bauchemie HandelsGmbH in 2008 to expand its portfolio of polymer dispersions to construction chemicals.
- Acquisition to strengthen a current market position. EMC Corporation, a leader in the information infrastructure market (in both hardware and software), acquired companies to strengthen its market position. It acquired Document Sciences Corporation to extend its enterprise content management leadership and RSA Securities to strengthen its position in unified storage and security management.
- Acquisition to diversify into a related market. AT&T acquired cable properties such as Falcon, Bresnan, Cablevision and Lenfest to gain access to the customer's home and bypass the Baby Bells, which had entered the long-distance telephone market.

A

In the pharmaceutical industry, the big pharma companies rely on the acquisition of innovative and smaller biotechnology companies to increase the number of products going through Phase 3 drug trials. This acquisition strategy has proven to be very useful in the launch of blockbuster drugs and in sustaining double-digit returns on innovation (ROI) in the pharmaceutical industry.

Further reading
McGrath, Michael E. (2000) *Product Strategy for High Technology Companies*, 2nd edn, New York, McGraw-Hill.

Adaptive design

In adaptive design, the principles of known and established solutions are kept and the way they are embodied in the item or process being designed is adapted to changed requirements. It may be necessary to undertake original designs of individual assemblies or components as part of the adaptation. Adaptive design is different from the design of entirely new products in which the working principle, model of shape, functionality, and technical shape are defined (Leondes, 2001). In adaptive designs, product's dimensions are adapted to the technical and technological possibilities for their manufacture (Leondes, 2001). The emphasis is on geometrical (strengths, stiffness etc.), production and material issues.

In drug development, adaptive design is a clinical trial design innovation (Chang et al, 2007). In clinical trial design, adaptive design refers to a design that allows adaptations or modifications to the trial design after its initiation without undermining the validity and integrity of the trial (Chow and Chang, 2006). There are many possible adaptive designs with different combinations of adaptations. However, clinical studies with adaptive designs are more complicated than those with classic designs. Also, practical aspects of adaptive designs (such as trial monitoring, data cleaning, and data analysis and reporting) must be considered prior to trial implementation. *Clinical trial simulation* is a tool to streamline the drug development process, increase the probability of success, and reduce the cost and time-to-market (Chang et al, 2007).

Adaptive designs in clinical trials can reduce the number of patients exposed to experimental therapies, decrease clinical development times, and improve product safety and efficacy (Chang et al, 2007).

Further reading
Chang, Mark; Susan Kenley; Jonca Bull; Yuan-Yuan Chiu; Wenjin Wang; Charles Wakeford; Katherine McCarthy. (2007). Innovative Approaches in Drug Development. *Journal of Biopharmaceutical Statistics*, (17): 775–789.
Chow, S. C., Chang, M. (2006). *Adaptive Design Methods in Clinical Trials*. Boca Raton, FL: Chapman & Hall/CRC, Taylor & Francis Group.
Leondes, Cornelius (2001). *Computer-Aided Design, Engineering, and Manufacturing: The Design of Manufacturing Systems*. CRC Press, Boca Raton, Florida

Adaptive learning
See **organisational learning**

Adoption rate
See **diffusion of innovations**

A

Agile development
Agile development is a people-centric process and recognises the value of team members' competencies in bringing agility to the software development process (Nerur and Balijepally, 2007). Providing team autonomy and building a diverse team are two key practices in agile development, although diversity brings conflict with it (Lee and Xia, 2010). Lee and Xia (2010) studied software development agility and found that team autonomy has a positive effect on response efficiency

and a negative effect on response extensiveness, and that team diversity has a positive effect on response extensiveness. These are the two main dimensions of agility (Lee and Xia, 2010).

Agile methodologies go by many names, including *extreme programming, feature-driven development and dynamic systems development.* Pair programming, refactoring, test-first design and simplicity are some of the practices common to extreme programming. In pair programming senior and junior developers are teamed together to encourage the adoption of *test-driven development (TDD).* This is a software development technique in which software developers write unit tests prior to integration testing, which is done before implementing the solution. Unit tests are code-level tests that developers write to prove that their code behaves as expected and provide immediate feedback. However, traditional practice in software development is not to write code in pairs and not to write unit tests before writing the code. TransCanada Pipelines successfully introduced extreme programming into its Greenfield project. Kitware, a small developer of commercial closed-source and open-source products, implements techniques from agile development and extreme programming in smaller projects as well as larger projects consisting of thousands of source files.

The value of agile development is that both requirements and solutions are co-developed through a collaborative process. Collaboration produces a feedback loop so that better solutions are found during the development process. Agile software development includes continual peer review of developed code, real-time interaction with users to evaluate ideas and implementations, and rapid development cycles measured in hours or days. Because solutions are user tested while still in development, and the solutions are improved based on user feedback, the agile development process is also iterative.

There is a concern about the scalability of agile development from small projects to large projects. Githens (2005) points to the agile development manifesto to allay such concerns. This is a statement of values and preferences: for individuals and personal interactions over processes and tools; for working on products over documentation; for customer collaboration over contract negotiation; and for responding to change over following a plan.

Agile development can lead to significant reductions in project time and higher rates of project success.

See also: **lean product development**

Further reading
Brandl, Dennis (2009) 'Agile Software Development', *Control Engineering*, 56(7), 18.
Githens, G. (2005) 'Agile and Lean Development: Old Wine in New Jugs?', *Visions*, 29(4).
Lee, Gwanhoo and Xia, Weidong (2010) 'Toward Agile: An Integrated Analysis of Quantitative and Qualitative Field Data on Software', *MIS Quarterly*, 34(1), 87–114.
Nerur, S., and Balijepally, V. (2007) 'Theoretical Reflections on Agile Development Methodologies', *Communications of the ACM*, 50(3), 79–83.

AK paradigm

See **growth theory**

Aligning innovation with market life cycle

Innovation plays a key role throughout the *market development life cycle*. Organisations must adapt their core competencies to ensure alignment with changing development requirements in three stages: the *technology adoption life cycle*, 'Main Street' and declining. Alignment between the core competencies of an organization and its market development life cycle is necessary to sustain profitable growth. Geoffrey Moore (2004) identified eight types of innovation: '**disruptive innovation, application innovation, product innovation, process innovation,** experiential innovation, **marketing innovation, business model innovation** and structural innovation'. Each stage of the life cycle requires that employers allocate appropriate resources to relevant innovation projects in order to compete for revenues and achieve the desired growth. The first three innovation types (disruptive, application and product) dominate the technology adoption cycle, leading to the creation of a new market opportunity. The next stage is 'Main Street', when innovations lose their leverage. Continued investments in disruptive, application and product innovations in the early Main Street stage lead to the acceleration of commoditisation. Clayton Christensen calls this phenomenon 'overshooting'.

A second suite of innovation types (process, experiential and marketing) can now be deployed to make incremental improvements. As the market enters the declining phase, companies can still utilise **business model** and structural innovation. For example, the automotive industry decommoditised roadside services and opened up new opportunities for structural and **business model innovations** as the market for maintenance was commoditised. However, many managers believe that such a market is at its end of life and prefer to divest the company to investors, who run the business as a *'cash cow'* rather than reinvest in the renewal of the business.

See also: **bowling alley, chasm**

Further reading
Moore, Geoffrey A. (2004) 'Darwin and the Demon. Innovating with Established Enterprises', *Harvard Business Review,* 82(7/8). 86–92.

Alliances

An alliance is a formal arrangement between two or more firms for the development of technology, product or markets. Alliances involve the exchange of information, hardware, **intellectual property** or enabling technology to access the resources and competencies held by other firms. Competencies acquired through alliances include new distribution channels, complementary knowledge to *co-develop new products* and shared production platforms. For instance, Volvo Cars created partnership networks to develop a new car, the Volvo C70. It established a *'transformation network'* and operated it across different organisational levels to secure the transformation and integration of knowledge into **innovation**. Alliances in media take advantage of the convergence of data and image technologies. For example, Optus, Nine and Microsoft announced plans for a major strategic alliance

A

to provide content and the delivery of communication services such as fixed and mobile phone to a personal computer and video and voice calls over the internet.

Through alliances companies can gain access to a particular resource, realise economies of scale, share the risks and costs of product and/or service development, accelerate learning, gain speed to market, achieve flexibility and realise **competitive advantage**. The disadvantages of alliances are that companies risk the loss of proprietary information; face increased management complexities and additional financial and organisational risks; risk becoming dependent on a partner for success; and suffer a partial loss of decision autonomy and organisational flexibility. The partners' cultures may also clash.

The technology-enabled ambient organisation is a new type of alliance with the capability to erode industry barriers and transform industry structures (Elliot 2006). Linnarsson and Werr (2004) found that the challenges of **radical innovation** could be reduced with *alliances for innovation*.

See also: **strategic alliances**

Further reading

Elliot, Steve (2006) 'Technology Enabled Innovation, Industry Transformation and the Emergence of Ambient Organizations', *Industry & Innovation*, 13(2), 209–25.

Harryson, Sigvald J, Dudkowski, Rafal and Stern, Alexander (2008) 'Transformation Networks in Innovation Alliances – The Development of Volvo C70', *Journal of Management Studies*, 45(4), 745–73.

Lazzarini, Sergio G. Brito, L., Luiz, Z. and Fabio, R. (2008) 'Conduits of Innovation or Imitation? Assessing the Effects of Alliances on the Persistence of Profits', *Academy of Management Proceedings*, 1–6.

Linnarsson, H. and Werr, A. (2004) 'Overcoming the Innovation–Alliance Paradox: A Case of an Explorative Alliance', *European Journal of Innovation Management*, 7, 45–56.

Alpha testing

Alpha testing is pre-production product testing. It provides a crucial 'first look' at the initial product design to see whether it meets performance specifications or falls short of them. It helps to find and eliminate obvious design defects or deficiencies early.

The alpha testing is usually done in-house, in an operating unit of a company or in a laboratory setting. In some cases it may be done with lead customers in a controlled environment. The testing must simulate the environmental conditions under which the product will actually be used. When the testing is done in-house, care must be taken to retain objectivity. The alpha test should not be performed by people who are also doing the development work.

Alpha testing is also used in services development. For example, CTC Communications Group used alpha testing to verify that switched long-distance and local voice services on its packet-based network were operational and that they worked. Motorola's Healthcare Communications Solutions group conducted alpha testing at Barnes-Jewish Hospital to verify that its DocLink system to reduce medical errors actually worked in a hospital setting.

See also: **beta testing, gamma testing**

Further reading

http://www.tmcnet.com/enews/122100d.htm (accessed on March 6, 2010).

Ambidextrous organisation

Ambidextrous organisations implement both **incremental** and **disruptive innovations** equally well. How do they manage this paradox of exploitation vs exploration? *Exploration* is associated with organic structures, loosely coupled systems, path breaking, improvisation, autonomy and chaos, as well as emerging markets and technologies. *Exploitation* is associated with mechanistic structures, tightly coupled systems, path dependence, routinisation, control and bureaucracy, as well as stable markets and technologies (He and Wong 2004). Ambidextrous companies use a different kind of structure and discipline in the ideation and early stages of development for each type of innovation. They set up different structures to manage different innovation projects in different ways.

In 'The Ambidextrous Organization', Charles O'Reilly III and Michael Tushman reported that ambidextrous organisations integrate independent project teams into the existing management hierarchy. A tightly integrated senior team ensures that the 'left hand knows what the right hand is doing'. Both the traditional business and new ventures report to the same senior leadership team, but each is managed under a different set of rules. O'Reilly and Tushman reported that ambidextrous structures were successful in achieving **breakthrough innovation** goals about 90 per cent of the time. Other structures, such as functional groups, were successful about 25 per cent of the time, whereas cross-functional teams and unsupported skunk works-style groups showed a dismal degree of success (O'Reilly and Tushman, 2004).

Insufficient implementation of the organisational structures that support ambidexterity is a **barrier to innovation**.

See also: **types of development teams**

Further reading:
He Zi-Lin and Wong Poh-Kam (2004) 'Exploration vs. Exploitation: An Empirical Test of the Ambidexterity Hypothesis', *Organization Science*, 15(4), 481–94.
O'Reilly III, Charles A. and Tushman, Michael L. (2004) 'The Ambidextrous Organization', *Harvard Business Review*, 82(4), 74–81.

Analytic hierarchy process (AHP)

The analytic hierarchy process (AHP) is a strategic-level decision support technique used in new product project screening and evaluation. It was developed in the 1970s by Thomas Saaty to gather expert judgements and make optimal decisions. At its simplest, AHP includes a goal, decision criteria and alternatives. When used as a full screening technique, it gathers managerial judgements and expertise to identify key criteria in the screening decision, obtain scores for each project and rank projects in order of desirability.

AHP is an algorithm, available as commercial PC software such as Expert Choice.

Further reading
Calantone, Roger J., Di Benedetto, Anthony and Schmidt, Jeffrey B. (1999) 'Knowledge Acquisition in New Product Screening with the Analytic Hierarchy Process', *Journal of Product Innovation Management*, 16(1), 65–76.

Application innovation

Technology-driven application innovation is ubiquitous. Innovative applications in mobile telephony, such as location awareness in mobile phones (Nokia introduced GPS into its phones in 2003) and touch technologies in the Apple iPad, enable innovative service firms to enhance the business value proposition in the Web 2.0 environment (Hsu, 2009).

Some application innovations originate from the development and adoption of new **platforms**. Roger Allan (2009) reported that automobile manufacturers and hardware and software suppliers used Genivi Alliance to develop and adopt an open-source in-vehicle infotainment (IVI) reference platform. Car manufacturers add their differentiated products and services (the consumer-facing applications and interfaces) to this common platform. The open-source platform reduces **time to market** and the total cost of ownership.

Further reading
Allan, Roger (2009) 'Alliance Launches Open-Source In-Vehicle Infotainment Development Platform', *Electronic Design*, 57(7), 19–20.
Hsu, Andrew (2009) 'Choosing a Touch Technology for Handheld-System Applications', *EDN*, 54(1), 40–44.

Architectural innovation

Architectural innovations change the architecture of a product without changes to its individual components. Henderson and Clark (1990) described architectural innovation as 'the reconfiguration of an established system to link together existing components in a new way'. In architectural innovation, the core design concept behind individual components remains the same. However, the individual components themselves may be amended, because a change in one component creates new interactions and new linkages with other components in the product architecture.

Henderson and Clark (1990) used a room air fan as an example to demonstrate the differences between architectural innovation, **incremental innovation** and **radical innovation**. The established technology is embodied in large, electrically powered fans, mounted to the ceiling. The motor is hidden from view and insulated to dampen the noise. For the makers of large, ceiling-mounted room fans, improvements in blade design or motor power would be incremental innovations. Central air conditioning of the room would add new technical disciplines and new interrelationships and would be a radical innovation. The introduction of a portable fan would be an architectural innovation. The architecture of the product would be quite different in a portable fan versus a room air fan, even though the primary components such as blade, motor and control system remain the same.

Further reading:
Henderson, Rebecca M. and Clark, Kim B. (1990) 'Architectural Innovation: The Reconfiguration of Existing', *Administrative Science Quarterly*, 35(1), 9–22.

Architecture

See **product architecture**

ARIZ

ARIZ (an abbreviation of the Russian for algorithm for inventive problem solving) is the central analytical tool of **TRIZ**, a problem-solving tool. It provides specific, sequential steps for developing a solution to complex, inventive problems. The first version of ARIZ was developed in 1968 and many modifications have been made, rendering it a precise tool for solving a wide variety of technical problems. The sequential steps of ARIZ include:

Step 1: Analysis of the problem. The process begins by making a transition from vaguely defined statements of the problem to a simply stated mini-problem without jargon or industry-specific terminology. For example, 'A technical system consisting of elements A, B and C has technical contradiction TC (state the contradiction). It is necessary to provide required function F (state the function) while incurring minimal changes to the system.'

Step 2: Analysis of the problem's model. This consists of drawing a simplified diagram modelling the conflict (technical contradiction) and an assessment of all available resources.

Step 3: Formulation of the ideal final result (IFR). This usually reveals contradictory requirements to the critical component of the system; that is, a physical contradiction and how it can be resolved.

Step 4: Analysis of the method that removed the physical contradiction. This step checks the quality of the solution; that is, has the physical contradiction been removed in the most ideal way?

Step 5: Utilisation of found solution. This step guides users through an analysis of the effects of the solution on adjacent systems. It also forces the search for applications for other technical problems.

Step 6: Checking of steps that lead to the solution. This compares the problem-solving process used to that suggested by ARIZ.

See also: **creative thinking methods**

Further reading
Introduction to TRIZ, http://www.triz.org/triz/40Ptriz.pdf (accessed on January 2, 2010). TRIZ in MEMS, http://www.ee.iitb.ac.in/~apte/CV_PRA_MEMS_PAP3.htm (accessed on January 2, 2010).

A-T-A-R model

A

The A-T-A-R model (awareness–trial–availability–repeat) is practised in consumer products marketing companies. They build awareness and trial through brand advertisement. Availability is determined by a company's channel strategy to distribute and deliver products to its target customers. Brand awareness is a key factor in the **concept testing** and selection phase of the **new product development process**.

The A-T-A-R model states the critical factors and shows their relationship to each other and to the sales and **profit** forecasts for new products (Crawford and Di Benedetto, 2008).

In the model:

Profits = Buying units × Percentage aware × Percentage trial
× Percentage available × Percentage repeat × Annual units bought ×
(Revenue per unit – Costs per unit)

Definitions of terms used in the ATAR model:

- *Buying unit* means the purchase point, which may be each person, household or department that participates in the decision.
- *Aware* means that someone in the buying unit hears about the existence of a new product.
- *Available* means the percentage chance that if a buyer wants to try the product, the effort to find it will be successful.
- *Trial* is variously defined. In packaged goods, it means to buy at least once, twice, three or more times; in durables it may mean to be happy with the product and/or make at least one recommendation to another buyer.

See also: **forecasting**

Further reading
Crawford, Merle and Di Benedetto, Anthony (2008) *New Products Management,* 9th edn, New York, McGraw-Hill.

Autonomous innovations

Autonomous innovations are constrained to a single portion of the **value chain**. They occur in relatively isolated 'pockets' and are not dependent on *simultaneous innovations* elsewhere in the value chain (Dorf, 1999). For example, a new specialised accounting software package can be installed and run without redesigning either the computer system it runs on, the books and records it documents, or the organisation that deploys the new system. A diffused, *decentralised innovation process* is more suitable for autonomous innovations than a structured process. In the Web 2.0 environment, YouTube, Flickr and Twitter are transforming the online participatory culture and creating new pathways for autonomous innovation.

China uses the term 'autonomous innovation' to refer to the development and use of domestic technology to manufacture products. Its national strategy is to become an *'innovation-oriented country'* by building its innovation capabilities through autonomous innovations. For example, China wanted to build its capabilities in telecommunications. The government gave priority access to capital to Chinese telecom companies developing and using local technologies. In 2008, it formed the Commercial Aircraft Corporation of China (CACC) to advance the civil aviation manufacturing sector to compete with Boeing and Airbus in global markets for large aircraft. CACC imports advanced manufacturing technologies as supplements to China's 'autonomous innovations'.

Further reading
Dorf, Richard C. (1999) 'Knowledge for the Technology Manager', in Technology *Management Handbook*, Vol. 49, Boca Raton, FL, CRC Press.

Autonomous team (team autonomy)

An autonomous team is a completely self-sufficient project team with overlapping skills. Autonomous teams collectively perform and deliver the *project charter* (such as bringing a **radical innovation** to the marketplace) while exercising a high level of discretion over the conduct of project work.

On a continuum of team autonomy, four team types can be identified: managed, autonomous, self-managed and self-led.

- *Managed teams* have no identity and are responsible solely for the execution of the team's work. They have no autonomy or decision-making power regarding work group goals or processes.
- *Autonomous teams* have a collective identity and are given the authority to make decisions about how to accomplish the work and assign team members to each task.
- *Self-managed teams* have a collective identity and autonomy with regard to work processes, selecting new team members, disciplining and terminating team members, and making compensation and other evaluation decisions for the team.
- *Self-led teams* are self-designing and exist in a state of complete autonomy, with the team maintaining control over all aspects of its work, team composition and purpose (Mumford and Mattson, 2009).

Improved team performance is not always correlated with team autonomy. It is also dependent on the nature of the work that the team performs.

See also: **types of development teams**

Further reading
Mumford, Troy V. and Mattson, Marifran (2009) 'Will Teams Work? How the Nature of Work Drives Synergy in Autonomous Team Designs', *Academy of Management Proceedings*, 2009, 1–6.

Axiomatic design (AD)

Axiomatic design (AD) is a system design methodology (Shu, 2001). Introduced by Nam P. Suh at MIT, it uses matrix methods to analyse the conversion of customer needs into functional requirements, design parameters and process variables. AD is an effective design methodology for achieving functional requirements in the early **concept definition, concept visualisation and concept testing phases** of the **new product development process**.

In axiomatic design, the overall design activity is divided into four domains:

- *Customer domain*, the benefits customers seek, e.g. the customer needs to preserve food.
- *Functional domain*, the functional requirements of the design solution, e.g. the customer selects cooling as a design solution to a problem to preserve food.
- *Physical domain*, the design parameters of the design solution, e.g. the customer decides to use the refrigerator as the design solution to store and preserve food.

A

- *Process domain*, the process variables, how a manufacturer manufactures the refrigerator.

Axiomatic design also reduces product development risk and development cost and speeds **time to market** by utilising two design axioms (Jeon *et al.*, 2008):

- *Independence axiom*: the independence of functional requirements must be maintained. Uncoupled design is considered the ideal design to satisfy the independence axiom.
- *Information axiom*: the information content of the design must be minimised.

Further reading
Jeon, S.-K., Shin, M.-K. and Park, G.-J. (2008) 'Design of the Occupant Protection System for Frontal Impact using the Axiomatic Approach', *Proceedings of the Institution of Mechanical Engineers – Part D – Journal of Automobile Engineering*, 222(3), 313–24.
Suh. Nam P. (2001) *Axiomatic Design: Advances and Applications*, Oxford, Oxford University Press. www.axiomaticdesign.com (accessed on November 4, 2009).

A

Balanced portfolio

See **portfolio management**

Balanced scorecard (BSC)

The balanced scorecard (BSC) was developed by Kaplan and Norton in 1992 as a performance measurement system to give strategic managers a fast and comprehensive view of their business and to measure organisational performance. In addition to traditional financial metrics, it included strategic non-financial performance measures to create a 'balanced' view of organisational performance (Kaplan and Norton, 2005).

The balanced scorecard views an organisation from four perspectives and helps managers develop metrics and analyse data relative to each of those perspectives. The four perspectives are:

1. Financial: how does the company appear to shareholders?
2. Customer: how do customers see the company?
3. Business process: at what business processes must the company excel?
4. Learning and growth: can the company continue to improve and create value?

The financial measures tell the results of actions already taken. These are complemented by the three sets of operational measures related to customer satisfaction, internal processes and the organisation's ability to learn and improve the activities that drive future financial performance.

The balanced scorecard has evolved considerably over time to become a strategic planning and management system, as in the case of National Insurance Company (Kaplan and Norton, 2007). The 'new' balanced scorecard helps managers plan and execute strategies by recording and analysing objectives, measures and targets as well as initiatives. Almost 50 per cent of Fortune 1000 companies use a balanced scorecard.

Further reading
Kaplan, Robert S. and Norton, David P. (2005) 'The Balanced Scorecard: Measures that Drive Performance', *Harvard Business Review,* 83(7/8), 172.
Kaplan, Robert S. and Norton, David P. (2007) 'Using the Balanced Scorecard as a Strategic Management System', *Harvard Business Review*, 85(7/8), 150–61.

Base of the pyramid (BoP) consumer

Base of the pyramid (BoP) consumers represent the poorest socioeconomic group in the world. They include more than 70 per cent of the world's population with annual per capita incomes below $3000.

The BoP consumer's primary goal is to secure affordable food, housing, health care, economic livelihood, education and other essential products and services. BoP consumers are different from consumers in the developed markets. So companies targeting BoP consumers need to experiment with new **business models** and new ways of addressing their 4As: awareness, availability, affordability and acceptability (Anderson and Markides, 2007):

1. *Awareness*: Use unorthodox advertising techniques to reach BoP consumers.
2. *Availability*: Restructure supply chains and distribution systems to adjust to poor infrastructure and underdeveloped distribution and collection systems.
3. *Affordability*: Offer goods and services at a price point and unit size that the BoP consumer can afford. Several mobile phone companies offer pre-paid cards in small denominations. Unilever and Procter & Gamble sell single-use sachets of products.
4. *Acceptability*: Adapt products and services to the cultural norms, aspirations and buying behaviours of BoP consumers. In China Haier pioneered ways to use its washing machines not only to launder clothes, but to clean vegetables and make cheese, while in India, Unilever modified the ingredients of its hair-care products to fit with the cultural preference for using a single soap for hair and body.

Altman *et al.* (2009) offer five lessons to expand business opportunities with BoP consumers:

1. The BoP consists of both consumers and producers.
2. Public–private partnerships open up opportunities at the BoP.
3. The BoP can drive innovation.
4. Respond to the BoP's needs with **design thinking**.
5. The BoP is a source of employees.

See also: **strategic innovation**

Further reading
Altman, David G., Rego, Lyndon and Ross, Peg (2009) 'Expanding Opportunity at the Base of the Pyramid', *People & Strategy*, 32(2), 46–51.
Anderson, Jamie & Markides, Costas (2007) 'Strategic innovation at the Base of the Pyramid', *Sloan Management Review*, 49(1), 83–8.
Dolan, Catherine and Scott, Linda (2009) 'The Future of Retailing? The *Apparajitas* of Bangladesh', *Retail Digest*, Summer, 22–5.
IFC and World Resources Institute (2007) *The Next 4 Billion*, report, Washington, DC, IFC/World Resources Institute.

Base of the pyramid (BoP) innovation

In 1998, Professors CK Prahalad and Stuart Hart published the article 'The fortune at the Bottom of the Pyramid: Eradicating poverty through profits'. It attracted the attention of big businesses who wanted to sell to the needs of two-thirds of the world's population. Hindustan Lever Limited (HLL) in India is one of the pioneering companies in BoP implementation. Its *Shakti programme,* where women in villages are trained as self-employed sales representatives (also called the 'Shakti Lady') became very successful. However, the model is not always sustainable on the *triple bottom line* of social, environmental and financial value.

Erik Simanis and Stuart Hart at Cornell University developed a second-generation approach to the *base of the pyramid protocol.* They defined a standard for *BoP enterprise* as:

● A business model of mutual value creation for community and company.
● A triple-bottom-line approach, with social, environmental and economic sustainability.
● The intention and potential to replicate and scale.

Whereas the first generation of BoP was easily reduced to a strategy of *selling to the poor,* second-generation BoP is about *business co-venturing* through deep dialogue, shared commitment and building local businesses based on the resources and capabilities of the community. The best example is Grameen Bank in Bangladesh, founded by Nobel Prize winner Mohammad Yunus in collaboration with local communities. For research and development professionals and engineers, the BoP can be an ideal incubator for technology innovations. A study prepared for the American Society for Mechanical Engineers, 'Engineering solutions for the base of the pyramid, noted cost-effective ways to increase access to food and clean water; effective sanitation; affordable energy, health care, education and transportation; and revenue-generating activities as major challenges for engineers when serving BoP consumers. Research and practice at the base of the pyramid and in social business raise many interesting questions and opportunities.

Further reading
Koch, Louise (nd) 'The business of development', in *International Development,* PSCA International, http://edition.pagesuite-professional.co.uk (accessed on December 7, 2009).

Benchmarking

Benchmarking is a management tool used for comparing the performance of an organisation against those companies recognised as leaders in its industry. It helps managers determine whether the company is performing efficiently, whether its costs are in line with those of competitors, and whether its internal processes need improvement. The idea behind benchmarking is to measure particular functions and activities of a company against an external standard, with the emphasis on how it can apply the same to achieve superior results.

Best practice

A best practice is a superior technique, methodology or tool that has proven to lead reliably to a desired result in a specific situation or context. It is usually recognised as 'best' by peer organisations in the industry and contributes to improved performance if replicated in an organisation. For example, conducting regular focus groups with customers is a best practice that organisations can replicate to incorporate the 'voice of the customer' early in the new product development process. Learning from best practices implies accumulating and applying knowledge about what is working, how and why in different situations and contexts to a continuing process of learning, feedback, reflection and analysis in the organisation.

Best practice is a widely used concept across various fields, including product development and innovation. CK Prahalad, however, argues that companies aiming for **competitive advantage** through **innovation** need to be focused on 'next practices', instead of 'best practices'.

See also: comparative performance assessment study, next practice

Further reading
Prahalad, C. K. (2010) 'Best Practices Get You Only so Far', Harvard Business Review, 88(4), 32–3.

Beta testing

Beta testing involves potential users of a product trying out the new product and reporting on their experience. The number of potential users who try out the product varies significantly from industry to industry. Beta testing is a regular part of new product development programmes in the computer industry, where a small number of potential adopters are invited to test the new product and that testing might be limited to a small number of complex products. It also features in the consumer products industry, where for example a new version of a shampoo is tested with a large number of respondents. A large sample size yields statistically valid results.

Beta testing is a very useful tool for validating the product concept and eliminating performance issues prior to market introduction. It also serves as an effective device for sales promotion. However, it possesses certain risks. For example, a poorly designed test can generate inaccurate performance data. It can also stimulate negative publicity and damage the relationship with potential customers. To be effective, the design of beta testing must be tailored to specific product and market conditions (Dolan and Mathews, 1993).

See also: alpha testing and gamma testing

Further reading
Dolan, Robert J. and Matthews, John M. (1993) 'Maximizing the Utility of Customer Product Testing: Beta Test Design and Management', Journal of Product Innovation Management, 10, 318–30.

Black-box design

Black-box design refers to the phenomenon whereby buyers merely specify performance requirements and suppliers or subcontractors are required to undertake the detailed designing and development of the component or the subsystem needed. Suppliers rely on their own technical expertise to perform the detailed design, engineering and manufacturing tasks to complete the black-box design. Although black-box designing reduces the complexity of a buyer's product-development tasks, it makes the buyer quite dependent on specific suppliers. From a *supplier-performance* viewpoint, the customer's dependence on technological or product design, or both, guarantees suppliers' steady demand while making it difficult for the *original equipment manufacturer (OEM) customer* to switch suppliers. This buyer dependence results in enhanced supplier performance in **incremental innovations**, as suppliers become more adaptive and collaborative through mutual reinforcement.

See also: **incremental innovation**

Further reading
Srinivasan, Raji and Brush, Thomas H. (2006) 'Supplier Performance in Vertical Alliances: The Effects of Self-Enforcing Agreements and Enforceable Contracts', *Organization Science*, 17(4), 436.

Blue ocean strategy

Challenging the conventional, competition-based approach to strategy, Chan Kim and Renee Mauborgne have argued that competing head on is not sufficient to sustain high performance and long-term profitable growth of companies. They suggest that companies can seize new profit and growth opportunities by creating blue oceans of new market space, where the objective is not to beat the competition, but to make it irrelevant. Blue ocean strategy (BOS) is based on the concept of **value innovation** developed by Kim and Mauborgne (the simultaneous pursuit of differentiation and low cost) to create a leap in value for both the company and its buyers. Blue Ocean Strategy achieves strategic alignment through three propositions:

1. **Value proposition**: 'The utility buyers receive from the product or service minus the price they pay for it.' To create a compelling reason for the target customers and non-customers to purchase the new offering, BOS raises selected value factors much above the industry standard or creates new ones that the industry did not think of before.
2. *Profit proposition*: 'The price of the offering minus the cost of producing and distributing it.' Lower cost is achieved by eliminating factors that add little value and reducing factors that the industry overdelivers.
3. *People proposition*: 'The readiness of employees to execute the new strategy voluntarily to the best of their abilities.' This is achieved by addressing the hurdles to adoption that may come from the company's employees, business partners or the general public.

Beyond the alignment of these propositions, six principles drive the successful formulation and implementation of BOS to minimise risks systematically while maximising opportunities. These six principles are:

1. Reconstructing market boundaries to attenuate search risk by creating new paths across alternative industries, strategic groups, buyer groups, complementary offerings, the functional-emotional appeal of an industry or time.

B

2. Focusing on the big picture to minimise strategic planning risk.
3. Reaching beyond existing demand by building on commonalities across non-customers to minimise scale risk.
4. Following the strategic sequence of utility, price, cost and adoption requirements to address **business model** risk.
5. Overcoming cognitive, resource, motivational and political hurdles to minimise organisational risk.
6. Building execution into strategy by practising fair process to minimise management risk.

The first four principles address the formulation of strategy, whereas the last two address its execution.

BOS uses a number of tools and analytical frameworks to suggest that **creativity** is a learnable process.

See also: **breakthrough growth, strategic innovation and value innovation**

Further reading
Kim, Chan and Mauborgne, Renee (2005) *Blue Ocean Strategy*, Boston, MA, Harvard Business School Press, www.blueoceanstrategy.com (accessed November 13, 2009).

Blue sky ideas

Blue sky or revolutionary ideas come from *thinking 'outside the box'* as contrasted with thinking 'inside the box'. This colloquialism sees the context of the idea-generation process as a space (or box) for the consideration of a stimulus. Thinking *inside the box* refers to a more pedestrian form of sense making. Making sense of things via fresh contexts and/or stimuli is necessary to break out of the box. Thinking outside the box refers to thinking differently, unconventionally or from a new perspective. The 3M Company credits much of its success to the practice of unrestricted, outside-the-box thinking.

Once the breakthrough ideas have been developed and commercialised, they evolve and expand and become refined in the process of **aligning innovation with life cycle**. 3M Post-it Notes™ with ultra-removable adhesive are an example of the *revolutionary/evolutionary cycle* of idea generation and implementation.

See also: **idea generation and enrichment**

Further reading
Wylant, Barry (2008) 'Design Thinking and the Experience of Innovation', *Design Issues*, 24(2), 3–14.

Bowling alley

The term 'bowling alley' refers to an *early growth stage strategy*, in the *market-development life cycle* proposed by Geoffrey Moore, which focuses on specific niche markets. In this phase of the life cycle, the innovation appeals to customers within narrowly defined market niches. These customers are conservative but open to new ideas, and are influential and active in the community. Companies can build a strong position by delivering clearly differentiated products to these niche customers, and can then use that *niche market* strength to move into neighbouring niche markets.

Success in the bowling alley is generally achieved by building product leadership via customer intimacy. Sales to niche markets generally provide high margins.

See also: **aligning innovation with life cycle**

Brainstorming

Brainstorming is the most widely used tool for stimulating **creativity** and innovative thinking in groups.

The brainstorming technique was first described in the early 1940s by Alex Osborn (Osborn, 1957) as an intervention in which individuals, groups and organisations adhere to a set of four rules while working in sessions designated to generate ideas. The four rules of brainstorming are: generate a lot of ideas, avoid criticising any of the ideas, attempt to combine and improve on previously articulated ideas, and encourage the generation of 'wild' ideas. Brainstorming rules facilitate the generation of a large number of ideas for a specific problem (Paulus and Brown, 2003). There have since been many modifications in format, each variation with its own name, such as nominal group technique, group passing techniques, team idea mapping, electronic brainstorming, directed brainstorming and so on. For example, in *nominal groups*, groups of individuals generate ideas in isolation. Putman and Paulus (2009) found that nominal groups generated more ideas and more original ideas and were more likely to select original ideas during the group decision phase than were the interactive groups during brainstorming. In *electronic brainstorming*, participants interact with each other using computers. Participants type their ideas into special computer software that collects the ideas and shares them with other members of the group (Dennis and Reinicke, 2004). Electronic brainstorming combines the best of both verbal brainstorming and nominal group brainstorming (Gallupe *et al.*, 1992).

There is ongoing research into the effectiveness of brainstorming to improve the quantity of ideas. Specifying the key elements of creativity needed for a particular context before idea generation and then tailoring interventions to match needs is one way to improve brainstorming (Litchfield, 2008).

B

See also: **idea generation and enrichment**

Further reading
Dennis, Alan R. and Reinicke, Bryan A. (2004) 'Beta versus VHS and the Acceptance of Electronic Brainstorming Technology', *MIS Quarterly*, 28(1), 1–20.
Gallupe, R. B., Dennis, A. R., Cooper, W. H., Valacich, J. S., Bastianutti, L. M. and Nunamaker, J. F. (1992) 'Electronic Brainstorming and Group Size', *Academy of Management Journal*, 35(2), 350–69.
Litchfield, Robert C. (2008) 'Brainstorming Reconsidered: A Goal-Based View', *Academy of Management Review*, 33(3), 649–68.
Osborn, A. F. (1957) *Applied imagination*, New York: Scribner's.
Paulus, P. B. and Brown, V. R. (2003) 'Enhancing Ideational Creativity in Hroups: Lessons from Research on Brainstorming', in P. B. Paulus and B. A. Nijstad (eds), *Group*

Creativity: Innovation through Collaboration (pp 110–36), New York: Oxford University Press.

Putman, Vicky L. and Paulus, Paul B. (2009) 'Brainstorming, Brainstorming Rules and Decision Making', *Journal of Creative Behavior*, 43(1), 23–39.

Brand-centred product development

Brands are companies' strategic **platforms** for interacting with their customers. Brand-centred product development refers to integrating the brand experience into the **product development process**. Brand's defining emotional and social benefits relate to how consumers feel about a brand or how the brand makes them feel while they're using it. Emotional experiences are intensely meaningful to customers. Focusing development on enhancing the emotional functionality of product thus creates stronger bonds with customers. The first and hardest lesson for product developers in brand-centred product development is to learn the dual focus on the physical product and its functionality, and the *customer emotional experiences* they want to create for their customers. *Apple* is well known for harvesting the economic value of emotional functionalities.

Harley-Davidson expanded from building motorcycles to licensing diverse products and experiences such as beer, clothing and a café that reinforce the brand experience. It likes to say that customers 'buy the lifestyle and get the bike for free'.

There are several different methods for understanding and codifying brands. The Brand Genetics™ Profile and the BrandAsset® Valuator system are two examples.

See also: **new product process**

Further reading
Briggs, Harvey (2004) 'The Value of Integrating the Brand Experience into the Product Development Process', *PDMA Visions*, 28(4), 8.

Brand community

A brand community is a community of admirers of a particular brand. Passionate admirers of a brand, brought together in a structured environment, can be a valuable source of innovation for firms. Brand communities are not geographically bound. Many such communities are brought together online and set up to interact socially in a structured manner. They generate new product ideas and improvements and share experiences with other members and with their favourite firm.

Füller (2008) described three basic characteristics of brand communities as:

- Shared consciousness. Members share a passion for the brand. They feel connected to other members of the community and share a demarcation from users of other brands.
- Common rituals and traditions. Members share experience with the brand's products. These shared product experiences become traditions for the community, as they represent a common meaning within the community.
- Members feel a sense of responsibility to the brand and the community and support one another in solving problems.

B

For example, ilounge.com is dedicated to the Apple iPod. At Niketalk, enthusiastic Nike fans create their own basketball shoe designs and develop new shoe features. Other well-known brand communities include 3Com Audrey, Apple Newton, Garmin and Mini Cooper.

Further reading
Füller, Johann, Matzler, Kurt and Hoppe, Melanie (2008) 'Brand Community Members as a Source of Innovation', *Journal of Product Innovation Management*, 25, 608–19.
Schau, Hope Jensen, Muñiz, Albert M. and Arnould, Eric J. (2009) 'How Brand Community Practices Create Value', *Journal of Marketing*, 73(5), 30–51.

Brand innovation

Marketers claim that brand innovation is the most effective and efficient way to drive *brand equity* and revenue. Brand equity refers to revenue that a product generates with a brand name, in comparison to the revenue that it generates with a private label. A brand sustains by capturing the expectations of its customers and then delivering on a set of promises for each of its target segments. Brand innovation occurs in process as well as product areas, and can be incremental as well as radical. All innovations use the brand to convey the jobs they do for their customers.

In his book *The Brand Innovation Manifesto*, John Grant recommends alternative tools to traditional advertising for building brands, such as customer relationship marketing, brand experience marketing and cultivation of customer communities. Obtaining insights into customers' relationship with the brand gives a clear and disciplined **platform** for innovation across the total *brand experience*, including service and the retail experience.

Strong brands have three *brand characteristics*: clarity, consistency and leadership. This means that the brand is clear about what it stands for, about its values, its purpose and its future direction; and that it communicates those clear values, both inside and outside the organisation. When all of a brand's elements are consistent, consumers can better understand what a brand stands for and what it does for them. For instance, Clorox Green Works became the leading green household care brand in the US by designing products to meet consumers' interest in ethical and 'green' products. The key to brand leadership is to have an impact on people's lifestyles, as demonstrated by brands such as Starbucks, iPod, Harley and eBay.

B

See also: **business, brand community, customer targeting, green design, target market**

Further reading
Grant, John (2006) The Brand Innovation Manifesto, London: John Wiley & Sons.
Jolly, Adam (2005) *From Idea to Profit: How to Market Innovative Products and Services*, London, Kogan Page.
Slotegraaf, Rebecca J. and Pauwels, Koen (2007) 'The Impact of Brand Equity and Innovation on the Long-Term Effectiveness of Promotions', *Journal of Marketing Research*, 45(3), 293–306.

Breakthrough growth

Breakthrough growth refers to major new market growth in mature markets that is exceptionally rewarding and profitable. To achieve breakthrough growth in existing companies, Govindarajan and Trimble (2006) prescribe three strategic actions:

- Strategic action 1: Manage the present while improving the current **business**.
- Strategic action 2: Abandon the past selectively.
- Strategic action 3: Create the future.

While managing the current business and maximising the profits from it is important, building for the future is even more important. Blue-chip companies like Sears and K-Mart in retailing and Kodak in photography lost market leadership because they failed to abandon the old business selectively and redefine their customer value proposition.

HCL Technologies is an example of an IT outsourcing company that created its future in a stagnant outsourcing market. It achieved breakthrough growth in the early 2000s by identifying large market opportunities; leveraging alliance partnerships; and creating exceptional value for its customers. For example, HCL understood that quality, trust and flexibility became increasingly important to IT customers during the 2001–03 recession and identified this as an unmet customer need and a large market opportunity. It therefore redefined its market and reverse engineered its **value proposition**. It then delivered outstanding quality and flexibility to frustrated corporate IT customers in order to achieve breakthrough growth. With trust, quality and flexibility as its core principles, HCL achieved a five-year compound revenue growth rate of 63 per cent in a stagnant outsourcing market (David G. Thomson, *BusinessWeek Online*, 16 November 2009).

Strategic innovation in the **business model** plays a central role in converting R&D into commercial value and achieving breakthrough growth.

See also: **blue ocean strategy**

Further reading
Govindarajan, Vijay and Trimble, Chris (2006) 'Achieving Breakthrough Growth: From Idea to Execution', *Ivey Business Journal*, 70(3), 1–7.

Breakthrough innovation

See **radical innovation**

Business

What is a business? In Peter Drucker's view, the first purpose of a business is to create a customer (i.e. to satisfy customer need) and the second purpose is innovation (Finklestein, 2009). A business is not defined from inside. It is defined from outside by the customer who buys a product or service from the business. In his discussion of **disruptive innovation**, Clayton Christensen reminds businesses to ask themselves a question: 'What jobs are customers hiring your products to do?' (Christensen, 2006). To satisfy the customer is the mission and purpose of every business.

Abell (1980) defined business according to three measures: scope; differentiated marketing; and competitive differentiation. These measures are viewed in three dimensions:

1. *Customer groups*: Describe who is being satisfied.
2. *Customer functions*: Describe what is being satisfied.
3. Technologies: Define technologies utilised to satisfy needs.

Value creation is fundamental to the operation of a business. A traditional measure of value creation is the return on investment (ROI) to shareholders. However, non-monetary value generation along social and ecological dimensions is also gaining the attention of policy makers and shareholders.

See also: **harmony**

Further reading
Abell, D. (1980) *Defining the Business: The Starting Point of Strategic Planning*, Englewood Cliffs, NJ: Prentice-Hall.
Christensen, Clayton M., Cook, Scott and Hall, Taddy (2006) 'What Customers Want from Your Products', http://hbswk.hbs.edu/item/5170.html (accessed on December 31, 2010).
Finklestein, Ron (2009) 'The Purpose of Business', http://www.yourbusinesscoach.net/Purpose-of-business.html (accessed on December 31, 2010).

Business angels

Business angels are private investors who provide capital for high-potential entre-preneurial start-ups in exchange for a share in the company. Business angels also contribute their expertise in business management and their personal network of contacts. Also called *angel or informal investors*, these individuals typically provide more business guidance than **venture capital** providers and play a major role in early-stage financing. Angel investors are considered key players in financing the high-growth start-up companies that are essential to regional economic develop-ment and are playing a significant role in the USA, UK and Sweden. The University of New Hampshire Center for Venture Research reported that angel investments in 2007 amounted to $26 billion. However, the market for business angel invest-ment suffers from a mismatch between investors and early-stage ventures.

Recently, there has been a trend towards forming *angel groups* or *angel networks* to pool the investment capital and expertise of a small group of business angels. While *business angel networks* (BANs) can act as a financial intermediary between investors and start-ups to remedy the problem of matching one to the other, Knyphausen-Aufseß and Westphal (2008) found that in Germany, most BANs do not mitigate the matching problem.

See also: **new ventures**

Further reading
European Trade Association for Business Angels, http://www.eban.org (accessed on December 9, 2009).

B

Knyphausen-Aufseß, Dodo Zu and Westphal, Rouven (2008) 'Do Business Angel Networks Deliver Value to Business Angels?', *Venture Capital*, 10(2), 149–69.

Business case

The business case is a compelling rationale that justifies investments in new products, product enhancements and major marketing programmes. It documents facts leading up to the investment request and presents the business, market and financial rationale behind the investment required for development. Knowledge-based business cases are written to match requirements to available resources. Resources include technical and engineering knowledge, time and funding. Knowledge-based business cases also include controls to ensure that sufficient knowledge has been attained at key phases within the product development process. The business case is baselined in the definition phase of the **new product development process** and evolves in the concept and feasibility phases. It continues its evolution across the **product life cycle** as market conditions change, competitors act, technologies evolve and people change jobs. The best business cases are written when the development team sees them as a tool for its own success.

Adams and Hublikar (2010) raised three business questions and answered them in the context of a business case as follows:

1. *Do we want to do this?* This question is answered by clearly defining project scope, strategic fit, market attractiveness, customer overview, competitive landscape and **value proposition**.
2. *Can we do this?* This question is answered by describing the rationale for the new product design, technical plan and project plan.
3. *Impact if we do this?* This question is answered by the financial plan and management approval.

See also: **concept definition**

Further reading
Adams, Dan and Hublikar, Sudhir (2010) 'Upgrade your New Product Machine', *Research Technology Management*, 53(2), 55–67.
Haines, Steven (2009) *The Product Manager's Desk Reference*, New York: McGraw-Hill.

B

Business charter

A business charter is a simultaneous expression of a company's vision and a capability road map for adapting the reality to that **vision**. It is what its customers, investors, suppliers and employees think the company is and what it is not. It is a living document that expresses both the nature of a business and the reality, and it should have built-in flexibility to reexamine and reconcile any differences between the two.

Reexamination of the business charter and the **core strategic vision** (CSV) often leads to out-of-the-box thinking and the development of new solutions. When market conditions change, a business can expand its charter. For example, one medical device manufacturer expanded into new markets after defining a CSV outside its existing business charter (McGrath, 2000).

Galunic and Eisenhardt (1996) defined the *divisional charter* of a multidivisional corporation as the businesses (i.e. product and market arenas) in which a division actively participates and for which it is responsible within the corporation. Multidivisional organisations can change charter strategically to adapt to rapidly coevolving markets and technologies. By effective recombination of divisional charters, multidivisional corporations can achieve high performance in competitive markets.

See also: **firm growth**

Further reading
Galunic, D. Charles and Eisenhardt, Kathleen M. (1996) 'The Evolution of Intracorporate Domains: Divisional Charter Losses in High-technology, Multidivisional Corporations', *Organization Science*, 7(3), 255–82.
McGrath, Michael E. (2000) *Product Strategy for Technology Companies*, 2nd edn, New York: McGraw-Hill.

Business intelligence (BI)

In 1958, IBM researcher Hans Peter Luhn defined business intelligence (BI) as 'the ability to apprehend the interrelationships of presented facts in such a way as to guide action towards a desired goal'. BI interprets data and transforms the information into insights. It can be used to guide strategy development and business decision making. BI is both a product and a process. The process consists of the methods used to develop intelligence. The product is the information that will allow organizations to predict the behaviour of their 'competitors, suppliers, customers, technologies, acquisitions, markets, products and services, and the general business environment' with a degree of certainty (Vedder *et al.*, 1999).

When effectively integrated into organisational processes, BI can advance corporate goals such as *sustainability management*, enhance sales outcomes, increase *customer satisfaction* and improve *return on investment*. Online companies such as Google, eBay and Amazon continually gain intelligence about their customers' experience and engagement, products, channels, partners, target markets and competitors. In the health-care industry BI can help health-care organisations to improve financial and operational performance and the quality of patient care.

Issues that have hampered widespread BI use are poor data quality, the complexity of BI tools, lack of BI skill sets within organisations and lack of management buy-in. Additionally, the use of BI remains limited to specialists and analysts. *Open source* offers a viable alternative to organisations looking for BI on a tight budget.

B

Further reading
McKendrick, Joe (2009) 'Business Intelligence Comes to the Masses', *Database Trends and Applications*, 23(1), 20.
Vedder, R. G., Vanecek, M. T., Guynes, C. S. and Cappel, J. J. (1999) 'CEO and CIO Perspectives on Competitive Intelligence', *Communications of the ACM*, 42(8), 108–11.

Business model

A business model 'describes the logic and principles' a firm uses to generate revenues (Hummel *et al.*, 2010). Firms build 'skills, capabilities, capital assets, intellectual property portfolios, skill sets as well as policies and procedures around their business models' (*ibid.*).

A business model consists of four interlocking elements: customer **value proposition**, profit formula, key resources and key processes. Kraemer *et al.* (2000) describe how Dell Computer's business model combined direct sales and build-to-order production. Dell established a strong customer value proposition with large customers and in the process built online infrastructure, to reach new customers at low marginal costs (*ibid.*). It utilised resellers and integrators as key resources to sell its products. It offered for sale non-Dell products such as software and peripherals to increase profits per transaction with individual buyers (*ibid.*). For instance, it became the second largest reseller of Hewlett-Packard printers (Schick, 1999). Dell's efficiencies in minimising inventory and increased speed in bringing new products to market came from its information technology-enabled processes (Kraemer *et al.*, 2000).

While a business model enables a firm to succeed, it can also be a constraint if the firm tries to change models or experiment with new ones (Hummel *et al.*, 2010). Hummel *et al.* (2010) suggest collaboration with a partner firm with a compatible business model as a way to understand the customer value proposition in new market spaces. The PepsiCo–Lipton Partnership is an example of this approach (Altman, 1999).

Further reading
Altman, R. (1999) 'Co-branding, Partnership, Joint Venture and Crossover: Tea Marketing Gains Sophistication', *Tea & Coffee Trade Journal*, 171(8), 58–62.
Hummel, Ed, Slowinski, Gene, Mathews, Scott and Gilmont, Ernest (2010) 'Business Model for Collaborative Research', *Research Technology Management*, 53(6), 51–4.
Kraemer, Kenneth L., Dedrick, Jason and Yamashiro, Sandra (2000) 'Refining and Extending the Business Model with Information Technology: Dell Computer Corporation', *Information Society*, 16(1), 5–21.
Schick, Shane (1999) 'The Great Sales Model Swap-Fest', *Computer Dealer News*, 2/21.

Business model innovation

Breakthrough, game-changing products usually require a new **business model**. Business model innovation can result from either successfully applying an existing **business model** to a new context or introducing a completely new business model for radically new products (http://www.investorglossary.com/business-model).

For example, the internet offered print publishers a new context in which to monetise their print assets. The publishing industry took this opportunity to build *new business models*. To meet the demand for digital content, Simon & Schuster teamed with multimedia start-up Vook to publish four hybrid video–e-book titles, while Disney Publishing Worldwide launched Disney Digital Books (disneydigital-books.com) (Milliot and Deahl, 2009).

In the services sector, McDonald's led the industrialisation of *services business model* by applying process standardisation and the mass-production techniques of the *manufacturing business model* to the food service industry (http://www.investorglossary.com/business-model).

Yahoo! became successful with the *online content provider business model* following the technological innovation of the World Wide Web (*ibid.*).

Apple built a ground-breaking business model that combined hardware, software and service. It essentially gave away low-margin iTunes music to lock in purchase of its high-margin hardware (iPod). The iPod model defined value in a new way and provided game-changing convenience to the consumer. Apple's true innovation in the iPod with the iTunes Store was to make downloading digital music easy and convenient.

Johnson *et al.* (2008) identified five strategic circumstances for business model innovation:

1. To address the needs of large groups of potential customers who are underserved. This may include reaching out to the **base of the pyramid consumer**.
2. To commercialise a new technology with a new business model (Apple and MP3 players).
3. To bring focus to a job to be done where there was no focus (FedEx).
4. To preempt new entrants such as low-end disrupters (Minimills).
5. To respond to a shifting basis of competition (Hilti).

See also: **strategic innovation**

Further reading
Johnson, Mark W., Christensen, Clayton M. and Kagermann, Henning (2008) 'Reinventing Your Business Model', *Harvard Business Review*, 86(12), 51.
Milliot, Jim and Deahl, Rachel (2009) 'Disney Try New Models', *Publishers Weekly*, 256(40), 6.

Business plan

A business plan is a document that describes a proposed new business or a new project within an existing business. A business plan for a new business seeks to capture the **vision** and describes the current status and expected needs of the markets, along with the projected results and financial needs for the new business. A business plan for a new project within an existing business describes the business idea, establishes its objectives and details the strategies, methods and processes that must be followed to achieve those objectives, including financial projections for the coming years.

B

Business planning for a new project begins at the front end by asking:

- Why do we want to do this project? Is the market large or rapidly growing?
- What products and services do we want to offer (what jobs we want to do for customers)? And why are they needed?
- How do these new products/services further our strategic objectives?
- Does our company have the people and organisational resources, capabilities and supply chains to do the job?

See also: **new product planning, product marketing plan**

Further reading
Sahlman, William A. (1997) 'How to Write a Great Business Plan', *Harvard Business Review*, 75(4), 98–108.

Business process innovation (BPI)

Business processes are the set of collaborative and transactional activities that deliver value to customers. Business processes occur within a complex sociotechnical system. Individuals working within a business have differing perceptions of business processes, depending on their role in the organisational sociotechnical system. So it can be argued that business processes are the result of cognitive and communicative interactions among individuals as they interpret, make sense of and coordinate actions (Lewis *et al.*, 2007) within the context of how organisational knowledge is created and used. Smith and Fingar (2003) describe three waves of fundamental paradigm shifts in *business process management*. The first wave began in the 1920s with non-automated processes; the second wave with manual reengineering of processes; and the third wave with a focus on how to implement changes to business processes without requiring companies to have significant technology knowledge.

Business process innovation (BPI) deals with redesigning how business processes are presented in an end-to-end way, and how technologies interact with them.

Lewis *et al.* (2007) combined the business process modelling and stakeholder analysis literature and proposed a four-step approach to business process innovation: engage process stakeholders; collect process data; explicate process knowledge; and design process innovations.

See also: **organisational knowledge-creation theory**

Further reading
Lewis, Mark, Young, Brett, Mathiassen, Lars, Rai, Arun and Welke, Richard (2007) 'Business Process Innovation Based on Stakeholder Perceptions', *Information Knowledge Systems Management*, 6(1/2), 7–27.
Smith, Howard and Fingar, Peter (2003) *Business Process Management: The Third Wave*, Tampa, FL: Meghan-Kiffer Press.
'The Essence of Business Process Innovation', http://www.1000ventures.com/business_guide/innovation_process.html (accessed on December 7, 2009).

B

Business process reengineering (BPR)

Business process reengineering (BPR) is a new paradigm of *organisational change* necessary to maintain flexibility and competitiveness in a global economy. BPR is defined as 'the fundamental rethinking and radical redesign of business processes to achieve dramatic improvements in critical, contemporary measures of performance, such as cost, quality, service, and speed' (Hammer and Champy, 1993). Effective reengineering of business processes begins with reconnecting with the purpose of the business. The fundamental question to ask is: 'Why does this process exist, and for whom?' or, as Champy (1996) asks: 'Why are we doing what we are doing?'

Reengineering depends on a company's strategic direction. For example, strategic focuses on price leadership, supply chain, high-quality production or micro-marketing require different reengineering strategies, each with a different outcome. So BPR requires a clear understanding of the problem space in order to define the scope of the project, the areas of the organisation most affected and the complexity of the problem. A typical reengineering problem space comprises of five interrelated factors at three managerial levels (strategic, managerial and operational). Grant (2002) describes the five interrelated factors as:

1. Processes: There are three types of processes: inter-organisational processes by which companies along the same value chain interact; interfunctional processes that overlap functional boundaries in the organisation; and inter-personal processes that occur between people within one functional group.
2. Technologies: Technologies allow companies to achieve performance gains. Technology includes hardware, tools, techniques and methods such as BPR, information engineering, joint application design, prototyping, computer-aided support systems, structured methods of development and object orientation.
3. People: In the people domain items for redesign include job titles and positions; team-based management techniques; team-based performance evaluation; individual and team compensation and reward; and individual and team authority and responsibility.
4. Communication: Redesign communication channels inside and outside the organisation to further mutual understanding in order to improve decision making, collaboration and the free exchange of information.
5. Structure: Structure defines the relationship between people and technology. Structural changes may include cross-functional teams, product teams and flattening of the organisational hierarchy.

BPR projects have a very high failure rate. Key factors in such failures are a lack of sustained management commitment and leadership, unclear definition of BPR projects, unrealistic scope and expectations, organisational resistance to change and inadequate resources.

Further reading
Grant, Delvin (2002) 'A Wider View of Process Reengineering', *Communications of the ACM*, 45(2), 85–90.
Champy, J. (1996) *Reengineering Management: The Mandate for New Leadership – Managing the Change to the Reengineered Corporation*, New York, HarperCollins.
Hammer, M. and Champy, J. (1993) *Reengineering the Corporation*, New York: HarperCollins.
Kesner, Richard M. and Russell, Bruce (2009) 'Enabling Business Processes through Information Management and IT Systems: The FastFit and Winter Gear Distributors Case Studies', *Journal of Information Systems Education*, 20(4), 401–5.

B

Buyer utility map

Which new product ideas to fund? This is the decision challenge faced by managers of innovation. Kim and Mauborgne (2000) propose three tools to identify which innovative ideas have real commercial potential: the 'buyer utility map', which indicates the likelihood that customers will be attracted to the new idea; the 'price corridor of the mass', which identifies what price will unlock the greatest number of customers; the 'business model guide', which offers a framework for how a company can profitably deliver the new idea at the targeted price.

The buyer utility map is a two-dimensional matrix that displays the six stages of the buyer experience cycle on one dimension and the six utility levers on the other (Kim and Mauborgne, 2000). A *buyers' experience* can be broadly broken down into a cycle of six stages: purchase, delivery, use, supplements, maintenance and disposal. At each stage a company can typically use six levers to unlock exceptional buyer utility: customer productivity, simplicity, convenience, risk, fun and image, and environmental friendliness.

By applying the buyer utility map, innovation managers gain initial insights into the unquestioned assumptions in their industry, how those assumptions detract focus from value creation and, more importantly, how they can raise buyer utility and deliver outstanding customer experiences.

See also: **blue ocean strategy**

Further reading
Kim, W. Chan and Mauborgne, Renee (2000) 'Knowing a Winning Business Idea When You See One', *Harvard Business Review*, 78(5), 129–38.
Kim, Chan and Mauborgne, Renee (2005) *Blue Ocean Strategy*, Boston, MA: Harvard Business School Press.

B

Cannibalisation

Cannibalisation is a distinct phenomenon of products and services and can be categorised into self-cannibalisation and cannibalisation by a competitor. Self-cannibalisation occurs when two or more of an organisation's products compete and take away market share from each other in a given product or service market. Competitor-driven cannibalisation occurs when a competitor launches substitutable products at a price lower than the market leader to take away sales from that market leader. The strategy of cannibalisation is common in the computer hardware and software, financial services, airline services, automotive, food and pharmaceutical industries.

In the computer industry new entrants routinely attack market leaders with new technologies. Personal computer (PC) companies used the cannibalisation strategy successfully against minicomputer companies. However, a market leader can proactively introduce a new technology ahead of the competition to discourage new entrants. Regular cannibalisation of its own products by the technology leader is successful when the underlying technology continuously advances, because the technology leader can pace the market. This is planned *self-cannibalisation*. However, the market leader risks loss of financial returns if the **launch strategy** of innovation is not in alignment with the life cycle.

The strategic challenge of planned self-cannibalisation for a manufacturing company is to choose between two launch strategies: sequential and simultaneous. A manufacturer of durable goods facing two customer segments with differing valuations of product quality can introduce two models (high end and low end) simultaneously or introduce only the high-end model first and then sequence it with the low-end model. Moorthy and Png (1992) showed that when cannibalisation is a problem and customers are relatively more impatient than the manufacturer-seller, the *sequential product introduction* strategy is better than the *simultaneous product introduction strategy*. In this example, if two models of digital cameras are introduced simultaneously, the lower-quality product would cannibalise demand for the higher-quality product.

Further reading
Mazumdar, Tridib, Sivakumar, K. and Wilemon, David (1996) 'Launching New Products with Cannibalization Potential: An Optimal Timing Framework', *Journal of Marketing Theory & Practice*, 4(4), 83–93.
McGrath, Michael E. (2000) *Product Strategy for Technology Companies*, 2nd edn, New York: McGraw-Hill.
Moorthy, K. Sridhar and Png, I. P. L. (1992) 'Market Segmentation, Cannibalization, and the Timing of Product Introductions', *Management Science*, 38(3), 345–59.

Taylor, Mark B. (1986) 'Cannibalization in Multiband Firms', *Journal of Consumer Marketing*, 3(Spring), 69–75.

Capability maturity model (CMM)

The capability maturity model (CMM) defines software developers' process maturity. Paulk *et al.* (1993) defined a software process as a 'set of activities, methods, practices, and transformations that people use to develop and maintain software and the associated products (e.g., project plans, design documents, code, test cases, and user manuals)'. The Carnegie Mellon University Software Engineering Institute (SEI) capability maturity model (CMM) organises software development process maturity into five levels: initial, repeatable, defined, managed and optimising (*ibid.*). However, it does not adequately address testing issues. The *testing maturity model (TMM)*, developed in 1996 at the Illinois Institute of Technology, is a complementary model to the CMM (Rana and Ahmad, 2005).

The CMM has been successfully used as a framework for improving the *software development process* utilizing *total quality management* protocols. Science Applications International (SAIC) and General Dynamics are examples of corporations that have successfully implemented the CMM.

Further reading

Fraser, P., Moultrie, J. and Gregory, M. (2002) 'The Use of Maturity Models/Grids as a Tool in Assessing Product Development Capability', 10th International Product Development Management Conference, Brussels, Belgium, pp. 244–9.

Paulk, Mark C., Curtis, Bill, Chrissis, Mary Beth and Weber, Charles V. (1993) 'Capability Maturity Model for Software, Version 1.1', Report CMU/SEI-93-TR-25, Software Engineering Institute, Pittsburgh, PA: Carnegie Mellon University, http://www.dfki.de/imedia/lidos/pbir/uCMU-SEI-93-TR-025.html (accessed on March 29, 2011).

Rana, Kibria Khalil and Ahmad, Syed Shams Uddin (2005) 'Bringing Maturity to Test', *Electronics Systems & Software*, 3(2), 32–5.

Cash flow

Cash flow is the actual amount of cash coming into or going out of a firm. The simplest definition and calculation for cash flow is:

$$\text{Cash flow} = \text{net income} + \text{depreciation} + \text{depletion} + \text{amortisation}$$

Cash flow before tax (CFBT) and cash flow after tax (CFAT) are common measures used to assess new product or new business opportunities in the **concept** phase of the **new product development process**. *Incremental cash flow* is the difference in cash flows resulting from the implementation of a new project, usually measured as the change in value of the firm before and after implementation of the new opportunity. The cash flow is negative in the *development phase* of the **new product process**. In the early launch phase, new revenues begin to flow back to the organisation. The generation of post-launch positive cash flow from the new opportunity results in the generation of a positive return on investment.

Cash flow is critical to maintain the normal operations of a business. When firms experience financial crisis, they slash product development budgets. In 2008 General Motors experienced severe cash-flow shortages due to depressed car sales. It cut its 2009 product development budget to save as much as $1.5 billion in cash. However, key to increasing long-term sales is to launch more successful new products and not fewer.

Rather than relying strictly on cash-flow projections based on **opportunity analysis** within the existing **business models**, technology managers should also consider the potential of **technological innovations** to create new market opportunities or to reshape existing markets.

Further reading
Dorf, Richard C. (1999) 'Tools for the Technology Manager' in *Technology Management Handbook*, Vol. 49, Boca Raton, FL, CRC Press.

Catalytic innovation

Catalytic innovations are disruptive innovations for social change (Christensen *et al.*, 2006). Catalytic innovations become successful when the features and functions of existing products are more than the capacity of consumers to absorb them. Catalytic solutions have lower quality and performance levels than existing products. They also cost less. Consumers consider them 'good buys'.

Catalytic innovations generate novel financial resources; utilise volunteer labour driven by passion for social change; and access intellectual capital in ways that are not attractive to incumbents.

In the health-care field, Minneapolis-based Minute-Clinic is a catalytic innovator. It located its clinics in CVS retail drugstores and other retail locations. It provides fast, affordable walk-in diagnosis and treatment for common health problems and offers vaccinations. Freelancers Union is another catalytic innovator that provides low-cost health insurance and other services to independent contractors, consultants and part-time and temporary workers in the New York area.

In the education field, the *community college model* is a catalytic innovation. Community colleges offer a low-cost alternative to four-year college and university courses and measure their success based on convenience of access to classes and percentage job placement of graduates. Micro-finance organisations such as the Grameen Bank are examples of catalytic innovation in economic development.

See also: **base of the pyramid consumer, base of the pyramid innovation**

Further reading
Christensen, Clayton M., Baumann, Heiner, Ruggles, Rudy and Sadtler, Thomas M. (2006) 'Disruptive Innovation for Social Change', *Harvard Business Review*, 84(12), 94.

Champions

See **product champion**

Change management

Change management is the process of managing individual, team and organisational change. Successful organisations adapt their **core competences,** launch new products and services and continually evolve in sync with a dynamic marketplace. However, managing change is not easy. It is more of an art than a structured process. The art lays in blending limited and flexible organisational structures with extensive communications; and in offering design freedom to developmental teams to improvise within current projects, leading to the launch of successful, multiproduct innovations (Brown and Eisenhardt, 1997). In the high-tech electronics and computer industries with short **product life cycles** and rapid technological change, the ability to change continually becomes a **core competence**. Continuous change includes management of both **incremental** and **radical innovations**. **Ambidextrous organisations** manage both incremental innovations and radical innovation projects effectively.

Continuous change is often played out through **product innovation** as firms are transformed by the act of continuously altering their products. Hewlett-Packard is a classic example of a company that changed rapidly through continuous innovations as it successfully moved from an instrument company to a computer company (Brown and Eisenhardt, 1997).

Esther Cameron and Mike Green (2004) describe four metaphors for *change leadership*: machine, political system, organism and flux and transformation:

1. *The machine metaphor* is symbolised by setting clear goals and implementing organisational structures in support of those goals. Overuse of rigid organisational structures (such as centralised planning and control) leads to a culture of **incremental innovations**.
2. In the *political system metaphor*, it is necessary to involve influential people when change is desired. For example, **product champions** play a much needed role in championing **radical innovation projects.** However, overuse of champions can be perceived as manipulation. Creation of both **champion** and *sponsor* roles within the **innovation process** can minimise the risk of the political system metaphor.
3. *The organism metaphor* recognises that people are involved in the change process and people need to support change actively. The organism metaphor runs the risk of moving too little and too late. Clearly defined organisational goals and chartering of innovation project teams with clearly understood goals can mitigate the downside risk of the organism metaphor.
4. *The flux and transformation model* reminds us that organisations and their people cannot be controlled unless they are ruled by fear. However, leading by fear is an **innovation killer.** To become innovative, leaders must encourage discussion of conflicts and tensions within their organisations and support learning behaviours, while balancing the need to be clear and provide unambiguous direction.

Successful leaders of change learn to implement a combination of the above four metaphors to achieve continuous **product innovations**.

Further reading
Brown, Shona L. and Eisenhardt, Kathleen M. (1997) 'The Art of Continuous Change: Linking Complexity Theory and Time-paced Evolution in Relentlessly Shifting Organizations', *Administrative Science Quarterly*, 42(1): 1–34.
Cameron, Esther and Green, Mike (2004) *Making Sense of Change Management: A Complete Guide to the Models, Tools and Techniques of Organizational Change*, London: Kogan Page.

Chasm

In his book *Crossing the Chasm*, Geoffrey Moore (2002) defines the chasm as the gap between the needs of two types of customers: visionaries and pragmatists. *Visionaries* are customers who embrace new high-technology products. Visionaries embrace 'leading-edge properties' that generate a lot of enthusiasm within the 'in crowd'. They seek radical, discontinuous change (that is, **radical innovations**). *Pragmatists* are more common mainstream customers who need to be convinced before they go and buy new technology products. They seek proven, continuous improvements (that is, **incremental innovations**).

Moore (2002) urges high-technology companies to focus on bridging the gap in needs between visionary and pragmatist customers; that is, to focus on crossing the chasm. To cross the chasm, a firm must change its corporate culture from one driven by technology to one focused on delivering solutions to customer problems.

Many small companies wish to cross the chasm and become successful, but do not have the resources or opportunities to cross it. Innovative owners of small start-up companies overcome this handicap by forming **alliances** with larger firms with global distribution channels or merge with dominant industry players.

See also: **aligning innovation with life cycle, mergers and acquisitions**

Further reading
Moore, Geoffrey A. (2002) *Crossing the Chasm: Marketing and Selling High-Tech Products to Mainstream Customers,* New York: HarperCollins.
Moore, Karl (2004) 'Forget Crossing the Chasm', *Marketing Magazine*, 109(12), 7.

Cloud computing

Cloud computing refers to the delivery of applications as services over the internet and to hardware and systems software located in the data centres that provide those services (Armbrust *et al.*, 2010). The National Institute of Standards and Technology (NIST) defines cloud computing as 'enabling convenient, on-demand network access to a shared pool of configurable computing resources (e.g., networks, servers, storage, applications, and services) that can be quickly released with minimal effort or service provider interaction' (Mell and Grance, 2009).

Developers with innovative ideas for new internet services can deploy their services without the capital or human expense to operate it (Armbrust *et al.*, 2010). There are three service models: cloud software as a service (SaaS), cloud platform as a service (PaaS) and cloud infrastructure as a service (IaaS) (Mell and Grance, 2009):

- *Cloud software as a service (SaaS)*. The consumer can access the provider's applications running on a cloud infrastructure from a web browser (e.g. web-based email). The provider controls the cloud infrastructure.
- *Cloud platform as a service (PaaS)*. The consumer can launch consumer-created or acquired applications using the provider's programming languages and tools. The provider controls the cloud infrastructure, but the consumer has control over the applications deployed and application hosting environment configurations.
- *Cloud infrastructure as a service (IaaS)*. The consumer has access to computing resources and the consumer can deploy and run operating systems and applications. The provider controls the cloud infrastructure, but the consumer has control over operating systems, storage and applications deployed.

Cusumano (2010) sees companies offering SaaS or cloud versions of their products (now web-based services) and allowing outside application developers to build and launch applications from the company's platforms. Microsoft and Amazon offer pay-as-you-go software licensing for Windows Server and Windows SQL Server and IBM has pay-as-you-go pricing for hosted IBM software (Armbrust *et al.*, 2010).

Further reading

Armbrust, Michael, Fox, Armando, Griffith, Rean, Joseph, Anthony D., Katz, Randy, Konwinski, Andy, Lee, Gunho, Patterson, David, Rabkin, Ariel, Stoica, Ion and Zaharia, Matei (2010) 'A View of Cloud Computing', *Communications of the ACM*, 53(4), 50–58.

Cusumano, Michael (2010) 'Technology Strategy and Management: Cloud Computing and SaaS as New Computing Platforms', *Communications of the ACM*, 53(4), 27–9.

Mell, Peter and Grance, Tim (2009) *The NITS Definition of Cloud Computing*, Version 15, 10-7-09, http://csrc.nist.gov/groups/SNS/cloud-computing/ (accessed on December 26, 2010).

Cluster analysis

To deliver superior customer value, organisations identify who their customers are and what those customers want; and they try to understand the key product attributes that underlie their customers' product preferences. Customer needs are heterogeneous. Approaches to understanding customers involve segmenting markets, identifying segments relevant to the organisation and uncovering product attributes that meet the needs of customers in the targeted segments. Cluster analysis is a statistical technique used to identify market segments. It gives management a data-driven view of customers.

In cluster analysis individuals are grouped into a relatively small number of homogeneous groups or benefit segments. The technique consists of initially clustering sampling units (customers, geographical areas and so on) on the basis of their sensitivities to various purchase attributes such as price, performance and marketing promotions (Sexton, 1974). For example, consumers that have similar utility values for selected attributes and their levels are clustered together, whereas consumers that have different utility values are classified into different clusters. Next, response functions are estimated for each cluster or segment. Aggregating the estimates from these segment response curves produces the estimate for the entire market (*ibid.*).

Canadian Imperial Bank Commerce (CIBC), a financial services company, used its customers' product holdings and demographic characteristics to further its understanding of their product and servicing needs. CIBC won customer loyalty by providing customised services to its custom clusters. Using cluster analysis, Proflowers.com, an e-business, gained understanding of the demographic profiles of its customers and used this understanding to form a strategic partnership with companies like Omaha Steaks and the Bombay Company.

Several commercial software packages are available to conduct cluster analysis. However, in their book *Advanced Data Mining Techniques*, David Olson and Dursun Delen (2008) posit that cluster analysis is computationally intensive and may be impractical in large datasets.

Further reading
Olson, David and Delen, Dursun (2008) *Advanced Data Mining Techniques*, New York: Springer.
Sexton Jr, Donald E. (1974) 'A Cluster Analytic Approach to Market Response Functions', *Journal of Marketing Research*, 11(1), 109–14.

Clusters

Theories of regional specialisation and clustering have existed for more than 100 years and date back to the *Marshallian district* (Marshall, 1889). Interest in clusters reemerged in part due to the published work of Michael Porter. Porter defined a cluster as a 'geographically proximate group of interconnected companies and associated institutions in a particular field, linked by commonalities and complementarities'. Why are clusters important to innovation? They provide a constructive and efficient forum for dialogue among related companies and their suppliers, governments, universities and research laboratories, as well as access to financial institutions. Such dialogue and access in geographical proximity are widely accepted as key enablers of innovation and creative capacities. The pioneering innovations of Silicon Valley's entrepreneurs are often attributed to its becoming a geographical cluster.

Many governments and public policy organisations have adopted clusters at city, state or country level or as a network of neighbouring countries as a public policy instrument to increase competitiveness, innovation and growth (OECD, 2001). Clusters generally include end-product or service companies; suppliers of specialised inputs, components, machinery and services; financial institutions; or firms in related industries (Porter, 2008). Also included in clusters are distribution channels for customers; producers of complementary products; specialised infrastructure providers; governments and other institutions providing specialised training, education, information, research and technical support (such as universities, thinktanks, vocational training providers); standards-setting agencies; and trade associations.

Some well-known industrial clusters in the US are the institutional furnishings cluster in Grand Rapids, Michigan and the California wine cluster (Porter, 2008).

Cluster building is a slow and tedious process and it takes time to transform competitive relationships into complementary relationships among industrial firms located within a geographical area.

See also: **regional innovation systems**

Further reading
Marshall, A. (1889) *Principles of Economics*. London: Royal Economic Society.
Organization for Economic Co-operation and Development (OECD) (2001) *Innovative Clusters: Drivers of National Innovation Systems*, Paris: Organization for Economic Co-operation and Development.
Porter, Michael E. (2008) 'Clusters and Competition' in *On Competition*, Boston, MA, Harvard Business School Publishing, pp. 213–21.

Co-creation

There is some confusion in the use of terms co-creation, **co-design** and **participatory innovation**. In their 2004 book *The Future of Competition: Co-Creating Unique Value with Customers,* C.K. Prahalad and Venkat Ramaswamy proposed: 'The meaning of value and the process of value creation are rapidly shifting from a product and firm-centric view to personalized consumer experiences. Informed, networked, empowered and active consumers are increasingly co-creating value with the firm.' Sanders and Stappers (2008) referred to co-creation as 'any act of collective creativity, i.e. creativity that is shared by two or more people'. However, the practice of collective creativity in design is not new. It was practised for nearly 40 years under the name **participatory design**. Also, **co-design** and co-creation are often treated synonymously. Co-design indicates the application of collective creativity throughout the design process. Co-design, in a broader sense, refers to the creativity of designers and people not trained in design working together at the front end of the design development process (Sanders and Stappers, 2008), whereas co-creation takes place throughout the development process.

Distributed co-creation is the process of opening up a firm's product-development process to new ideas from the outside, such as from suppliers, independent inventors, university laboratories and customers. Companies adopt distributed co-creation in three ways:

1. They capture value from the co-created product or service by merchandising ideas rated as good from the network (examples are LEGO and Threadless in the USA and Missha, a co-created cosmetic brand, in South Korea).
2. They capture value by providing a complementary product or service (Red Hat sells technology services to users of Linux).
3. They benefit indirectly from the co-creation process through an enhanced brand or strategic position.

For Ramaswamy and Gouillart (2010), the co-creation approach serves the interests of all stakeholders by focusing on their experiences and how they interact with one another. In the co-creative approach, stakeholders benefit from 'improved experiences, increased economic value (higher earnings, the acquisition of skills, opportunities to advance), and increased psychological value (greater satisfaction, feelings of appreciation, higher self-esteem)' (Rangaswamy and Gouillart, 2010). Consumer product, fashion, and technology sectors are utilizing co-creation to develop solutions to customers' problems or creating new markets for their products and services. Also, computer companies such as Hewlett-Packard and Apple's iPod are including parts invented and manufactured by their suppliers (Bughin and Johnson, 2008). Eli Lilly licenses and sells products that other companies develop. The Best Buy Company is harnessing the expertise and creativity of its employees, vendors, and customers by instituting results-oriented learning environment (Congemi, 2009). In India, consumers are creating animated commercials for DoCoMo brand using its microsite www.create. docomo.com (Sharma and Mazumdar, 2009).

Maddock and Vitn (2010) pointed that co-creation can succeed between profit-driven organizations; also between profit-driven organizations and cause-driven ones. For example, Procter & Gamble (P&G) and Clorox created an environment friendly bag sold under the Glad-brand name (Maddock and Vitn, 2010). Wal-Mart and Sierra Club co-created public sustainability pledges. These are lifestyle pledges that Wal-Mart employees make to each other, and they had a marked effect on Wal-Mart employee morale and health. Duke Energy and the Environmental Defense Fund teamed up to achieve Duke Energy's goal of 'decarbonizing' its operations by 2050 (Maddock and Vitn, 2010).

See also: **outside innovation**

Further reading
Bughin, Jacques, Chui, Michael and Johnson, Brad (2008) 'The Next Step in Innovation', *McKinsey Quarterly*, June, http://www.mckinseyquarterly.com/next_step_in_open_innovation_2155 (accessed on September 16, 2009).
Congemi, John (2009). New Role. *Training*, Jun 2009, Vol. 46 (5): 4–4.
Maddock, G. Michael; Vitn, Raphael Louis (2010). *Co-Creation Innovation.* BusinessWeek.com, 11/3/2010, p. 5–5.
Prahalad, C. K. and Ramaswamy, Venkat (2004) *The Future of Competition: Co-Creating Unique Value with Customers,* Boston, MA: Harvard Business School Press.
Ramaswamy V, Gouillart F. (2010). *Build the Co-creative Enterprise.* Harvard Business Review, 88 (10): 100–109.
Sanders, Elizabeth B.-N. and Stappers, Pieter Jan (2008) 'Co-creation and the New Landscapes of Design', *CoDesign*, 4(1), 5–18.
Sharma, Amit and Rakhi Mazumdar (2009). Companies in India involve consumers in co-creation of brands. *Economic Times, The* (India), 12/29/2009.

Co-design

Co-design is the development of coherent product or service concepts in collaboration with industry partners in a firm's **value chain**. Development teams visualise co-design concepts and illustrate them with concept sketches, prototypes and **scenario-based designs**. Co-designing requires creative initiative on the part of the entire design team: researchers, designers, clients and the people who will ultimately benefit from the co-designing experience (Sanders and Stappers, 2008).

Co-designing threatens the existing power structures in hierarchical organisations because it requires the relinquishing of control of product design and development to potential customers, consumers or end-users. Development managers

in non-learning organisations have difficulty giving up decision authority to co-
design teams. Non-learners also feel threatened by new ways of doing business.

See also: **participatory design, co-creation**

Further reading
Burr, Jacob and Matthews, Ben (2008) 'Participatory Innovation', *International Journal of
Innovation Management,* 12(3), 255–73.
Sanders, Elizabeth B.-N. and Stappers, Pieter Jan (2008) 'Co-creation and the New
Landscapes of Design', *CoDesign,* 4(1), 5–18.

Cognitive modelling

The Product Development and Management Association (PDMA) defined cognitive modelling as 'a method for producing a computational model for how individuals solve problems and perform tasks' (http://www.pdma.org/npd_glossary.
cfm). Cognitive models are of two types: the computational process model, which
captures internal computational (mechanistic) processes that generate cognitive
behaviour; and the mathematical model, which measures behavioural parameters
(such as recall rates, response time or learning curves) and relates them through
mathematical equations.

Olson and Olson (1990) attributed the growth of cognitive modelling in human–
computer interaction to Card, Moran and Newell's (1983) work on cognitive engineering models.

Cognitive models improve the user interface to minimise interaction errors or
interaction time by anticipating users' behaviour. Cognitive models for games and
animation explore the interface between computer graphics and artificial intelligence.

Further reading
Card, S.K., Moran, T.P. and Newll, A. (1983) *The Psychology of Human–Computer
Interaction,* Hillsdale, NJ: Lawrence Erlbaum Associates.
Funge, John (2000) 'Cognitive Modeling for Games and Animation', *Communications of
the ACM,* 43(7), 40–48.
Olson, Judith Reitman and Olson, Gary M. (1990) 'The Growth of Cognitive Modeling in
Human–Computer Interaction Since GOMS', *Human–Computer Interaction,* 5(2/3), 221.

Co-ideation

Burr and Matthews (2008) defined co-ideation as: 'the generation of ideas for
product and service business opportunities based on the understanding of
people's practices in collaboration with industry partners and the people studied'.
This definition recognises that design ideas for new business opportunities and
improvements to existing systems stem from a 'meeting of minds', with different
perspectives brought together to solve a clearly defined set of problems.

Burr and Matthews organised a workshop within the idea-generation phase of
the **new product development process** of Danfoss, a wastewater-treatment
company. The co-ideation workshop had two parts. In the first part, participants
used game pieces to discuss the instrumentation of future water-treatment plants.
In the second, they went to a wastewater-treatment plant to illustrate where the

new components should be placed. The application of new solutions in a real context encouraged the groups to enrich their original ideas (*ibid.*). Plant operators responded favourably to product ideas, while business unit engineers were sceptical. The mixed support for the implementation of new ideas at Danfoss is a phenomenon normal to non-learning organisations. The new solutions disrupt the status quo.

The **lead user** research by Eric von Hippel advocates the inclusion of lead users early in the **innovation process**. He illustrates the benefits of lead users using 3M as an example. The Danfoss example illustrates that ordinary users are also capable of contributing to the generation of new business opportunities.

Further reading
Burr, Jacob and Matthews, Ben (2008) 'Participatory Innovation', *International Journal of Innovation Management*, 12(3), 255–73.
Von Hippel, E. (2005) *Democratizing Innovation*, Cambridge, MA, MIT Press.

Collaborative innovation

Collaborative innovation is 'the product of a joint effort by two or more organizations' (Gilson *et al.*, 2009). In collaborative innovation transactions occurring within the boundaries of firms (that is, the manufacture of key parts or components, assembly of final products and research leading to new products) move across boundaries. The shift in organisational boundaries necessitates organised exchanges of information between collaborating firms to mitigate the risk of vulnerability to each other's performance. Gilson *et al.* (2009) label the governance of this novel form of innovation '**contracting for innovation**'.

The development of the Boeing 787 aircraft is an example of collaborative innovation. Innovation in the design and manufacture of the wing was dependent on the design and manufacture of the fuselage. The two designs had to fit together. So the innovation of the 787 became a collaborative and iterative process. An iterative process differs from the traditional contract process. In traditional supply chain contracts, the downstream customer sets specifications for parts and contracts with upstream suppliers to produce them.

Collaborative innovation occurs under different circumstances. It can be governed effectively by '**contracting for innovation**' practices.

Further reading
Gilson, Ronald J., Sabel, Charles F. and Scott, Robert E. (2009) 'Contracting for Innovation: Vertical Disintegration and Interfirm Collaboration', *Columbia Law Review*, 109(3), 431–502.

Collaborative new product and process development (NPPD)

Collaborative new product and process development (NPPD) differs from *Stage Gate™ new product development (NPD)*. The Stage Gate™ system recognises that **product innovation** is a process and applies process-management methods to control variation within it. It exerts control by dividing the NPD process into a number of stages or work stations. For example, a five-stage system consists of concept development and testing, business case, product development, testing and validation, full production and market launch stages. Between each stage there is a control point or a gate. Gates are go/no-go checkpoints. Gates allow projects meeting specified criteria to move from one stage to the next while 'killing' bad projects or recycling them. The Stage-Gate process improves performance, enhances efficiency and reduces new product cycle time (Cooper, 1994). However, its rigid controls also act as **innovation killers** by negatively influencing learning and harming new product development. Thus Stage Gate™ firms often modify the traditional Stage Gate™ process (Cooper, Edgett and Kleinschmidt, 2002).

In contrast, the goal of collaborative NPPD is to integrate **product innovation** and supply chain **process innovation**. Collaborative NPPD emphasises the integration of internal and external stakeholders early into the **new product development process** through the **product life cycle**. Internal stakeholders (that is, marketing, product engineering, manufacturing, sales and distribution, and customer service and field support teams) and external stakeholders (that is, customers, product engineering specialists, development partners, suppliers and regulators) become full partners in the NPPD **design–build–test** (support) team (Swank, 2006).

Interorganisational collaboration in product development across the **value chain** and through the product life cycle is complex. The complexity of this integration of stakeholders from suppliers to customers can be overcome with the use of web-based software technologies. *Collaborative product commerce (CPC)* is such a software technology to integrate the products, processes and resources of different enterprises using web technologies (Kim *et al.*, 2006).

To be effective for new product development, the collaborative NPPD process should be flexible. **Radical innovation** project teams need flexibility, while **incremental innovation** project teams can become successful under structured development processes such as the Stage Gate™ process.

Further reading
Cooper, Robert G. (1994) 'Third-Generation New Product Processes', *Journal of Product Innovation Management*, 11(Jan.), 3–14.
Cooper, Robert G., Edgett, Scott J. and Kleinschmidt, Elko J. (2002) 'Optimizing the Stage-Gate Process: What Best Practices Companies Are Doing – Part I', *Research-Technology Management*, 45(Sept.–Oct.), 21–7.
Ettlie, John E. and Elsenbach, Jorg M. (2007) 'Modified Stage-Gates Regimes in New Product Development',
Journal of Product Innovation Management, 24(1), 20–33.
Kim, Huyn, Kim, Hyung-Sun, Lee, Joo-Haeng, Jung, Jin-Mi, Lee, Jae Yeol and Do, Nam-Chul (2006) 'A Framework for Sharing Product Information across Enterprises', *International Journal of Advanced Manufacturing Technology*, 27(5/6), 610–18.
Swink, Morgan (2006) 'Building Collaborative Innovation Capability', *Research Technology Management*, 49(2): 37–47.

Collective innovation

Collective innovation is a combination of 'private investment' and 'collective action' models that encourages and supports innovation (von Hippel and von Krogh, 2003). Private firms and individuals spend private resources to create 'public good' in collective innovations.

Open-source software projects (such as Linux, MySQL and Apache) are such a combination. Von Hippel and von Krogh (2003) term this compound the 'private–collective' innovation model. Other examples of private–collective innovations are user-generated media products, drug formulas and sport equipment designs (Stuermer *et al.*, 2009). Two key characteristics of private–collective innovations are non-rivalry and non-exclusivity in consumption *(ibid.)*.

Private–collective innovation is similar to **open innovation** in that firms are open to source ideas, knowledge and technology from outside. However, private–collective innovation is different from open innovation in that software developers forfeit their intellectual property rights, while firms in open innovation retain such rights.

Further reading
Stuermer, Matthias. Spaeth, Sebastian and von Krogh, Georg (2009) 'Extending Private–Collective Innovation: A Case Study', *R&D Management*, 39(2), 170–91.
von Hippel, Eric and von Krogh, Grog (2003) 'The Private-Collective Innovation Model in Open Source Software Development: Issues for Organization Science', *Organization Science*, 14, 209–23.

Co-location

Co-location of research and development (R&D) staff is 'the positioning of departments and offices of R&D personnel in proximity to each other' (Xie *et al.*, 2003). Co-location stimulates knowledge dissemination both within R&D and between R&D and other functional groups (such as marketing, engineering, finance and manufacturing) (Song *et al.*, 2007). Exchange of knowledge within and across different divisions and functions and spontaneous communication and exchange of ideas are critical to innovation. Co-location helps in the alignment of innovation goals across multiple divisions and functions.

Co-location fosters the integration of teams and departments. It raises the probability that innovation teams incorporate user needs and technology needs, in a balanced way, into new product designs. Ford Motor Company co-located its key suppliers with the Ford engineering team at Automobile Alliance International, a joint Ford and Mazda assembly plant in Flat Rock, Michigan. Ford engineers led the design effort of the 2005 Mustang, and suppliers implemented their decisions.

Some organisations invest in *computer-mediated communication (CMC)* and prefer this over face-to-face communications. Effective knowledge dissemination requires a balanced investment in co-location and information technologies (Song *et al.*, 2007).

Further reading
Song, Michael, Berends, Hans, van der Bij, Hans and Weggeman, Mathieu (2007) 'The Effect of IT and Co-location on Knowledge Dissemination', *Journal of Product Innovation Management*, 24(1), 52–68.
Xie, Jinhong, Song, Michael and Stringfellow, Anne (2003) 'Antecedents and Consequences of Goal Incongruity on New Product Development in Five Countries: A Marketing View', *Journal of Product Innovation Management*, 20(3), 233–50.

Commercialisation

Commercialisation denotes *commercial product development* (Di Norcia, 2005). It is the third step in a three-phase 'RT&D' process, involving *scientific research* (R), *technological innovation* (T) and *commercial product development or commercialisation* (D) (*ibid.*). Commercialisation activities in the **new product development process** include production, launch, ramp-up, marketing, sales, customer service, training and supply chain development. *Commercialisation of innovation* may be led by inventors, entrepreneurs or third-party developers who license innovation. In the outdoor industry (e.g. the kayaking, canoeing and rafting industry), *user innovators* create new materials and equipment and commercialise their own innovations or participate in the commercialisation process of other firms (Hienerth, 2006).

Commercialisation of innovation is a priority in firms such as GE, BMW, Nestlé and Samsung. These firms work across functional and divisional boundaries, and empower employees with an entrepreneurial mindset to achieve success (Von Krogh and Raisch, 2009). Firms that achieve growth through internal innovations (that is, *organic growth*) fund breakthrough ideas for commercialisation. The commercialisation of **radical innovations** is heavily dependent on the corporate **innovation culture** (Tellis *et al.*, 2009).

Further reading

Di Norcia, Vincent (2005) 'Intellectual Property and the Commercialization of Research and Development', *Science & Engineering Ethics*, 11(2), 203–19.

Hienerth, Christoph (2006) 'The Commercialization of User Innovations: The Development of the Rodeo Kayak Industry', *R&D Management*, 36(3), 273–94.

Tellis, Gerard J., Prabhu, Jaideep C. and Chandy, Rajesh K. (2009) 'Radical Innovation across Nations: The Preeminence of Corporate Culture', *Journal of Marketing*, 73(1), 3–23.

Von Krogh, Georg and Raisch, Sebastian (2009) 'Focus Intensely on a Few Great Innovation Ideas', *Harvard Business Review*, 87(10), 32.

Commitment

Commitment in new product development (NPD) refers to two types: resource commitment and emotional commitment. Resource commitment is management's promise to invest resources in a project. In emotional commitment, people invest their ego and passion into a project even in the face of uncertain outcomes.

Uncertainty and ambiguity characterise **radical innovation** projects. Organisations must institute mechanisms to develop commitment, even in the presence of uncertainty and risk, to succeed with radical innovation.

Commitment strategies improve competitive position by influencing the future behaviour of a firm's rivals. Both market leader and follower can use two commitment strategies alternately or in combination, investing in R&D and delegating quantity decisions to managers. These two commitment strategies, 'cost-reducing R&D' and 'delegation to managers', influence the market leader and follower differently due to different interaction effects. By leveraging the differences in commitment strategies, *follower companies* can overcome the advantage of *first-mover companies in the market* and even obtain a higher profit than the market leaders (Kopel and Löffler, 2008).

See also: **first-mover advantage**

Further reading
Githens, Greg (2004) 'Radical Innovation Success Starts with Asking the Question – Why Are We Doing This Project?' *Visions*, 28(3), 21.
Kopel, Michael and Löffler, Clemens (2008) 'Commitment, First-Mover, and Second-Mover Advantage', *Journal of Economics*, 94(2), 143–66.

Communities of practice (CoP)

Lave and Wenger (1991) first used the term 'communities of practice' (CoP) to describe 'a set of relations among persons, activity and world, over time and in relation with other tangential and overlapping communities of practice'.

A community of practice is a group of people with common characteristics (similar interests, expertise, roles and goals), who interact with each other in a physical or virtual space and possess a common domain, practice or set of practices (Koliba and Gajda, 2009). CoPs connect practitioners to share ideas and solve problems. They are valuable to organisations because they build the **social capital** necessary for knowledge creation, sharing and use (Lesser and Prusak, 2000). CoPs transcend organisational boundaries (Wenger, 1998). While they are powerful sources of knowledge, they are also limited by the world view of the 'in crowd'.

Further reading
Koliba, Christopher and Gajda, Rebecca (2009) '"Communities of Practice" as an Analytical Construct: Implications for Theory and Practice', *International Journal of Public Administration*, 32(2), 97–135.
Lave, J. & Wenger, E. (1991) *Situated Learning: Legitimate Peripheral Participation in Communities of Practice*, New York: Cambridge University Press.
Lesser, E. & Prusak, L. (2000) 'Communities of Practice, Social Capital and Organizational Knowledge,' in E. L. Lesser, M. A. Fontaine & J. A. Slusher (eds), *Knowledge and Communities*, Boston, MA: Butterworth Heinemann, pp. 123–32.
Wenger, E. (1998) *Communities of Practice: Learning, Meaning, and Identity*, Cambridge: Cambridge University Press.
Wenger, E. C. and Snyder, W. M. (2000) 'Communities of Practice: The Organizational Frontier', *Harvard Business Review*, 78(1), 139.

Community innovation survey (CIS)

In Europe, community innovation survey (CIS) results are used extensively in setting innovation policy. The CIS has revealed that European innovation took place across a broad cross-section of industrial sectors. This knowledge shifted *European innovation policy* from a narrow to a broad view of innovation. European researchers utilise CIS data in academic studies and it is also incorporated into the *European Innovation Scoreboard (EIS)*.

CIS visualises the **innovation process** as a 'step-wise activity' leading to discrete and well-defined (mostly tangible) innovative outputs from quantifiable inputs. CIS2 and CIS3 define innovative firms as those introducing either a product or a process (technological) innovation. This definition reflects a manufacturing bias. CIS3-rated **technological innovation** intensity for service sectors is lower than for manufacturing industries. This reflects the differences in business practices of service firms and technology-intensive firms. For example, service firms conduct more training, undergo more organisational changes and devote fewer resources to R&D and intellectual property than do high-technology firms. Service firms also interact less with universities, research institutions and other science and technology (S&T) institutions and rely more on suppliers for technology innovations.

See also: **innovation governance**

Further reading
Evangelista, Rinaldo (2006) 'Innovation in the European Service Industries', *Science and Public Policy*, 33(9): 653–68.
Nauwelaers, Claire and Wintjes, Rene (2008) *Innovation Policy in Europe: Measurement and Strategy*. Cheltenham, Edward Elgar.

Comparative performance assessment study (CPAS)

The Product Development and Management Association (PDMA) conducted three longitudinal studies (in 1990, 1995 and 2003) on best practices in new product development. These studies are commonly referred to as comparative performance assessment studies (CPAS).

The 1990 CPAS study found that only 54.5 per cent of responding firms had a well-defined new product development (NPD) process, and only 56.4 per cent of firms had a specific new product strategy.

The 1995 study identified NPD practices with a high degree of success as use of formal NPD processes and strategies; measuring NPD outcomes; use of cross-functional development teams; use of a variety of qualitative market research (e.g. voice of the customer, customer visits and beta-testing techniques); and use of engineering design tools and computer simulations.

The 2003 study reinforced the 1995 findings. It also identified differentiators between the 'best' performing companies and the 'rest' as a focus on the 'soft' tools and processes; encouraging and enabling functional and senior managers to achieve effective cross-functional integration; and willingness to continually experiment with new technology-based tools.

The new product success rate remained relatively stable at under 60 per cent during 1990–2003. However, the 2003 study reported a success rate of 75.5 per cent for the 'best'-performing companies and 53.8 per cent for the 'rest'. Another critical finding is that the success rate of incremental innovation projects for the 'rest' is 66 per cent, while the success rate of radical innovations for the 'best' performers is 65 per cent. This finding dispels the notion that radical innovation projects are riskier than incremental innovation projects, especially in the case of 'best'-performing firms. This finding has serious implications for the management of innovation.

See also: **innovation mindset**

Further reading
The PDMA Foundation's 2004 Comparative Performance Assessment Study (CPAS), http://www.pdma.org/shop_pdma_description.cfm?pk_store_product=25 (accessed with purchase on March 16, 2010).

Competitive advantage

In his 1990 book *The Competitive Advantage of Nations,* Michael Porter proposed that nations and firms create competitive advantage and that it is not inherited. He theorised that competitive advantage comes from innovation. Once a firm achieves competitive advantage, it can sustain it through continuous improvements (that is, through *sustaining innovations*).

Competitive advantage occurs at the level of the firm and the industry level. Industry-level competitive advantage occurs when clusters of firms within industries collaborate and innovate via vertical or horizontal relationships.

Firm-level competitive advantage occurs when the value a firm creates for its customers exceeds its cost of creating it. To achieve competitive advantage, firms choose from three generic strategies: cost leadership, differentiation and focus. A focus strategy includes cost focus, differentiation focus and cost and differentiation focus. Wal-Mart is a cost leader with everyday low prices. Apple Computer is a differentiator with exceptional customer experiences (e.g. the iPod).

Cost leadership is critical for success when a **dominant design** emerges in the industry. The television industry is an example. Cost advantage in the television industry requires efficient picture tube facilities, a low-cost design and automated assembly. In addition, global scale allows multinational corporations to leverage that scale in amortising R&D costs.

All employees contribute to sustaining competitive advantage at the level of the firm.

See also: **competitive strategy**

Further reading
Porter, Michael E. (1985) 'Competitive Strategy: The Core Concepts' in *Competitive Advantage: Creating and Sustaining Superior Performance,* New York: Free Press, pp. 11–25.
Porter, Michael E. (1990) *The Competitive Advantage of Nations,* New York: Free Press.

Competitive strategy

Competitive strategy can be viewed as a set of offensive and defensive actions taken by a firm from a group of strategic dimensions. Examples of strategic dimensions are brand definition, customer channel, technology push vs market pull, market leadership (first mover or fast follower), technology leadership, vertical integration, cost positioning, pricing policy and customer service policy.

Michael Porter created four generic competitive strategies (cost leadership, market differentiation, innovation differentiation, market focus) as a framework for industrial firms (Porter, 1998). He argued that firms need to adopt one or more of his strategies to be successful.

In cost leadership strategy firms improve their competitive position by lowering production and marketing costs. Emerson Electric, Texas Instruments, Black and Decker and DuPont are examples of cost leaders.

Implementation of market **differentiation** strategies in an electronic market enables firms to charge premium prices for unique products and services (Koo *et al.,* 2004). Differentiation strategies based on innovation can create a dynamic market environment in which competitors become powerless to predict, plan and react. Innovative capabilities can also be leveraged. For example, Amazon leveraged its IT capabilities and built an efficient warehousing and delivery system to distance itself from competitors.

See also: **first-mover advantage**

Further reading
Koo, Chul Mo, Koh, Chang E. and Nam, Kichan (2004) 'An Examination of Porter's Competitive Strategies in Electronic Virtual Markets: A Comparison of Two On-line Business Models', *International Journal of Electronic Commerce*, 9(1), 163–80.
Porter, Michael E. (1998) *Competitive Strategy: Techniques for Analyzing Industries and Competitors*, New York: Free Press.

Complex product systems (CoPS)

Complex product systems (CoPS) are large-scale technology systems such as the enterprise resource planning (ERP) system. CoPS are high-cost, engineering-intensive products, adaptive systems or networks. The term 'complex' reflects complexity from the perspective of the supplier rather than as perceived by users of the system. Until recently, CoPS research was supplier oriented. Knowledge concerning the capabilities required to deploy CoPS within a user organisation was sparse. This is a critical gap in knowledge, because Web 2.0 organisations use such high-technology products and systems to support their primary service or productive activities. In internet-based businesses such as Amazon, effective utilisation of CoPS becomes a core competence. For Amazon, it also leads to competitive advantage (as, Amazon offers ERP for sale via its Cloud).

The core process of CoPS innovation depends on organisational environment, innovation strategy and resource supply. It is a user-producer–driven, flexible and craft-based **innovation process**. The utilisation of CoPs is not 'the end unto itself', however. Disruptive e-businesses such as eBay use it to build their innovation capabilities.

Further reading
Cai, Xiang, Jianyuan, Song and Ying, Chen (2009) 'Empirical Study on Innovation Process for Complex Product System in China' in Hu Shuhua and Hamsa Thota (eds), *Proceedings of the 4th International Conference on Product Innovation Management*, Wuhan, China: Hubei People's Press, Part 1, pp. 13–17.
Hobday, Mike (1998) 'Product Complexity, Innovation and Industrial Organization', *Research Policy*, 26, 689–710.

Computer-aided design (CAD)

Joseph Frisch, a professor emeritus of mechanical engineering at the University of California at Berkeley, pioneered computer-aided design (CAD). CAD systems use surrounding surfaces to define solid objects and mathematics to define boundary surfaces. The original two-dimensional CAD technology evolved into modern three-dimensional (3D) modelling.

Engineers utilise 3D CAD in 3D digital prototyping and share product drawings and detailed data with dispersed product development teams. CAD technology enabled Sikorsky Aircraft's helicopter development team to build a helicopter by integrating components from six continents. The Sikorsky team used computer-aided, three-dimensional, interactive application (CATIA) software to develop virtual helicopter designs.

CAD technology is also used in bioengineering and life sciences. BioCAD applications differ from traditional engineering applications. In the traditional engineering geometrical shapes are uniform and designed using CAD and computer-aided manufacturing applications. In biomedical applications such as cardiovascular anatomical design, vascular anatomies are patient specific. Their shapes are not uniform and cannot be created by a sequence of digital counterparts of manufacturing operations (Pekkan *et al.*, 2008).

BioCAD uses free-form CAD surfaces and not closed-form mathematical definitions to allow for the direct manipulation of surfaces necessary in surgical applications. BioCAD is an example of how mechanical engineering has cross-pollinated with bioengineering and other life sciences to achieve further innovation.

Further reading
Pekkan, Kerem, Whited, Brian, Kanter, Kirk, Sharma, Shiva, De Zelicourt, Diane, Sundareswaran, Kartik, Frakes, David, Rossignac, Jarek and Yoganathan, Ajit P. (2008) 'Patient-Specific Surgical Planning and Hemodynamic Computational Fluid Dynamics Optimization through Free-Form Haptic Anatomy Editing Tool (SURGEM), *Medical & Biological Engineering & Computing*, 46(11), 1139–52.
Lee, Doheon, Kim, Sangwoo and Kim, Younghoon (2007) 'BioCAD: An Information Fusion Platform for Bio-network Inference and Analysis', *BMC Bioinformatics*, Supplement 9(8), 1.

Computer-enhanced creativity

Computer-enhanced creativity is the use of specially designed computer software to enable users to become more productive and innovative. Web-enabled software tools enable new discovery and innovation processes for individuals, groups and social environments. Some examples are exploratory research (e.g. Google), enabling collaboration (e.g. Wikipedia and Wikimedia in social creativity) and providing rich history keeping (TRIZ problem solving) (Shneiderman, 2007), **visualisation** and composition. General-purpose commercial creativity tools support **brainstorming**, **idea generation** and knowledge elicitation with concept maps. In a special issue of the *International Journal of Human–Computer Studies* devoted to computer support for creativity, Lubart (2005) distinguished between four types of support: support the creative process as a 'nanny'; facilitate communication among collaborators as a 'pen-pal'; enhance the creative process as a 'coach'; and cooperate in the actual production of creative ideas as a 'colleague'.

Further reading
Lubart, T. (2005) 'How Can Computers Be Partners in the Creative Process', *International Journal of Human–Computer Studies*, 63, 365–9.
Shneiderman, Ben (2007) 'Creativity Support Tools: Accelerating Discovery and Innovation', *Communications of the ACM*, 50(12), 20–32.

Concept

A new product concept is a statement about anticipated product features that will yield selected benefits relative to other products or problem solutions already available. So, a new product *concept statement* is a claim of proposed satisfaction. This is open to interpretation in relation to claims and promises about features and benefits, made by the producer and understood by the consumer. New products fail when a disconnect exists between the interpretations of producers and those of consumers.

An example of a concept statement that states a difference and how that difference benefits the customer or end-user is: 'This new refrigerator is built with modular parts; consequently, the consumer can arrange parts to best fit a given kitchen location and then rearrange them to fit another location' (Crawford and Di Benedetto, 2008).

The *concept-development process* typically includes the **front end of product development**, idea selection and concept and technology development. Other activities include setting final specifications, project planning, **business case**, benchmarking of competitive products, **concept visualisation** and **model-based design** (prototyping).

See also: concept definition, concept generation, concept selection, concept testing

Further reading
Crawford, Merle and Di Benedetto, Anthony (2008) *New Products Management,* 9th edn, New York, McGraw-Hill.

Concept definition

A concept has a defined form (that is, a written and visual form) with features and customer benefits, combined with a broad understanding of the technology required. Concept definition occurs at the **front end of the innovation** process. It consists of several activities: development of a **business case**; risk assessment of technology and markets; and determination of detailed product specifications (for incremental products). At times, an overlap occurs between concept definition and other front-end activities. For example, a food manufacturing firm identifies non-fat potato crisps as an opportunity in the **opportunity analysis** step. The firm funds research to develop specific types of non-fat molecules once the idea has been selected for development (Koen, 2003).

Further reading
Koen, Peter A. (2003) 'Front End of Innovation: Effective Methods, Tools and Techniques in Proceedings of Managing the Front End of Innovation', PDMA and IR Conference, May, Boston, MA.

Concept testing

A concept test is a research procedure to determine whether the intended user really needs the proposed product. (Does it solve a customer problem? Is there a market demand for it?)

The purpose of concept testing is to select the 'best' design, positioning, pricing and manufacturing for a new product (Dahan and Mendelson, 2001).

Concept testing reduces uncertainty about customers, markets, technologies and new product designs. New product development success depends as much on the elimination of poor concepts as on the selection of winning ones. A concept can be presented for testing in the form of written descriptions, a rough sketch, a detailed drawing or a 3-D model, or even in the form of a physical prototype. Respondents evaluate the concepts and/or give preferences (Schoormans *et al.*, 1995). The market research representative contacts the consumer by telephone and conducts an interview. During the interview, the consumer is asked about his or her reaction to the materials and the likelihood that he or she would buy the product described.

Properly executed concept tests predict trial rates if concepts are not radically different from products already in the marketplace. Standard concept tests give discouraging results for major innovations. Allocation of concept testing resources across new product development projects is an ongoing challenge for NPD managers. Despite some drawbacks for really new products, consumer research is a valid tool in the development process.

See also: **conjoint analysis, idea generation and enrichment**

Further reading
Dahan, Ely and Mendelson, Haim (2001) 'An Extreme-Value Model of Concept Testing', *Management Science*, 47(1), 102.
Schoormans, Jan P., Ortt, Roland J. and de Bont, Cees J. P. (1995) 'Enhancing Concept Test Validity by Using Expert Customers', *Journal of Product Innovation Management*, 12, 153–62.

Concept visualisation

Concept visualisation includes methods to elaborate qualitative concepts, ideas, plans and analyses. In the visualisation, an observer can build a mental model. Multidimensional visualisations and prototypes display research findings via dynamic and interactive graphics. Choosing an appropriate representation is critical to comprehensive appreciation of the data, thus benefiting subsequent analysis, processing or decision making.

The US Navy uses a visualisation system to view, test and manipulate new ship designs, reducing the time and cost of building new vessels.

Further reading
http://www.visual-literacy.org/periodic_table/periodic_table.html (accessed on October 25, 2009).
http://www.siggraph.org/education/materials/HyperVis/visgoals/visgoal2.htm (accessed on October 25, 2009).
'Navy's Ship Design Center Selects SGI Onyx Visualization', *Imaging Update*, March 2004, 15(3).

Concurrent engineering (CE)

Concurrent engineering (CE) compresses project schedules by developing the product and all its associated processes at the same time. Multifunctional teams work on multiple aspects of a new product simultaneously. Marketing and R&D

team members focus on integrating customers' needs and marketing strategies into design decisions. Associated processes include manufacturing, customer service, field service and distribution.

Three types of multifunctional teams used in CE projects are programme management, technical and design-build teams. Different CE projects operate with different focuses, such as quality, speed or cost. CE projects with a focus on design qualities use formal presentations and periodic reviews in project management. Projects emphasising development speed utilise frequent, informal communications. Programmes addressing design quality require extended product definition and performance testing. Efforts to reduce development time involve small, informal teams led by design engineers and managers. Aggressive product cost goals necessitate intensive interaction between product designers and manufacturing personnel. Highly innovative projects require **early supplier involvement** and collaborative engineering problem solving.

See also: **Design for X (DFX) tools, types of development teams**

Further reading
Swank, Morgan L., Sandvig, Christopher J. and Mabert, Vincent A. (1996) 'Customizing Concurrent Engineering Processes: Five Case Studies', *Journal of Product Innovation Management*, 13, 229–4.

Conjoint analysis

Conjoint analysis (CA) originated in psychological research. In product development, it is commonly used in the analysis of consumer preferences. Conjoint results indicate the likely success of product line extensions.

Using conjoint analysis, companies form benefit segments and make design trade-off decisions among various features (Moore *et al.*, 1999) to develop optimal product configurations (for example, designs forecast to maximise sales or profits for a given competitive setting). Sometimes organisations fail to optimise features due to the inherent limitations of testing. For instance, Colgate introduced body washes with a unique, no-leak 'Zeller valve' cap that enabled bottles to be stored cap side down. The increase in cost was only a few percentage points. However, subsequent market research suggested that consumers' 'true willingness to pay' for body washes did not justify this improvement (Hauser and Toubia, 2005).

Further reading
Hauser, John R. and Toubia, Olivier (2005) 'The Impact of Utility Balance and Endogeneity in Conjoint Analysis', *Marketing Science*, 24(3): 498–507.
Moore, William L., Louviere, Jordan J. and Verma, Rohit (1999) 'Using Conjoint Analysis to Help Design Product Platforms', *Journal of Product Innovation Management*, 16(1), 27–39.
Pullman, Madeline E., Moore, William L. and Wardell, Don G. (2002) 'A Comparison of Quality Function Deployment and Conjoint Analysis in New Product Design', *Journal of Product Innovation Management*, 19, 354–64.

Connect and develop (P&G model for innovation)

In the twentieth century, companies gained **competitive advantage** by keeping idea sourcing and technology development the exclusive domain of R&D departments. Most companies still embrace the *invention model*. In 2000, Procter & Gamble (P&G) realised that its *'invent-it-in-house'* model was not enough to sustain the desired top-line growth (Huston and Sakkab, 2006). A paradigm shift was occurring, and P&G recognised that internal and external sourcing of ideas, co-development with partners and co-invention with customers are now key success factors.

P&G discovered that innovation was increasingly taking place at small and medium-sized entrepreneurial companies. It also knew that most of its best innovations came from connecting ideas across internal businesses. After studying the performance of a small number of products it had acquired beyond its own labs, it concluded that external connections would produce highly profitable innovations. P&G therefore focused on external connections for future growth.

This model became successful and came to be known as the 'connect and develop' model of innovation.

Further reading
Huston, Larry and Sakkab, Nabil (2006) 'Connect and Develop: Inside Procter & Gamble's New Model for Innovation', *Research Technology Management,* 84(3), 58.

Consumer innovativeness

Consumer innovativeness is the predisposition of some consumers to adopt new products and innovations before other consumers in a given market segment. Innovative consumers adopt innovations earlier than other members of their system. They are *early adopters*. Several factors influence consumer innovativeness and new product adoption behaviour. Tellis *et al.* (2009) describe seeking new and different new products for variety, stimulation or risk as factors that influence adoption. **Lead users** influence the purchasing behaviour of other consumers. Consumers also learn by observing others within their social networks (that is, by *modelling*). Many consumers resist change, while others are nostalgic about products in the past. Frugal customers purchase new products only at discount prices.

Further reading
Tellis, Gerard J., Yin, E. and Bell, S. (2009) 'Global Consumer Innovativeness: Cross-Country Differences and Demographic Commonalities', *Journal of International Marketing,* 17(12), 1–22.

Consumer needs

See **customer needs**

Contracting for innovation

Contracting for purchasing is old, but contracting for innovation is new. Contracting for innovation triggers collaboration across organisational boundaries. It offers an alternative to the traditional vertical integration of companies.

Innovation contracts differ from traditional purchasing contracts. Contracts for innovation cannot specify, with certainty, the exact nature of the product to be produced or its precise performance characteristics. So contracts for innovation interweave explicit and implicit terms that respond to uncertainty in the **innovation process**. Buyer and supplier firms collaborate and iterate throughout the development process.

Traditional purchasing contracts are explicit and subject to renegotiation, thus anticipation of advantageous future dealings encourages opportunism (Gilson *et al.*, 2009), whereas collaborative innovation raises switching costs for buyers, which deters buyer opportunism.

Innovative industries are practising contracting for innovation, a practice that is ahead of the theory.

Further reading
Gilson, Ronald J., Sabel, Charles F. and Scott, Robert E. (2009) 'Contracting for Innovation: Vertical Disintegration and Interfirm Collaboration'. *Columbia Law Review*, 109(3), 431–502.

Contradiction matrix

Genrich Altshuller (1969), the father of the *theory of inventive problem solving (TRIZ)*, found that technological systems evolve according to predictable patterns. Patterns of inventive solutions cross domains of both knowledge and industry. Altshuller found 'nesting' patterns in mechanical, chemical and electrical systems. He found 39 universal characteristics of technical systems that generate contradictions and 40 principles to overcome them.

The contradiction matrix plots the 39 engineering parameters as 'undesired effects' on the x-axis, and 'features to improve' on the y-axis. The matrix's cells contain any of the 40 inventive principles that should be considered for improvement of one of the engineering parameters without degradation of another one.

Variations of the original contradiction matrix are widely available. TRIZ allowed Crayola® to develop a new crayon product that does not write on walls.

Further reading
Altshuller, G. S. (1969) 'Algorithm of Inventions', Moscow, Moscowskiy Rabochy (in Russian).
Product examples, Innovation-TRIZ, http://www.innovation-triz.com/triz/examples.html (accessed on March 31, 2011).

Convergent thinking

Discussion of convergent and divergent thinking followed Guilford's 1950 article on creativity published in *American Psychologist*. Divergent thinking quickly became equated with creativity and convergent thinking became a necessary evil. However, creative production requires both divergent and convergent thinking (Cropley, 2006).

Convergent thinking derives a single answer to a clearly defined question. There is no ambiguity; the answer is either right or wrong. Convergent thinking is most effective where a ready-made answer exists and is available for recall.

Convergent thinking is generally performed late in the idea-generation and enrichment phase of the **new product development process**. It helps funnel ideas generated through divergent thinking into a small group of ideas that can be developed further.

Further reading
Cropley, Arthur (2006) 'In Praise of Convergent Thinking', *Creativity Research Journal*, 18(3), 391–404.
Guilford, J. P. (1950) 'Creativity', *American Psychologist*, 5, 444–54.

Co-opetition

Co-opetition is the practice of collaboration with rivals for joint technology development. Choi (2005) defined co-opetition as 'the situation where a group of competitors cooperate in activities associated with creating mutual benefits while, at the same time, they compete with each other in activities associated with dividing up the benefits'. There are three categories: channel co-opetition, marketing co-opetition and R&D co-opetition (Choi, 2005).

- *Channel co-opetition* denotes a situation where one competitor becomes the other competitor's buyer or supplier, or one competitor uses the other's distribution channels or production facilities while competing in an end-product market.
- *Marketing co-opetition* occurs where direct competitors cooperate in joint marketing efforts (e.g. brand alliances, joint promotion, bundling etc.) while competing in the same market.
- *R&D co-opetition* occurs when direct competitors collaborate in research and development activities (e.g. joint new technology development, joint industry standard development etc.) while competing in other activities.

The incentive for collaboration is the mutual sharing of development costs and project risks. Gnyawali and Park (2009) studied the relationship between co-opetition and **technological innovation** in small and medium-sized enterprises (SMEs) and found that a co-opetition strategy helped SMEs to pursue technological innovations. Co-opetition may pose new challenges to regulators concerned with the enforcement of antitrust laws, however. While co-opetition among SMEs may not result in monopolistic markets, among large national or multinational competitors it may become problematic (Gnyawali and Park, 2009).

Further reading
Choi, Pilsik (2005) 'The Nature of Co-Opetition: Literature Review and Propositions', *AMA Winter Educators' Conference Proceedings*, 16, 105–6.
Gnyawali, Devi R. and Park, Byung-Jin (Robert) (2009) 'Co-opetition and Technological Innovation in Small and Medium-Sized Enterprises: A Multilevel Conceptual Model', *Journal of Small Business Management*, 47(3), 308–30.
Yu, Larry (2008) 'Co-opetition Without Borders', *MIT Sloan Management Review*, 49(2), 8–9.

Core competence

A core competence is a difficult-to-duplicate collection of capabilities. Core competences complement each other to create enduring **competitive advantage**. Core competences comprise a firm's internal capabilities (that is, its engineering, design, research and development, marketing and management and manufacturing capabilities) plus its capabilities to leverage external resources (for instance its innovation ecosystem). Honda converted its core competences in small engines to extend product lines to motorcycles, cars, lawnmowers, outboard motors and home generators.

The core competences form the boundaries of the **core strategic vision (CSV)**. Companies revise their CSV in response to revisions to the business strategy or **business model**. The CSV also changes with **technological innovations** or **open innovation**. Companies acquire new technologies in response to changes in CSV; they may scale back investments in old competences or divest them entirely. In 1991, for instance, Compaq changed its CSV from 'the first to incorporate high-performance leading-edge technology into its products' to 'lead in PCs and servers by pricing competitively and controlling costs'. It developed new competences in low-cost manufacturing and reduced R&D spending from 6 per cent to 2 per cent of revenues. In another example, Qualcomm changed its CSV in 1999 to become 'the wireless industry's best research lab'. It therefore licensed its mobile telecommunications product manufacturing technology that it no longer required.

In their path-breaking *Harvard Business Review* article 'The Core Competence of the Corporation, C. K. Prahalad and Gary Hamel *(*1990) stimulated much attention on the importance of organisations' core competences. Those core competences should never be outsourced. In the electronics industry, brand-name manufacturers focus on core competences, especially **product innovation**, marketing and other activities related to brand development, while using specialised suppliers for non-core functions such as manufacturing (Gereffi *et al.*, 2005). They can capture higher value (gross profit) from R&D (product innovation) than can contract manufacturers and component suppliers (Shin *et al.*, 2009).

Further reading

Gereffi, G., Humphrey, J. and Sturgeon, T. J. (2005) 'The Governance of Global Value Chains', *Review of International Political Economy*, 12, 78–104.

McGrath, Michael E. (2000) *Product Strategy for Technology Companies,* 2nd edn, New York: McGraw-Hill.

Pinchot, Gifford and Pellman, Ron (1999) *Intrapreneuring in Action: A Handbook for Business Innovation,* San Francisco, CA: Berett-Koehler.

Prahalad, C. K. and Hamel, Gary (1990) 'The Core Competence of the Corporation', *Harvard Business Review*, 68(3), 79–91.

Shin, Namchul, Kraemer, Kenneth L. and Dedrick, Jason (2009) 'R&D, Value Chain Location and Firm Performance in the Global Electronics Industry', *Industry & Innovation*, 16(3), 315–30.

Core purpose

The core purpose is the organisation's reason for being. In his book *Good to Great*, Jim Collins writes that purpose is not a description of the organisation's output or its target customers. It is the meaning people attach to the company's work. In great companies people talk about their achievements, not the organisation's output. A great purpose captures the soul of the organisation.

An entrepreneurial owner or chief executive officer (CEO) is responsible for infusing his or her company with its own purpose and translating that purpose into action through organisational mechanisms. The simpler the purpose statement, the better the understanding, adoption and alignment of it throughout the organisation.

Companies' core purposes differ. Collins and Porras describe the core purposes of several organisations. 3M's core purpose is solving unsolved problems innovatively. HP's is to make technical contributions for the advancement and welfare of humanity. Sony's is to experience the joy of advancing and applying technology for the benefit of the public, and Disney's core purpose is to make people happy (Collins and Porras, 1996). These companies have clear and compelling purpose statements.

Further reading
Collins, James C. (2001) *Good to Great: Why Some Companies Make the Leap... And Others Don't*, New York: Random House.
Collins, James C. and Porras, Jerry (1996) 'Building Your Company's Vision', *Harvard Business Review*, 74(5), 65–77.

Core strategic vision (CSV)

Every company has a core strategic vision (CSV), which addresses three issues: What are the strategic objectives? What value does the firm deliver to customers? How does the firm differentiate itself in the marketplace? *Market vision* helps companies to differentiate themselves from the competition. This is a clear and specific mental model or image of a desired and important product-market for a new advanced technology (Reid and De Brentani, 2010).

A framework for the CSV has two purposes: to ensure that the strategic vision is aligned and to identify any changes necessary to achieve it. The CSV has six boundaries that can be contracted or expanded to achieve alignment: **core competences** (value), financial plan (economic model), **business charter**, technology strategy or trends, **product strategy** and market trends or **competitive strategy.**

Michael McGrath (2000) cites Cisco as an example of a company with a well-written CSV: 'Cisco expects to continue to be the world leader in networking for the rapidly growing Internet market, especially for enterprise networks, service providers, and small to medium-size businesses. It expects to maintain this leadership position by providing end-to-end network solutions using its competitive advantage of being the technology leader. Technology leadership in this rapidly changing environment will be maintained aggressively by development, acquisition of technology and **alliances**.'

Further reading
Cohen, Shoshanah and Roussel, Joseph (2005) *Strategic Supply Chain Management: The Five Disciplines for Top Performance*, New York, McGraw-Hill.
McGrath, Michael E. (2000) *Product Strategy for Technology Companies*, 2nd edn, New York, McGraw-Hill.
Reid, Susan E. and De Brentani, Ulrike (2010) 'Market Vision and Market Visioning Competence: Impact on Early Performance for Radically New, High-Tech Products', *Journal of Product Innovation Management*, 27(4), 500–18.

Core values

A company's core values are its essential tenets. Collins and Porras (1996) suggest that successful companies do not change their core values in response to changes in markets; they change the markets in which they compete. The core values motivate people to work for a purpose higher than a pay cheque. For example, ethical behaviour, integrity and mutual respect inspire trust and enhance employees' job satisfaction, which creates the environment and support for innovation. Core values are limited in number from three to five. Collins and Porras (1996) cite Merck, Disney and Sony as examples of companies that use core values to guide actions and decisions:

1. *Merck core values:* corporate social responsibility; unequivocal excellence in all aspects of the company; science-based innovation; ethics and integrity; and profit, but profit from work that benefits humanity (i.e. put patients first).
2. *Disney core values:* no cynicism; nurturing and promulgation of 'wholesome American values'; creativity, dreams and imagination; fanatical attention to consistency and detail; and preservation and control of the Disney magic.
3. *Sony core values:* elevation of the Japanese culture and national status; being a pioneer, not following others, and doing the impossible; encouraging individual ability and creativity.

Without core values an organisation can lose its moral compass. In the aftermath of the global financial crisis, for instance, the US Securities and Exchange Commission (SEC) charged investment bank Goldman Sachs for misleading investors (Gandel and Altman, 2010).

Further reading
Collins, James C. and Porras, Jerry (1996) 'Building Your Company's Vision', *Harvard Business Review*, 74(5), 65–77.
Gandel, Stephen and Altman, Alex (2010) 'The Case against Goldman Sachs', *Time*, 175(17), 30–37.

Corporate model of innovation

Engel and Freeman (2007) describe three destination states in the corporate model of innovation: 'go to market', spinout or 'go to grave' (that is, termination of innovation). In their article, they refer to existing firms that are older and usually larger as corporations. Corporations innovate top down or bottom up. Hierarchical structures such as the executive committee recognise and evaluate corporate opportunities (that is, **opportunity identification and analysis**). Some corporate boards make decisions about **strategic innovation** initiatives such as **open innovation** or **business model innovation** and communicate decisions downward. In other companies, invention begins at the bottom or in the middle. Engineers or marketing managers initiate the invention process and persuade those above them in the hierarchy to support it. Sometimes new technology is acquired by mergers and acquisitions or by licensing.

Internal frictions characterise the corporate model. Frictions resulting from the misalignment of goals impede **technology transfer**, because they cause a disconnect in the allocation of resources and slow down the **innovation process**. Most importantly, corporate models misalign organisational incentives such that some parties (corporate equity owners and senior managers) win while others (inventors and innovators) lose. In addition, the corporate governance structure imposes the interests of owners on management. Such organisational design factors impede an **innovative climate** and become barriers to **product and process innovations**.

Frustrated entrepreneurs in hierarchical organisations leave them to start and build nimble companies that succeed while their larger but slower competitors are constrained. In the corporate model of innovation, innovation is a means to an economic end. Termination of innovation for a business is a write-off, while the death of an innovation for an inventor is a personal disaster (Engel and Freeman, 2007). In the **entrepreneurial model of innovation** entrepreneurs' streams of discovery and invention are personal. The entrepreneurial model supports **breakthrough innovation**, whereas the corporate model of innovation encourages **incremental innovation.**

Further reading
Freeman, John and Engel, Jerome S. (2007) 'Models of Innovation: Startups and Mature Corporations', *California Management Review*, 50(1), 94–119.

Corporate social networks

Social networks, which are informal networks, exist throughout organisations. Corporate social networks must be shaped and cultivated in order to discover opportunities they present and exploit them. In the corporate model of innovation, large companies innovate within hierarchical structures (that is, silos), but are inadequate at collaboration across divisions to build new businesses. Business unit boundaries exist because they are efficient structures for executing **strategy**. However, the focus and efficiency that silos bring come at the cost of cross-divisional collaboration and hamper the realisation of **radical innovation** opportunities.

Organisations must manage informal, interdivisional networks to promote innovation throughout the innovation process. *Idea brokers* are excellent sources of ideas for productive collaboration. They maintain broad networks and draw connections between technologies, market offerings and unmet needs that might otherwise remain undiscovered.

As an example, Dow Chemical built a *corporate social network* to connect its employees with alumni and retirees. Mercedes-Benz used its Generation Benz community to solicit opinions about the Porcelain White Edition interior for its SL-Class cars (Levack, 2009). Organisations strapped for resources can leverage social networking sites such as Facebook, LinkedIn and Twitter for **crowd sourcing** and use social media outlets as alternatives to setting up their own social networks.

Further reading
Kleinbaum, Adam M. and Tushman, Michael L. (2008) 'Managing Corporate Social Networks', *Harvard Business Review*, 86(7/ 8), 26.
Levack, Kinley (2009) 'Tough Times Call for Social Measures: Marketing with Social Computing Tools', *E Content*, 32(1), 10–11.

Corporate venturing (CV)

When established firms invest in and/or create new businesses, it is called corporate venturing. Firms may invest in internal venturing (creating a new business within the parent company), external venturing (investing in the founding and/or growth of businesses outside the parent company's operations) or joint ventures (co-investing with another organisation to create new, external businesses). Intel, 3M, GE, Motorola and Microsoft use CV initiatives as key components of **strategic management**. DuPont Corporation's Development Department and Ralston Purina's New Venture Division promoted new ventures internally. General Electric's Business Development Services, Inc. invested directly in start-up companies (Gompers, 2002).

Corporate venturing efforts appear to be as successful as those backed by independent venture organisations (Gompers, 2002). However, some companies struggle to achieve long-term growth and corporate renewal with CV. This happens when there is managerial uncertainty over how CV might be operationally linked to the firm's overall strategic process and agenda.

Further reading
Covin, Jeffrey G. and Miles, Morgan P. (2007) 'Strategic Use of Corporate Venturing', *Entrepreneurship Theory and Practice*, 31(2), 183–207.
Gompers, Paul A. (2002) 'Corporations and the Financing of Innovation: The Corporate Venturing Experience', *Federal Reserve Bank of Atlanta Economic Review*, 87(4), 1–17.

Cost innovation

Cost innovation is the ability to exploit low costs in radically new ways. Cost innovations in research and development, marketing and customer service deliver more **customer value** for less. Ming Zeng and Peter Williamson (2007) found that leading Chinese firms are pursuing cost innovation strategies to penetrate global markets, which is disruptive to global competition.

The Chinese apply cost innovation to create the new 'value–cost equation' in three categories: high technology at low prices; product variety at mass-market prices; and speciality products at low prices. For example, in the high-tech-at-low-price category, computer manufacturer Dawning put supercomputer technology into ordinary servers used in IT networks. It used its own design and arranged a large number of ordinary processors on a special motherboard to achieve high performance at low cost. Its competitors, especially the dominant players, achieve superior performance using state-of-the-art chip design, building the highest performance only into their most expensive products. Dawning achieved competitive advantage without using state-of-the-art technologies. This is the new game of the 'high-value-low'cost' equation.

The Dawning example reminds us of Clayton Christensen's disruptors in *The Innovator's Dilemma*.

Further reading
Christensen, Clayton M. (1997) *The Innovator's Dilemma: When New Technologies Cause Great Firms to Fail*, Boston, MA, Harvard Business School Press.
Zeng, Ming and Williamson, Peter J. (2007) *Dragons at Your Door: How Chinese Cost Innovation Is Disrupting Global Competition*, Boston, MA, Harvard Business School Press.

Crawford Fellow

The title of 'Crawford Fellow of Product Innovation' is given by the Product Development and Management Association (PDMA) in recognition of its founding president, C. Merle Crawford from the University of Michigan. This prestigious award was created on August 16, 1991 by unanimous vote of the PDMA's board of directors. It recognises individuals who make 'superior and unique contributions to advance the state of professionalism in the field of new products management through direct contributions of knowledge, service, practice, and stature in the field'. In addition, the PDMA board specified, 'This will be a highly selective honor.' Merle Crawford was the first recipient in 1991. The award was later given to Thomas P. Hustad (1993), Robert G. Cooper (1998) and Abbie Griffin (2009).

Thomas P. Hustad was the co-founder of the two oldest executive programmes in new product development, both beginning in the 1970s. The oldest was at York University in Toronto, with Charles S. Mayer, and the second was at the University of Michigan, with Merle Crawford. Tom taught in the Michigan programme for 20 years and the York programme for about 10 years. He currently teaches courses in new product management and marketing at Indiana University. He was the fourth president of the PDMA (1981) and managed the association's office for over 15 years, as well as being founder of the *Journal of Product Innovation Management* (*JPIM*) and serving as its second editor from 1985 to 2000.

Robert G. Cooper has been the leading author of articles investigating the correlates of new product success and failure in the *Journal of Product Innovation Management*. His book *Winning at New Products* can be found on the shelves of many new products professionals. Bob introduced the term 'Stage Gate' process to the new product management vocabulary.

Abbie Griffin became involved with PDMA shortly after completing her doctoral studies at MIT and was the principal researcher for its second comprehensive best practices survey. Abbie has co-authored PDMA handbooks and toolbooks and succeeded Tom Hustad as editor of *JPIM* (1999–2004). She has had a notable influence on professional practices in new product management.

Creative destruction

Creative destruction is a process articulated by Joseph Schumpeter, a legend in innovation research. In his early works he stressed the importance of **discontinuities** of the **innovation process**. He called it *industrial mutation*, 'because it revolutionizes the economic structure from within, destroying the old one, and creating the new one'. Schumpeter argued that a system may fully utilise its possibilities and gain best advantages in the short term, but in the long term be inferior to a system that does not gain the best advantage at any given point in time. Schumpeter argued that the lack of the best opportunities in the short term may be a condition for the level or speed of long-term performance. Proponents of *Schumpeterian theories* view **innovation** as the destruction of contemporary frames of thought and action, which lead to the creation of new goods or quality of goods; development of new methods of production; establishment of new markets; utilisation of new supply sources; or industrial reorganisation (Lassen and Nielsen, 2009). According to this view WalMart is a force of creative destruction in retailing.

Further reading
Lassen, Astrid Heidemann and Nielsen, Suna Lowe (2009) 'Corporate Entrepreneurship: Innovation at the Intersection between Creative Destruction and Controlled Adaptation, *Journal of Enterprising Culture*, 17(2), 181–99.
Schumpeter, Joseph A. (1943) *Capitalism, Socialism and Democracy*, New York, Routledge.
Utterback, James M. (1994) *Mastering the Dynamics of Innovation: How Companies Can Seize Opportunities in the Face of Technological Change*, Boston, MA, Harvard Business School Press.

Creative leadership

By definition, **creativity** is the ability to create something unique and appropriate. It is necessary to **entrepreneurship** and to maintain companies' performance after they have reached a global scale. However, the lack of a theory of novelty prevails. Jim March, professor emeritus at Stanford University, proposed three conditions for novelty: slack, hubris and optimism. Slack refers to the availability of sufficient time and resources for exploration. Hubris refers to encouraging managers to take risks. Optimism refers to organisational vision, which gives a compelling reason to achieve the future it articulates.

Amabile and Khaire (2008) offer six guidelines for creativity leadership:

1. *Enhance creativity* by asking inspiring questions and allowing ideas to bubble up from the workforce.
2. *Enable collaboration* by redefining 'outstanding performers' as someone who helps others succeed. Managers can also use metaphors, analogies and stories to help teams conceptualise together.
3. *Enhance diversity* by getting people with different backgrounds and ability to work together. Managers can encourage individuals to obtain diverse experiences by opening up to external creative contributors.
4. *Map the creative process* by mapping the stages of creativity and then supporting their different needs. For example, managers should avoid process management at the **fuzzy front end**, allow sufficient time and resources for research and manage the hand-off to **commercialisation**.
5. *Accept results and failures*. The creative process deals with uncertainty. Managers should accept the value of uncertainty at the **front end** and the inevitability of failure. Failures in the creative process are learning opportunities and learning from failures can lead to the next 'big' idea.
6. *Challenge to achieve*. Managers can motivate people with intellectual challenge.

According to Peter Chernin (2002), chairman and CEO of FOX Network in Los Angeles, an outward focus on and a clear sense of the needs, desires and lives of other people is essential to creative leadership. He also suggests that creative leadership in the television industry is built on the ability to incorporate the perspective of others into truly original and interesting work (Chernin, 2002).

Further reading
Amabile, Teresa M. and Khaire, Mukti (2008) 'Creativity and the Role of the Leader', *Harvard Business Review*, 86(10), 100–9.
Chernin, Peter (2002) 'Creative Leadership', *Vital Speeches of the Day*, 68(8), 245.

Creative problem solving

See Osborn–Parnes creative problem solving (CPS) model

Creative thinking methods

Creative thinking methods in **product innovation** can be classified into four categories (Li *et al.*, 2007):

1. *Problem-oriented creative design*: The purpose of problem-oriented creative design is to improve performance, solve problems affecting the primary function and move towards an idealised solution. For example, the **contradiction matrix** and inventive principles of *TRIZ* solve system conflicts.
2. *Function-oriented creative design*: New product or system designs make use of function-oriented creative designs. These methods include the *concept fan, FBS mapping* and searching for new principles from other scientific fields.
3. *Product-oriented creative design*: The purpose of product-oriented creative design is to explore and improve products' secondary functions while the primary functions remain adequate. The creative template is a suitable method to implement this design strategy. For example, the component control template creates a link between a component within the product configuration and a component in the external environment (outside the product) and eliminates negative configuration factors.
4. *Form-oriented creative design*: The main purpose of form-oriented design is to meet customers' cultural and emotional needs and develop novel designs by association of ideas or other creative (non-logical) methods.

Further reading
Li, Yan, Wang, Jian, Li, Xianglong and Zhao, Wu (2007) 'Design Creativity in Product Innovation', *International Journal of Advanced Manufacturing Technology*, 33, 213–22.

Creativity

Creativity is the main driver of innovation. It is a mental process involving the construction of new ideas, concepts, approaches or actions by individuals or teams. This can cause the development of previously unknown artifacts, insights or interpretations. It can also involve the recombination or modification of what is already present. Chief executive officers consider creativity as necessary to advance innovations and sustain **competitive advantage**.

Psychologist Teresa Amabile (1983) identified three psychological components of enterprise creativity: domain-relevant skills, creativity-relevant skills and task motivation. Her research showed that intrinsic motivation positively influenced creativity, while extrinsic motivation did not.

Creativity is also positively associated with the use of team structures. Rosabeth Moss Kanter (1983) found that organisations working through teams function effectively in uncertain assignments, overlapping territories and uncertainty in resources associated with innovation.

The enemies of creativity are homogeneity, conformance, standardisation and repetition.

See also **creativity in R&D**

Further reading
Amabile, Teresa (1983) *The Social Psychology of Creativity*, New York, Springer-Verlag.
Bröckling, Ulrich (2006) 'On Creativity: A Brainstorming Session', *Educational Philosophy and Theory*, 38(4), 513–21.
Kanter, Rosabeth Moss (1983) *The Change Masters: Innovation for Productivity in American Corporations*, New York, Simon & Schuster.
Kelley, Tom (2001)*The Art of Innovation*, New York, Doubleday.

Creativity in R&D

Frank Andrews (1975), a pioneer in R&D creativity research, found that social and psychological factors affect the translation of creative talent into innovative performance. Amabile and Gryskiewicz (1987) categorised the qualities of creative R&D scientists into five categories:

1. *Intrinsic motivation*: Intrinsically motivated scientists are self-driven, motivated by the work itself, attracted by the challenge of the problem, committed to the idea and are not motivated by extrinsic rewards such as money, recognition or other external directives.
2. *Ability and experience*: Creative scientists possess outstanding problem-solving abilities and apply realistic tactics to creative thinking. They are bright and are experts in their knowledge domains.
3. *Risk orientation*: Creative scientists are pioneering, flexible, open to doing things differently, attracted to challenge and they take risks.
4. *Social skill*: Creative scientists have rapport with others, are good listeners, are open to others' ideas and have political savvy.
5. *Other qualities*: Individual scientists have persistence, curiosity, intellectual honesty and are unbiased by preconceptions about the problem. As a group, they have positive group synergy.

Further reading
Amabile, Teresa M. and Gryskiewicz, Stan S. (1987) 'Creativity in the R&D Laboratory', Technical Report 30, Greensboro, NC, Center for Creative Leadership.
Andrews, Frank M. (1975) 'Social and Psychological Factors which Influence the Creative Process', in Jacob W. Getzels and Irving A. Taylor (eds), *Perspectives in Creativity*, Piscataway, NJ, Transaction Publishers, p. 117.

Creativity measurement

Ellis Paul Torrance legitimised the study of creativity with his *Torrance Tests of Creative Thinking (TTCT)*. TTCT measure creativity and employ divergent thinking in assessment. Creativity measures are fluency (lots of ideas), flexibility (many different ideas), originality (unique ideas) and elaboration (detailed ideas).

Creativity assessment methods continue to grow. Other creativity measurement methods are association tests and attitude and interest scales. Creativity can also be measured though biographical inventories, because creativity is an expression of personality. Another measurement system is nomination by teachers, peers or supervisors. Researchers continue to expand their understanding of creativity and measurement.

Creativity researchers suggest that assessors of creativity should determine which combination of assessment methods fulfils the purpose and select multiple indicators for each 'assessment need'.

Further reading
Torrance, E. P. and Hall, L. K. (1980) 'Assessing the Further Reaches of Creative Potential', *Journal of Creative Behavior*, 1, 1–19.
Venable, B. (1994) 'Philosophical Analysis of Creativity', full text from ERIC, http://www.eric.ed.gov/ERICDocs/data/ ericdocs2sql/ content_storage_01/ 0000019b/80/13/52/9c.pdf (accessed on February 25, 2010).

Creativity techniques

Creativity and innovation in product design and engineering can be stimulated with practical, creative thinking techniques. Specific creative techniques available to develop innovative and practical solutions to problems include the following:

1. *Mismatch of scale*: This is a 'What if...?' question-and-answer session. 'What if the product in question is much bigger or smaller? How would the problem be changed?' For example, a drug made of large molecules in tablet form can be formulated into powder form with micro molecules.
2. *Word association*: Word association or *linguistic analysis* is an easy-to-use technique. It states a problem and then takes several words at random from the dictionary. For example, a hospital has a storage security problem for a medical device. This problem can be solved by picking two words at random from the dictionary, such as 'burglar' and 'tree', and taking a word path such as 'burglar' 'criminal' 'scene of crime' 'fingerprints' (unique) = biometric solution (fingerprint recognition unlocks treatment device).
3. *Bionics*: Bionics uses nature, because nature is a rich source of comparable problems and solutions. For example, 'How does the human body solve a fluid-transport problem?' may be the question asked to solve a fluid-transport problem in engineering. The answer leads to an innovative solution that is unlikely to be developed by a classically trained engineer.
4. *Synectics*: The term synectics means 'bringing different things into unified connection'. The process involves a team leader working with the team to develop a unified statement of the problem and then leading the team to explore an entirely different, seemingly unrelated area of interest. Free thoughts create totally different viewpoints and create possible solutions for the stated problem.
5. *Morphological analysis*: Fritz Zwicky developed morphological analysis, in which the product is considered an object with different dimensions. The first step involves identifying the most salient dimensions (usually three or four)

of a product; the next the creation of a list of attributes relating to various dimensions. Computer programs are available to order and evaluate a large number of attributes occurring in morphological analysis, including BRIAN, a general-purpose, creative problem-solving tool, and MORPHY, a program specific to the evaluation of new product market opportunities.

See also: **lateral thinking, TRIZ**

Further reading
Hockey, John (2004) 'Practical Creativity', *Medical Device Technology*, 15(1), 20–23.
Proctor, R. A. (1989) 'Innovations in New Product Screening and Evaluation', *Technology Analysis & Strategic Management*, 1(3), 313.

Cross-functional interface management

Product innovation involves cross-functional team collaboration from different functions (such as R&D, engineering, marketing, manufacturing, finance and customer service). Legitimate conflicts arise in cross-functional product innovation teams due to this functional diversity (that is, tensions arise between explicit and implicit beliefs, values and practices). The lack of shared knowledge and experience in a cross-functional team also complicates communication, a conflict that is compounded in *virtual product development teams*. Effective cross-functional interface management is essential for enterprise competitiveness.

The five conflict management styles for solving cross-functional interface issues are:

1. *Confrontation*: collaborative problem solving by developing solutions through discussion, customer interviews and so on.
2. *Give and take*: negotiating a set of features to move the project forward.
3. *Withdrawal*: team members with unpopular positions back out of the decisions.
4. *Smoothing*: appeasing team members who are strongly committed to specific product features.
5. *Forcing*: decision making by the project manager when the team cannot.

Confrontation and give and take are integrative problem-solving styles that provide better conditions for innovation. Withdrawal, smoothing and forcing are dysfunctional styles and are detrimental to innovation outcomes.

Information and communication technologies (ICT) facilitate conflict management by offering a systematic means of communication. However, ICT may exacerbate conflict due to misinterpretations or lack of empathy among virtual team members.

See also: **research and development**

Further reading
Crawford, Merle and Di Benedetto, Anthony (2008) *New Products Management*, 9th edn, New York, McGraw-Hill.

Crowdsourcing

Jeff Howe (2006) of *Wired* magazine defined crowdsourcing as the outsourcing of functions traditionally performed within the boundaries of the firm (e.g. design and development). For instance, a smartphone ground travel planner application, for platforms like the iPhone, was developed by 9292 using crowdsourcing (Staal, 2010). The management of 9292 decided to engage talent through a public contest for college students in the Netherlands. It created a special site for developers to start using dummy examples of 9292 datasets. It also wrote a set of developer's documentation for that purpose (*ibid*.).

Within a few hours of launching the application in 2009, it reached the no.1 position for downloads of paid applications in the App Store. 9292 gained from letting a young programmer show it that a graphic way to build and use planners leads to visually appealing solutions, attracts a younger generation and highlights new avenues of development for the organisation (*ibid*.). The crowdsourcing contest opened up 9292 to outside development and turned the company's paradigm upside down (*ibid*.).

Crowdsourcing is also known as interactive value creation, meaning customer and user involvement throughout the development process (Ogawa and Piller, 2006). Because with crowdsourcing companies are asking users early on about their purchase decisions, the risk of product failure is minimal.

Frank Piller (*ibid*.) recognised t-shirt manufacturer Threadless for successfully outsourcing its entire development process to users and customers, from design to evaluation and selection of products for sale. At Threadless users judge weekly between 400 and 600 new designs on a scale from zero (I do not like this design) to five (I love this design) (*ibid*.). On average, 1500 users score each design. An acceptable score corresponds to three or above. Customers can also opt in to purchase the design (*ibid*.). Through crowdsourcing as its business model, Threadless uses customer and user capabilities to carry out all aspects of the innovation process (*ibid*.).

See also: **idea generation and enrichment**

Further reading
Howe, Jeff (2006) 'The Rise of Crowdsourcing', *Wired*, June, http://www.wired.com/wired/archive/14.06/crowds.html (accessed on December 27, 2010).
Ogawa, Susumu and Piller, Frank T. (2006) 'Reducing the Risks of Product Development', *MIT Sloan Management Review*, http://sloanreview.mit.edu/the-magazine/articles/2006/winter/47214/reducing-the-risks-of-new-product-development/ (accessed on December 9, 2009).
Staal, Gert H. (2010) 'Crowdsourcing Is Used by Dutch Internet Company 9292 to Create a New Travel Application', *Visions*, 34(2), 25–7.

Customer-centric enterprise

A customer-centric enterprise aspires to put the customer at its centre. It builds processes and systems and organises its resources to serve customers as individuals in the best way possible and the most efficient manner. The customer-centric enterprise incorporates both cost efficiency and rapid response to customer needs.

A customer-centric enterprise recognises that customers have alternative choices to purchasing goods and services. Customers might want mass-customised products because they fit better (clothing), compromise and purchase an established product at a lower cost and marginal fit (clothing) or order a customised product with extra features and at a higher price. The **customer value** reflects the price customers are willing to pay for customised solutions.

A company expresses its customer-centric behaviour through **mass customisation**, **personalisation** and **crowdsourcing** strategies. Sports goods maker Adidas, apparel manufacturer Land's End and Threadless in the fashion industry are examples of customer-centric enterprises (Tseng and Piller, 2003).

Further reading
Tseng, Mitchell M. and Piller, Frank T. (2003) 'Heading Towards Customer Centric Enterprises' in Mitchell M. Tsend and Frank T. Piller (eds), *The Customer Centric Enterprise*, New York: Springer, pp. 3–16.

Customer-driven innovation

Customer-driven innovation turns customers into collaborators. Customers play a key role in innovation efforts and are deeply involved in the firm's product and service development.

The customer-driven architecture of a mobile device may begin with a series of questions about the limitations of existing technology, the evolving needs of users of mobile devices, how to fulfil chosen user needs and how the new device may change user behaviour.

The implementation of customer-driven architecture in a telecom company will differentiate it by putting users' needs first. A customer-driven telecom company's customers will be in charge of significant daily activities such as communication, entertainment and personal finances. The Apple iPhone is a customer-driven innovation.

Further reading
'The Customer-Driven Architecture: Interactive Financial Services Technology in the iPhone Era', *M2PressWIRE*, June 12, 2009.

Customer emotional clusters

Marketers create exceptional **customer experiences** with effective management of customer emotions. Four emotional clusters affect customers' commitment to a company: the destroying cluster, the attention cluster, the recommendation cluster and the advocacy cluster. The destroying cluster drives customers away by evoking negative emotions. It lowers a firm's value. The attention, recommendation and advocacy clusters add to a firm's value.

Customers feel positive emotions when they are stimulated, interested, energetic, indulgent and experimental. They start building loyalty when they feel valued, cared for and trusted. Loyal customers recommend the products and services of companies they love to their friends. In the advocacy cluster, loyal customers become advocates for the company.

Customer emotional clusters should not be confused with industrial **clusters**.

See also: **cluster analysis, customer experience**

Further reading
Shaw, C. (2007) *The DNA of Customer Experience,* London: Palgrave Macmillan.

Customer experience (CE)

Customer experience (CE) is the sum of all the interactions a customer has with an organisation compared to his or her expectations. It occurs at various connection points. Consumers touch an organisation while shopping for products, receiving them or consuming them. CE is a combination of the organisation's actual perform-ance and the resulting senses stimulated and emotions evoked. It is positive if the experience equals or exceeds expectations. Positive interactions will underpin the subsequent relationship with the firm and the customer's desire to return, spend more and recommend. Negative interactions weaken it. CE is structurally determin-istic: every firm delivers the customer experience that it is organised to provide. CE is not just practical, it is also emotional, and sometimes even irrational.

Technological innovations and mergers in the telecommunications industry disrupted customer loyalties to land line-based companies. Mobile telecommuni-cations operators such as Verizon utilise online billing services to enable consumers to take control of their own expenses. This contributes to a positive CE.

See also: **customer scenario mapping, profit**

Further reading
Goncalves, Alexis P. (2007) 'Connecting to Innovate Capabilities in Services', paper presented at 5th Annual Front End of Innovation PDMA & IIR Conference, Boston, MA, May.

Customer feedback management

Customer feedback management implies collecting and analysing data from various sources on an ongoing basis. Customer feedback is an essential component of the **new product development (NPD) process** and an indispensable tool for organisations with customer-driven and customer-focused NPD systems. Companies use surveys to solicit feedback in the **concept-testing** phase of the NPD process or post-**launch** analysis of customer satisfaction. It is a proactive approach to achieve the desired **customer experience** levels to improve business performance.

In a networked world, customer feedback management is complex, as customers speak through horizontal networks and new media such as communi-ties, forums and blogs. Successful companies continually improve their NPD processes and customer feedback mechanisms to keep pace with an increasingly connected world. Effective customer feedback management contributes to a posi-tive customer experience.

See also: **profit**

Further reading
Davies, Jim, Maoz, Michael and Thompson, Ed (2008) 'Manage the Customer Experience to Improve Business Performance', Gartner Research, 10 July, ID Number G00159629.

Morello, Diane (2008) 'The Hyperconnected Enterprise: Anticipating the Next Wave of Business', Gartner Research, 22 February, ID Number G00154926.

Customer needs

John Berrigan and Carl Finkbeiner (1992) define customer needs as internal conditions that influence the use of a particular product. Abbie Griffin (1996) defines customer needs as 'problems that a product or service solves and the functions it performs'.

New product development organisations have the opportunity to build product **platform** architecture based on customer needs. Planning tools for developing platforms to support customer needs are product planning, differentiation planning and commonality planning. Platform planning involves identifying market segments and what customers want in each market segment; then designing the product architecture to simultaneously deliver different products while sharing many parts and production processes. Identifying common parts/platform/ modules in the early stages of design lowers cost and design cycle time (Stone *et al.*, 2008).

Further reading
Berrigan, John and Finkbeiner, Carl (1992) *Segmentation Marketing, New Methods for Capturing Business Markets*, New York, Harper Business.
Griffin, Abbie (1996) 'Obtaining Customer Needs for Product Development' in Kenneth B. Kahn, George Castellion and Abbie Griffin (eds), *The PDMA Handbook of New Product Development*, Hoboken, NJ, John Wiley & Sons.
Stone, Robert B., Kurtadikar, Ravi, Villanueva, Nicholas and Arnold, Cari Bryant (2008) 'A Customer Needs Motivated Conceptual Design Methodology for Product Portfolio Planning', *Journal of Engineering Design*, 19(6), 489–514.

Customer scenario mapping

Customer scenario mapping is a method used in **customer experience** design. It involves business analysts and one or more consumers working together to map out a scenario that would save time, be more convenient or allow the user to make a better purchasing decision when buying a product or service. In a mapping session the analyst and customer walk through a typical purchasing scenario step by step.

Citigroup learned from its scenario-mapping programme that customer scenarios provide multiple benefits. Scenario mapping is a precise way to segment customers. Co-designing with customers also yields reliable results compared to surveys. The **lead users** provide new product and service ideas. Finally, customers are adept at telling the company what the **customer experience** metrics should be. Business leaders become more sensitive to **customer needs** when they become involved in the customer scenario-mapping process.

Further reading
Goncalves, Alexis P. (2007) 'Connecting to Innovate Capabilities in Services' in *Proceedings of the 5th Annual Front End of Innovation Conference*, Boston, May.
Zager, Masha (2004) 'Mapping the Customer Point of View', *Rural Telecommunications*, 23(4), 22–7.

Customer targeting

Customer targeting is the practice of developing products for a specific segment of the target customer base. Product managers seeking to enhance an existing product line or develop new products use customer targeting in support of a long-term branding effort.

Targeted **product innovation** is a common practice in consumer and financial services industries. In financial services, Barclays Stockbrokers and Citibank use customer targeting. E-businesses using software such as Omniture's SiteCatalyst applications continually assess the effectiveness of their segmentation and customer targeting.

See also: **differentiation, target market**

Customer value

Woodruff (1997) defined customer value as a customer's perceived preferences for and evaluation of products under actual use conditions. Market-oriented organisations recognise that customer value can only be defined by customers themselves. They also recognise that a focus on customer value leads to a customer-centric organisation, positive business performance and increased customer retention.

There are five types of customer value: functional, social, emotional, epistemic and conditional. In a business-to-business context, economic value comes from fulfilling economic needs at minimal transaction costs. Social value comes from satisfaction with the relationship compared with other alternatives. In business markets, suppliers can provide a strong customer value proposition by discovering and articulating those elements that their customers value most. For example, the supplier of an electronic medical record (EMR) module for anaesthesia documentation in surgical rooms provided seamless, coordinated pre-operative and intra-operative documentation to prevent operating room delays. The result was more revenue-producing time for surgeons, anaesthesia providers and the hospital.

Further reading
Flint, Daniel J., Woodruff, Robert B. and Fisher Gardial, Sarah (2002) 'Exploring the Phenomenon of Customers' Desired Value Change in a Business-to-Business Context', *Journal of Marketing*, 66(4), 102–17.
Woodruff, Robert B. (1997) 'Customer Value: The Next Source of Competitive Advantage', *Journal of the Academy of Marketing Science*, 25(2), 139–53.

Cyclical economy

Economies move in cycles, which may be upward or downward. A downward-cyclical economy moves to progressively worse states until it reaches a trough period, and then moves directly to a boom period. The exact opposite happens in the upward cycle.

At a national level, governments can play a counter-cyclical role in mitigating the cyclical nature of the economy. For example, the Chinese government introduced a stimulus package in 2008 to stabilise the domestic market until the materials and auto-export industries re-emerged from global recession. The US also launched an economic stimulus package in 2009.

At the organisational level, some organisations deliver stable performance through economic cycles. Such companies supply products and services people want and use, dominate their markets with strong brand names and maintain strong cash flows and balance sheets.

Procter & Gamble and Johnson & Johnson are two cycle-proof companies. P&G's Crest toothpaste and Tide detergent products are always in demand. Johnson & Johnson's products are part of people's daily living (Tylenol, Aveeno, Band-Aid and Johnson's Baby Powder are well-known brands).

The increasing share of services in the gross domestic product (GDP) of nations and technology-enabled supply chains has made the global economy more resilient to economic swings in the twenty-first century. Both downward and upward cycles are modulated by bold market and financial interventions from national governments.

Further reading
Lin, Diane (2009) 'The China Syndrome', *Money Management*, 23(23), 25–6.
Steidtman, Carl (2007) 'The Case for Optimism: One Last Time', *Gourmet Retailer*, 28(5), 75–6.

Cyclical innovation model (CIM)

The cyclical innovation model (CIM) is an *innovation system* that generates, diffuses and utilises technology. The CIM views **innovation processes** as running interactions between changes in markets, product and services, technology and science (Van Der Duin *et al.*, 2007). Innovative companies integrate changes in all four aspects (Berkhout *et al.,*, 2006).

In the CIM, innovation is a systematic *circle of change*. The system has four creative 'nodes of change': scientific exploration, technological research, product development and market transitions (Van Der Duin *et al.*, 2007). Between these nodes are 'cycles of change' ensuring that the dynamic processes in the nodes inspire, correct and supplement each other. The CIM presents the inspiration for human **creativity** and is a necessary condition for **sustainability** (Van Der Duin *et al.*, 2007). In the view of this model, **regional innovation systems** need to act as integrated, multicultural systems in order to become innovative.

Further reading
Berkhout, A. J., Van der Duin, P., Hartmann, L. and Ortt, R. (2006) 'Innovating the Innovation Process', *International Journal of Technology Management*, 34(3/4), 390–404.
Van Der Duin, Patrick, Ortt, Roland and Kok, Matthijs (2007) 'The Cyclic Innovation Model: A New Challenge for a Regional Approach to Innovation Systems?', *European Planning Studies*, 15(2), 195–215.

C

Data mining

Data mining, also known as *data discovery* or *knowledge discovery*, is the process of analysing large amounts of collected data and extracting meaningful patterns and useful information from that. It uses machine learning, statistical and **visualisation** techniques to discover knowledge from data. Frawley *et al*, described data mining as 'the nontrivial extraction of implicit, previously unknown, and potentially useful knowledge from data' (Frawley *et al*., 1992).

In **product innovation**, data mining is commonly used to classify relationships between **products** offered for sale and **target markets**. *Cluster analysis, decision trees and visualisation* are data-mining techniques utilised in product innovation.

Companies such as Wal-Mart, Amazon and American Express use data-mining techniques. Wal-Mart uses its customer knowledge to support lean inventories. It also shares point-of-sale data with suppliers as part of its vendor-managed inventory programme. Amazon recommends extra books to customers who purchased a particular book to increase sales per customer. Its title purchase recommendation begins with a leading statement such as 'Customers who purchased this title also purchased the following books ...' American Express uses its customer knowledge to **target customers**, for example, offering them its high-end products.

With the increasing adoption of *grid and* **cloud computing** technologies, Talia and Trunfio (2010) expect web-based knowledge services such as *distributed data-mining systems* to become as ubiquitous as public water, gas and electricity utilities. Widespread use of data mining also raises ethical and privacy concerns, especially in the health-care sector. For example, some pharmacies and insurance companies provide data on drug prescribing by individual physicians to data-mining companies, which sell the information to pharmaceutical companies (Loria, 2010). Pharmaceutical companies, in turn, make use of this knowledge to direct sales presentations to targeted physicians to increase the sales of profitable drugs.

Further reading
Frawley, W., Piatetsky-Shapiro, G. and Matheus, C. (1992) 'Knowledge Discovery in Databases: An Overview', *AI Magazine*, 13, 213–28.
Loria, Keith (2010) 'Contemporary Topics in Health Care: Data Mining', *PT in Motion*, 2(4), 24–31.
Silver, Michael, Sakata, Taiki, Su, Hua-Ching, Herman, Charles, Dolins, Steven B. and O'Shea, Michael J. (2001) 'Case Study: How to Apply Data Mining Techniques in a Healthcare Data Warehouse', *Journal of Healthcare Information Management*, 15(2), 155–64.
Talia, Domenico and Trunfio, Paolo (2010) 'How Distributed Data Mining Tasks Can Thrive as Knowledge Services', *Communications of the ACM*, 53(7), 132–7.

Data visualisation

Data visualisation is the process of converting numerical data into visual form to enable understanding, **creativity** and inference of new meaning from complex data. It leverages humans' advanced visualisation abilities to see patterns, spot trends and identify outliers (Heer *et al.*, 2010).

All visualisations contain a set of mappings between data properties and visual attributes. Customised visualisations can be constructed by varying these visual attributes, such as position, size, shape and colour. Examples of visual data representation techniques are time-series data, statistical data and maps, hierarchies and networks. *Time-series data* are sets of values changing over time. This is one of the most common forms of recorded data in domains such as finance, science and public policy (*ibid.*). *Concept maps* are a representation tool used as an adjunct to learning and study in science education, engineering, mathematics, psychology and health fields. They are a 'tool of the mind' (Jonassen and Marra, 1994). In addition, concept maps can be designed to become valuable tools in qualitative social research (Wheeldon and Faubert, 2009).

Data visualisation is emerging as a powerful tool for artists and designers when contextualising and interpreting complex data (Popova, 2009). Organisations can use data visualisation to determine which patterns and elements within the consumer behaviour data sets are significant to sustain **business** over time and then to construct a narrative to tell a compelling story.

Further reading
Heer, Jeffrey, Bostock, Michael and Ogievetsky, Vadim (2010) 'A Tour Through the Visualization Zoo', *Communications of the ACM*, 53(6), 59–67.
Jonassen, D. and Marra, R. M. (1994) 'Concept Mapping and Other Formalisms for Representing Knowledge,' *ALT-J*, 2(1), 50–56.
Popova, Maria (2009) 'Data Visualization: Stories for the Information Age', *BusinessWeek Online*, August 13, 7.
Post, Frits H., Nielson, Gregory M. and Bonneau, Georges-Pierre (2003) *Data visualization: The State of the Art*, Norwell, MA, Kluwer.
Wheeldon, Johannes and Faubert, Jacqueline (2009) 'Framing Experience: Concept Maps, Mind Maps, and Data Collection in Qualitative Research', *International Journal of Qualitative Methods*, 8(3), 68–83.

Delphi method

D

Delphi is an interactive forecasting method, developed by the RAND Corporation in the 1950s, for gathering and refining the opinions of experts in order to gain consensus. It uses a structured group interaction process with 'rounds' of opinion collection and feedback. Multiple rounds lead to a convergence and then stabilisation of group opinion. The stabilised group opinion may be agreement, disagreement or a combination.

Linstone and Turoff (1975) wrote an authoritative book on the Delphi method; other useful reviews include Okoli and Pawloski (2004) and Keeney *et al.* (2006). The optimum number of Delphi panellists has not been determined. It needs to be large enough to have a representative pooling of views and is between 10 and 50 members in the final round of the process. There are also no generally accepted criteria for the selection of experts.

See also: **forecasting**

Further reading

Keeney, S., Hasson F. and McKenna H. (2006) 'Consulting the Oracle: 10 Lessons from Using the Delphi Technique in Nursing Research', *Journal of Advanced Nursing*, 53(2), 205–12.

Linstone, H. and Turoff, M. (1975) *The Delphi Method: Techniques and Application*, Reading, MA, Addison-Welsey.

Okoli, C. and Pawloski S. (2004) 'The Delphi Method as a Research Tool: An Example, Design Considerations and Applications', *Information and Management*, 42, 15–29.

Demand

There is uncertainty about demand for new products at the time of **product launch**. Two objectives of a new product **launch strategy** are to stimulate and forecast demand accurately. Demand for new products can be generated in three ways. The first is trial and repurchase: consumers try the product first (that is, they make a trial purchase); if satisfied with the experience, they become regular consumers of the product. The second is customer migration: customers discontinue purchase of an existing product in favour of the new product. The third is the adoption and **diffusion of innovation**.

There are many techniques to forecast demand for new products and services. *Efficient consumer response (ECR), vendor-managed inventory (VMI)* and *collaborative planning, forecasting and replenishment (CPFR)* have transformed the supply and demand planning functions in consumer and retailing industries (Kiely, 2004). The *market-based forecasting method* is useful to estimate future demand for health-care services. It uses informed judgement along with analysis of historical data to analyse future demand scenarios (Beech, 2001). In the telecommunications industry, Rao and Wester (1989) used *simulated test markets* to produce robust estimates of demand for broadband services. The uncertainty of adverse market outcomes (such as lack of adoption of a new drug by physicians and patients, or market withdrawal due to adverse side effects) make demand forecasting more difficult in the pharmaceutical industry.

See also: **risk-quantification techniques**

Further reading

Beech, A. J. (2001) 'Market-based Demand Forecasting Promotes Informed Strategic Financial Planning', *Healthcare Financial Management*, 55(11), 46–54.

Guiltinan, Joseph P. (1999) 'Launch Strategy, Launch Tactics, and Demand Outcomes', *Journal of Product Innovation Management*, 16, 509–29.

Kiely, Dan (2004) 'The State of Pharmaceutical Industry Supply Planning and Demand Forecasting', *Journal of Business Forecasting Methods & Systems*, 23(3), 20–22.

Rao, Murlidhar and Wester, Gregory E. (1989) 'Consumer Research and Demand Forecasting for Wideband Telecommunications Services: Some Perspectives', *Advances in Consumer Research*, 16(1), 619–28.

D

Democratisation of innovation

The distributed nature of information and communications technologies (ICT) has enabled the creation of a new paradigm of economic and cultural production that can be described as the **democratisation of innovation**. In his book of the same name, Eric Von Hippel (2005) documented the impact of users on **product innovation**. According to him, a majority of the novel products he studied were first prototyped by users, whom he calls **lead users.**

Von Hippel's research is notable because it clearly documented the emergence of **user-centric innovation** as a powerful source of new **product innovation** for manufacturers. The shift from manufacture-centric innovation, a cornerstone of manufacturers during the nineteenth and twentieth centuries, to user-centric innovation in the twenty-first century requires the overhaul of a company's hierarchical structures, its resource-allocation policies, organisational processes and work routines. Manufacture-centric innovation supports a culture of **incremental innovations** while discouraging the learning behaviours and risk taking required in radical new product development. In contrast, **customer-centric enterprises** have the ability to drive incremental innovations from within and radical **new product development** from outside with lead users and others.

The principle behind the democratisation of innovation is gaining acceptance in information technology products such as software as well as in services and physical products (*ibid.*).

See also: **user innovations**

Further reading
Von Hippel, E. (2005) *Democratizing Innovation*, Cambridge, MA, MIT Press.

Design

The meaning of design has evolved from a **product development** focus to being broad and vague, encompassing all domains of communicative and social action (Meier, 2001). There is no commonly agreed definition for *product design* due to the breadth of the topic. Luchs and Swan (2011) proposed a product design definition as 'the set of properties of an artifact, consisting of the discrete properties of the form (i.e. the aesthetics of the tangible good and/or service) and the function (i.e. its capabilities) together with the holistic properties of the integrated form and function.' They defined product design process as 'the set of strategic and tactical activities, from idea generation to commercialization, used to create a product design' (Luchs and Swan, 2011).

D

Design covers many areas of human activity. There are two primary schools of design thought: positivistic and constructivistic (Cross, 2007). In the positivistic approach, design problems and processes are organised into categories or category systems. In the constructivistic approach, design problems are considered 'wicked problems'; that is, they are too complex and diffuse to be analysed by categories or a finite number of dimensions (Rittel, 1972). In the constructivistic approach, the design process consists of illumination of the problem and solution spaces, and iterative alignment of those spaces (Lindberg *et al.*, 2010). In illumination, designers explore the problem space with their intuitive understanding. They make use of user cases or **scenarios** to aid in their understanding of the problem. They also explore alternative ideas in multiple dimensions. In iterative alignment, designers transform ideas into **prototypes**, which facilitate communication within the design team as well as with users and other key stakeholders in the value chain.

A **robust design** blends form and function; quality and style; and art and engineering. The range of leading design applications includes goods, services, architecture, graphic arts and offices. Many organisational factors such as governance structures, resource allocations and decision authorities influence *product design*. The contribution of design to the success of **new product development** is underappreciated. Also there is benign neglect of product design in marketing inquiry (Luchs and Swan, 2011). The lack of research on product design practice is addressed in the special issue on product design research and practice in the May 2011 edition of the *Journal of Product Innovation Management* (Volume 28, Number 3).

See also: **prototype as design, scenario-based designs**

Further reading
Cross, N. (2007) 'From a Design Science to a Design Discipline: Understanding Designerly Ways of Knowing and Thinking', in R. Michel (ed.), *Design Research Now – Essays and Selected Projects*, Basel, Birkhäuser, pp. 41–54.
Huang, George Q. (1996) 'GIM: GRAI Integrated Methodology for Product Development' in George Q. Huang (ed.), *Design fox X: Concurrent Engineering Imperatives*, London: Chapman & Hall, pp. 153–72.
Lindberg, Tilmann, Noweski, Christine and Meinel, Christoph (2010) 'Evolving Discourses on Design Thinking: How Design Cognition Inspires Meta-disciplinary Creative Collaboration', *Technoetic Arts*, 8(1), 31–7.
Luchs, Michael and K. Scott Swan (2011) 'Perspective: The Emergence of Product design as a Field of Marketing Inquiry', *Journal of Product Innovation Management*, 28 (3): 327–45
Meier, C. (2001) *Designtheorie – Beiträge zu einer Disziplin*, Frankfurt a.M., Anabas.
Rittel, H. W. J. (1972) 'On the Planning Crisis – Systems Analysis of the First and Second Generation', *Bedrifsøkonomen*, 8, 390–96.

Design anthropology

Design anthropology provides a comprehensive understanding of users and contexts of use. It selectively applies anthropological theory to challenge existing conceptualisations of products, services, technology, users and uses (Nafus and Anderson, 2006).

Large corporations employing anthropologists in design include Intel, Whirlpool and IBM. The need for a deeper understanding of the use of digital technology (e.g. laptop computers) led Intel to conduct **ethnography research** in broadband-connected American homes (Melhuish, 2006). At Whirlpool, designers utilise contextual usage information in the **ideation generation and enrichment** and **concept definition** steps in the **new product development process.**

How can design anthropologists understand home settings and interactions, especially when their personal observations may be limited to a few hours or days? This is a valid question. They must take precautions to make sure that they do not reach too many conclusions from minimal experiential observations or memories. When done correctly, anthropology draws novel conceptual boundaries that provide a glimpse into how other cultures and people might see the world and brings unarticulated conceptual distinctions into focus. The difficulty

D

often encountered with the use of anthropology in innovation relates to its lack of pay-off in innovation projects.

Further reading
Melhuish, Clare (2006) 'Interior Insights: Design, Ethnography and the Home, Royal College of Art, London, 24–25 November 2005', *Anthropology Today*, 22(1). 24–5.
Nafus, D. and Anderson, K. (2006) 'The Real Problem: Rhethorics of Knowing in Corporate Ethnographic Research,' Ethnographic Praxis in Industry Conference, American Anthropological Association.

Design–build–test cycle

The design–build–test cycle is a repetitive problem-solving cycle in the **new product development process.** It generates insights into the connection between specific design parameters and customer attributes (Wheelwright and Clark, 1992). In the *design phase,* a product development team frames the problem, establishes goals for the problem-solving process and generates alternatives to explore the relationships between design parameters and specific customer attributes. In *the build phase,* the development team builds working models of alternative designs and tests them. In the early build phase, the team may build models using **computer-aided design (CAD)** with plastic, clay or soft materials. In the later build stage, the development team builds prototypes that are close to the actual product. *The test phase* consists of testing working models, prototypes or computer-generated images. In some cases, testing includes full-scale system evaluation.

Rapid modelling and prototyping allow a more flexible and responsive new product development process (Ulrich and Eppinger, 2004). For instance in the video gaming industry, game designer Mark Cerny champions the building and testing of at least four prototypes of a game in the pre-production process of development. According to Cerny, this practice allows artists, audio engineers and programmers to work quickly without worrying about debugging the system (Adams, 2004).

See also: **rapid prototyping**

Further reading
Adams, Ernest (2004) *Fundamental of Game Design*, Berkeley, CA, New Riders.
Ulrich, Karl T. and Eppinger, Steven D. (2004) *Product Design and Development*, New York, McGraw-Hill/Irwin.
Wheelwright, Steven C. and Clark, Kim B. (1992) *Revolutionizing Product Development*, New York, Free Press.

D

Design-driven innovation

There are two schools of thought about how to create successful new products. One focuses on fulfilling articulated customer needs with new products and services. Clayton Christensen asks: 'What problems are we solving for customers?' Eric Von Hippel advocates the **democratisation of innovation** and champions the use of **lead users** in new product development. Concepts such as **co-creation, user-centric innovation** and **user-centred design** have been thoroughly vetted by academic researchers and industry practitioners.

The second school of thought believes that an exclusive focus on users and **customer needs** leads to **incremental innovations**. It also believes that a focus on technology-driven **radical innovations** misses the opportunity to develop radically new markets. Gerald Zaltman and Lindsay Zaltman (2008) identified deep metaphors as a way to communicate unarticulated consumer needs.

Roberto Verganti also belongs to the second school of thought. However, unlike the Zaltmans, he proposes the creation of new markets, with new meanings (Verganti, 2009). New products that create new meanings build strong brands and sustainable growth. Verganti calls his approach '**design-driven innovation**'.

Design-driven innovation is based on the idea that each product holds a particular meaning for consumers (Claudio *et al.*, 2010). According to Verganti, companies can create compelling, emotional experiences for their customers by understanding the emotions that drive their purchasing decisions. However, customers cannot articulate their emotions, so who can? Verganti's response is 'interpreters', with whom he recommends companies collaborate. *External interpreters* include designers, firms in other industries, suppliers, schools, artists and the media (Verganti, 2008). Companies can leverage their relationships with interpreters by listening to them to gain knowledge, develop unique proposals based on that knowledge, and use the power of interpreters to influence customers (Verganti, 2009). Italian manufacturers Alessi and Artemide leverage external interpreters well.

Further reading
Claudio, Dell'Era, Marchesi, Alessio and Verganti, Roberto (2010) 'Mastering Technologies in Design-Driven Innovation', *Research Technology Management*, 53(2), 12–23.
Verganti, Roberto (2008) 'Design, Meanings, and Radical Innovation: A Metamodel and a Research Agenda', *Journal of Product Innovation Management*, 25, 436–56.
Verganti, Roberto (2009) *Design Driven Innovation: Changing the Rules of Competition by Radically Innovating What Things Mean*, Boston, MA, Harvard Business School Press.
Zaltman, Gerald and Zaltman, Lindsay (2008) *Marketing Metaphoria: What Seven Deep Metaphors Reveal about the Minds of Consumers*, Boston, MA, Harvard Business School Press.

Design for all (inclusive design)

Design for all, also called the *inclusive design* approach, refers to the designing of products, services or environments that meet the needs of the broadest range of consumers (that is, products that work for people irrespective of age or ability). It is a novel approach to using design as a tool for delivering on equality and inclusivity.

D

In design for all, the issue is not necessarily providing increased functionality but rather ensuring that the functionality is accessible to anyone who wishes to use the product. An example would be a company that creates an inclusive design that accommodates the needs of senior citizens and disabled people, as well as the young and more able. The Apple iPod and Ford Focus are two good examples of products made with inclusive designs. Both have simplicity and useful, easy-to-use features (Hosking and Sinclair, 2006).

Design for all enables seniors and disabled people to live more normal lives without the stigma and social cost attached to products made specifically for them.

Youth dominates Western culture. Marketers actively promote youthful attributes while implying a loss of beauty, energy or health with advancing age. This cultural bias can again be seen in the Apple iPod and Ford Focus, which despite their inclusive design, shied away from making any claims or references to inclusivity.

Further reading
Hosking, I. and Sinclair, K. (2006) 'Whose Design Is It Anyway? (Product Inclusive Design)', *Engineering Management*, 16(2), 10–13.
Marshall, R., Case, K., Porter, J. M., Sims, R. and Gyi, D. E. (2004) 'Using HADRIAN for Eliciting Virtual User Feedback in Design for All', *Proceedings of the Institution of Mechanical Engineers – Part B – Engineering Manufacture*, 218(9), 1203–10.

Design for assembly (DFA)

Design for assembly (DFA) simplifies the product by reducing the number of separable parts. Three common DFA methods are *Hitachi, Lucas* and *Boothroyd-Dewhurst* (Huang, 1996). The Boothroyd-Dewhurst DFA methodology provides three criteria for adding a part to the product during assembly: Does the part move relative to all other parts? Must the part be of a different material than all other parts? Must the part be separate from all other parts already assembled? If the answer is yes to any of these questions, then the part must be a separate item. This DFA assessment procedure leads to ideas about how the product may be simplified.

The Hitachi DFA method is enhanced by the additional consideration given to all parts in identifying candidates for elimination. The evaluation is based on completing the evaluation sheet in the same order as the envisaged assembly sequence. The evaluation sheet consists of nine main columns which, when completed for each part, will lead to the total assembly analysis. There are five scoring headings: assembly time, assemblability, assembly cost ratio, part count design efficiency and simplicity factor.

Lucas DFA evaluation encompasses a functional analysis, a handling (of feeding) analysis and a fitting analysis, and the resulting penalty factors are entered into an evaluation sheet called an assembly flowchart. The penalty factors are manipulated into three 'assemblability' scores, which are compared to values established for previous designs (such as design efficiency, feeding/handling ratio and fitting ratio).

DFA methods are useful for assembly at a desktop. Typical products benefiting from DFA analysis are tape recorders, video recorders or many car subassemblies, such as alternators or water pumps.

Further reading
Huang, George Q. (1996) 'Design for Manufacture and Assembly' in George Q. Huang (ed.), *Design fox X: Concurrent Engineering Imperatives*, London, Chapman & Hall, pp. 41–71.

DFX (Design for excellence or Design for X)

Design for excellence (DFX) has two definitions based on an expanded view of the product development process. The first is design for X, where X may be any phase of a product's life cycle. The second definition is design for excellence, which places the focus on quality and customer satisfaction (Maltzman *et al.*, 2005).

A generic definition of design for X is making decisions in product development related to products, processes and plants (Huang, 1996). It rationalises the design of products, associated processes and systems. It starts with product development goals and relates product design decisions and the **new product development process** to associated process activities. Huang and Mak (1998) consider DFX and **business process reengineering** as the two key business processes for transforming sequential engineering to **concurrent engineering**.

See also: **design for X (DFX) tools**

Further reading
Huang, George Q. (1996) 'Design for Manufacture and Assembly' in George Q. Huang (ed.), *Design fox X: Concurrent Engineering Imperatives*, London, Chapman & Hall, pp. 1–11.
Huang, G. Q. and Mak, K. L. (1998) 'Re-Engineering the Product Development Process with "Design for X"'. *Proceedings of the Institution of Mechanical Engineers – Part B – Engineering Manufacture*, 212(4), 259–68.
Maltzman, Richard, Rembis, Kevin M., Kevin M. Donisi, Kevin M., Farley, Matthew, Sanchez, Roberto C. and Ho, Augustine Y. (2005) 'Design for Networks The Ultimate Design for X', *Bell Labs Technical Journal*, 9(4), 5–23.

Design for manufacture (DFM)

Design for manufacture (DFM) is one of the main approaches to implementing **concurrent engineering (CE)** during the **concept** and **design** phases of the **new product development process**. DFM considers production process requirements and the reasonableness of the assembly and testing process, while taking into account after-sales service requirements. In the electronic products industry, 70 per cent of the costs are decided at the design stage. The cost impacts of design defects can be a hundred times greater in the back end of manufacturing.

DFM application is common in the electronics industry. The DFM of a printed circuit board (PCB) refers to the board-level circuit module design for manufacturing. Its focus is the PCB processing capacity and efficiency; that is, requirements for size and shape, operating capabilities of the manufacturing equipment with respect to maximum and minimum processing sizes, and PCB performance requirements. The PCB working panel utilisation rate can be improved from 58 per cent to 83 per cent through *DFM optimisation design* (Wang, 2009).

However, DFM is not successful in other manufacturing industries due to 'knowledge silos' and competing priorities (such as the design for function versus design for manufacture). Brainboxes, a designer and manufacturer of professional serial devices, overcame organisational hurdles with a highly formalised project management system that enforces consideration of design-for-manufacture issues throughout the **product life cycle** (Wheatley, 2010).

Further reading
Wang, Wenli (2009) 'Excellence Design for Manufacturability of Electric Product' in Hu Shuhua and Hamsa Thota (eds), *Proceedings of the 4th International Conference on Product Innovation Management*, Wuhan, China, Hubei People's Press, Part 1, pp. 1–6.

D

Wheatley, M. (2010) 'Fit for Purpose (Design-for-Manufacture)', *Engineering & Technology*, 5(5), 54–6.

Design for manufacturing and assembly (DFMA)

Design for manufacturing and assembly (DFMA) encourages dialogue during the early stages of design between designers and manufacturing engineers, sales and marketing, quality assurance, customers, procurement, suppliers and any other stakeholder with a role in determining the final product costs (Boothroyd *et al.*, 2002). Parts count reduction is one of the basic principles of DFMA. For instance, an appliance manufacturing company in Hungary reduced the number of parts in its coffee percolator design by 42 per cent (*Design News*, 1994).

DFMA deployment is a rigorous process. Its implementation leads to simplification and standardisation of design and manufacturing processes. It also facilitates the integration of function and form and optimisation of the work flow. For example, integrating the sheet metal parts assembly into an integral machined part in an aircraft door panel assembly led to a decrease in **product development** cost and time (Selvaraj *et al.*, 2009). Companies that successfully implement DFMA report material and labour reductions of 20 per cent to 50 per cent (Shipulski, 2009). Cross-functional teamwork is critical to achieving the benefits of DFMA.

See also: **design for assembly (DFA), design for manufacture (DFM)**

Further reading
Boothroyd, Geoffrey, Dewhurst, Peter and Knight, Winston (2002) *Product Design for Manufacture and Assembly*, New York, Marcel Deckker.
'DFMA cuts appliance parts 42%', *Design News*, 26 September 1994, 50(18), 51.
Selvaraj, P., Radhakrishnan, P. and Adithan, M. (2009) 'An Integrated Approach to Design for Manufacturing and Assembly based on Reduction of Product Development Time and Cost', *International Journal of Advanced Manufacturing Technology*, 42(1/2), 13–29.
Shipulski, Mike (2009) 'Resurrecting Manufacturing', *Industrial Engineer*, 41(7), 24–8.

Design for reuse

New product companies reduce cost and cycle time with design for reuse. Design for reuse began in the 1940s with Cambridge University's reuse of subroutines in computer software code. The steps for software reuse are finding, understanding, adapting and integrating (Retkowsky, 1998). Common methods for design reuse are **computer-aided design (CAD)** models, *FBS (function, behaviour, structure) models* and *semantics-based representations*. These methods are commonly used in the design phases, such as the conceptual design stage (Qiu *et al.*, 2004) and the detailed design stage (Jensen, 1998).

There are two types of reuse: product reuse and process reuse. In *product reuse*, products or product parts are reused with or without modification in other products. *Process reuse* is sometimes called the '*generator approach*' because one process used to develop a successful class of products is reused to develop other, similar products. Design knowledge reuse (that is, reuse of knowledge from previous completed or dormant projects in a current or active project) is common practice in the architecture, engineering and construction (AEC) industry (Demian and Fruchter, 2006).

Design reuse is also common practice among manufacturers of printed circuit boards. One manufacturer made printed circuit boards with almost 90 per cent design reuse. While the focus on cost, efficiency and quality drives design reuse, its negative impact on the novelty of new products is not given adequate attention. John E. Ettlie and Matthew Kubarek (2008) reported that in small firms, the tipping point was 43 per cent of reuse, beyond which the novelty of new products suffered. For manufacturing companies, the tipping point was lower and novelty suffered after 33 per cent design reuse. In spite of its potential negative impact on product novelty, design for reuse is a valuable tool to increase design efficiency in **mass customization** (Jin *et al.*, 2008).

Further reading
Demian, Peter and Fruchter, Renate (2006) 'An Ethnographic Study of Design Knowledge Reuse in the Architecture, Engineering, and Construction Industry', *Research in Engineering Design*,16(4), 184–95.
Ettlie, John E. and Kubarek, Matthew (2008) 'Design Reuse in Manufacturing and Services', *Journal of Product Innovation Management*, 25(5), 457–72.
Jensen, T. (1998) 'A Taxonomy for Design Reuse Systems: Proposing a System for Formalized Online Knowledge Capturing', *Proceedings of the Engineering Design Conference (EDC'98)*, Brunel University, Uxbridge, June, pp 483–94.
Jin, Bo, Teng, Hong-Fei, Wang, Yi-Shou and Qu, Fu-Zheng (2008) 'Product Design Reuse with Parts Libraries and an Engineering Semantic Web for Small and Medium-Sized Manufacturing Enterprises', *International Journal of Advanced Manufacturing Technology*, 38(11/12), 1075–84.
Qiu, L. R., Liu, H. and Gao, L. P. (2004) 'A Multi-agent System Supporting Creativity in Conceptual Design', *Proceedings of the 8th International Conference on Computer Supported Cooperative Work in Design (CSCWD 2004)*, Xiamen, China, May, pp 362–70.
Retkowsky, F. (1998) 'Software Reuse from an External Memory: The Cognitive Issues of Support Tools', *Proceedings of the Tenth Workshop on Psychology of Programming Interest Group (PPIG)*, Open University, UK.

Design for X (DFX) tools

Manufacturing organisations widely use 'design for X' (DFX) tools to delineate customer needs early in the **new product development (NPD) process** (Huang, 1996). Some DFX tools used in manufacturing include **DFA (design for assembly), DFM (design for manufacture), DFMA (design for manufacture and assembly) and DFX (Design for X or design for excellence)**. Comprehensive DFX programmes transform the product development process from a problem-prone sequential engineering environment to a problem-free **concurrent engineering** environment (Huang and Mak, 1998).

Customer value chain analysis (CVCA) is an original DFX tool used in the product-definition phase of the NPD process. CVCA enables design teams to understand key customers and their relationship to each other and the product. This allows them to define the product relative to their company's market and product objectives as well as **customer needs** (Donaldson et al., 2006). Knowledge gained about customers and their needs becomes an input to later DFX methodological tools, such as **quality function deployment (QFD)** and **failure modes and effects analysis (FMEA).** Rose and Stevels (2000) applied CVCA to *design for environment* to develop *environmental value chain analysis (EVCA).*

D

See also: **voice of the customer**

Further reading

Donaldson, Krista, Ishii, Kosuke and Sheppard, Sheri (2006) 'Customer Value Chain Analysis', *Research in Engineering Design*, 16(4), 174–83.

Huang, G. (1996) *Design for X: Concurrent Engineering Imperatives,* London, Chapman & Hall.

Huang, G. Q. and Mak, K. L. (1998) 'The DFX Shell: A Generic Applying 'Design for X' (DFX) Framework for Tools', *International Journal of Computer Integrated Manufacturing,* 11(6), 475–84.

Maltzman, R., Rembis, K. M., Donisi, M., Farley, M., Sanchez, R. C. and Ho, A. Y. (2005) 'Design for Networks – The Ultimate Design for X', *Bell Labs Technical Journal*, 9(4), 5–23,

Rose, C. M. and Stevels, A. (2000) 'Lessons Learned from Applying Environmental Value Chain Analysis to Product Take-back', 7th CIRP – Life Cycle Engineering Conference, Tokyo.

Design thinking

Design thinking is the study of the cognitive processes manifested in design action (Cross *et al.*, 1992). In design thinking, designers consider how people interact with products, with other people and with their surroundings (Brown, 2009). Martin (2009) described four components of design thinking as a 'complete understanding of the user, creative resolution of staff tensions, fostering a collaborative business culture and continuously modifying and enhancing ideas and solutions'. Design firm IDEO developed and commercialised its design thinking approach with Amtrak, Intel and Marriott (Brown, 2009) with notable success by complementing monodisciplinary thinking with interdisciplinary creative work (Lindberg *et al.*, 2010).

The process of design thinking evolved in four phases: the 'intuitive and artistic' designer, the rationalist and logical designer, the designer with bounded rationality and the reflective practitioner (Bousbaci, 2008). Design thinking encourages team-based learning, in which multiprofessional teams develop a shared understanding about the problem and its possible solutions. The 'problem space' and 'generative processes' are concepts from Newell and Simon's *problem-solving model* (Newell and Simons, 1992). The problem space is creating a representation for the problem. Thomas Edison envisioned how people would like to use what he made and engineered his inventions towards that insight. He was a pioneering practitioner of design thinking (Brown, 2008).

Further reading

Bousbaci, Rabah (2008) '"Models of Man" in Design Thinking: The "Bounded Rationality"', *Design Issues*, 24(4), 38–52.

Brown, Tim (2008) 'Design Thinking', *Harvard Business Review,* 86(6), 84.

Brown, Tim (2009) *Change by Design: How Design Thinking Transforms Organizations and Inspires Innovation,* New York, HarperBusiness.

Cross, Nigel, Dorst, Kees and Roozenburg, Norbert N. (1992) 'Preface', in Nigel Cross, Kees Dorst and Norbert N. Roozenburg (eds), *Research in Design Thinking*, Delft, Delft University Press.

Lindberg, Tilmann, Noweski, Christine and Meinel, Christoph (2010) 'Evolving Discourses on Design Thinking: How Design Cognition Inspires Meta-disciplinary Creative Collaboration', *Technoetic Arts*, 8(1), 31–7.

Martin, Roger (2009) *The Design of Business: Why Design Thinking Is the Next Competitive Advantage*, Boston, MA, Harvard Business School Publishing.

Newell, Allan and Simon, Herbert A. (1972) *Human Problem Solving*, Englewood Cliffs, NJ, Prentice-Hall.

Development

See **new product development process**

Development funnel

Product development is a funnel-shaped business process. The funnel narrows as an idea progresses through various stages of the **new product development process**. The product development funnel may be viewed as the front end of the new product development process. A generic NPD process has six stages: **idea generation and enrichment**, idea screen, business analysis, development, test and validation and **commercialisation.** The development funnel consists of the first three stages.

Idea generation is critical to new product development. Organisations generate a large number of ideas with help from experts, consumers and their own employees. However, they have limited resources and can use only a limited number of these ideas, so they weed out impractical or weak ideas. However, selecting strong ideas that reflect consumers' needs and preferences is a challenge. One way to screen ideas is to ask experts to judge them (Urban and Hauser, 1993). Another approach is to test ideas within the organisation for the size of the opportunity, level of strategic fit and project feasibility. Technical feasibility is the most frequently used criterion for idea-screening purposes, but intuition also plays a key role (Hart *et al.*, 2003).

Some ideas fall outside the scope of the innovation team, but they may be significant business opportunities. Such ideas are *opportunity innovations* (Russell and Tippett, 2008). In strategic fit screening, ideas are evaluated against the **product innovation charter (PIC).** In business analysis, companies use sales criteria such as sales in units and margins (profit levels) and market potential to ensure that forecasted sales and profit projections meet financial objectives (Hart *et al.*, 2003).

See also: **business case, front end of innovation, idea management**

Further reading
Hart, Susan, Jan Hultink, Erik, Tzokas, Nikolaos and Commandeur, Harry R. (2003) 'Industrial Companies' Evaluation Criteria in New Product Development Gates', *Journal of Product Innovation Management*, 20(1), 22–36.
Russell, Richard K. and Tippett, Donald D. (2008) 'Critical Success Factors for the Fuzzy Front End of Innovation in the Medical Device Industry', *Engineering Management Journal*, 20(3), 36–43.
Urban, G. L. and Hauser, J. R. (1993) *Design and Marketing of New Products*, Upper Saddle River, NJ, Prentice Hall.

Differentiation

Differentiation refers to the relative comparison that distinguishes one product from another. Chamberlin (1965) defined product differentiation as 'distinguishing the goods or services of one seller from those of another on any basis that is pertinent to the buyer and leads to a preference'. There are two types of product differentiation: horizontal and vertical. *Horizontal differentiation* occurs within the same quality product group; *vertical differentiation* takes place in terms of different quality levels (Conrad, 2006).

Product differentiation and pricing define the relative positioning of competitors in a market (McGrath, 2000). Differentiation also segments the market. In horizontal differentiation, a marketer maximises profit by modifying a standard product so that it meets the needs of target customers or satisfies the preferences of the **target market**. In vertical differentiation, a marketer chooses a trade-off 'in higher prices for better quality or a lower price for the lower quality' (Conrad, 2006).

Horizontal product differentiation is common in markets using consumer testing (such as banks and investment firms). Based on their consumer test results, banks may offer different menus of accounts or focus on different types of lending businesses. They may attract a different mix of retail or wholesale funds or they may expand into non-bank products such as insurance or mutual funds (Allen and Gale, 2000). Investment firms differ in their equity and advisory services. Hoppe and Lehmann-Grube (2008) found that banks and investment firms charge higher prices for highly differentiated products because they learn that captive customers will pay premium prices for them.

Product differentiation is becoming a common practice in the pharmaceutical industry to improve profitability and replenish product pipelines (Dubey and Dubey, 2009).

See also: **competitive strategy, customer targeting**

Further reading
Allen, F. and Gale, D. (2000) *Comparing Financial Systems*, Cambridge, MA, MIT Press.
Chamberlin, Edward H. (1965) *The Theory of Monopolistic Competition*, Cambridge, MA, Harvard University Press.
Conrad, Klaus (2006) 'Price Competition and Product Differentiation When Goods Have Network Effects', *German Economic Review*, 7(3), 339–61.
Dubey, Rajesh and Dubey, Jayashree (2009) 'Pharmaceutical Product Differentiation: A Strategy for Strengthening Product Pipeline and Life Cycle Management', *Journal of Medical Marketing*, 9(2), 104–18.
Hoppe, Heidrun C. and Lehmann-Grube, Ulrich (2008) 'Price Competition in Markets with Customer Testing: The Captive Customer Effect', *Economic Theory*, 35(3), 497–521.
McGrath, Michael E. (2000) *Product Strategy for High Technology Companies*, 2nd edn, New York, McGraw-Hill.

D

Differentiation strategy

Differentiation strategy positions a product in the market. The purpose is to create **customer value** and achieve a lasting performance advantage (Hill, 1998). Firms differentiate themselves from competition with innovative products and services, leading-edge technology, superior quality, differentiated brand positioning and excellence in service (Frambach *et al.*, 2003). Business strategists use Michael Porter's *five forces framework* to develop differentiated products and services aligned with business strategy (Porter, 2008).

McGrath (2000) identified 12 generic differentiation strategies in the high-technology industry:

1. Differentiation using unique features, e.g. Sharp ViewCam with a 4 ft colour LCD panel.
2. Differentiation by measurable customer benefits, e.g. Genentech TPA blood clot remover.
3. Differentiation through ease of use, e.g. Apple McIntosh user-friendly interface.
4. Differentiation by improved productivity, e.g. Lotus 1-2-3 Notes.
5. Differentiation by protecting the customer's investment, e.g. Digital Equipment's VAX computers.
6. Differentiation through lower cost of product failures, e.g. Stratus computers.
7. Differentiation with higher-performance products, e.g. Hewlett-Packard (HP) OmniBook computer.
8. Differentiation by unique, fundamental capabilities, e.g. Polaroid instant camera.
9. Differentiation through design, e.g. Apple iMac.
10. Differentiation by total cost of ownership, e.g. Microsoft Windows 2000.
11. Differentiation based on standards, which is more avoiding a negative consequence of a non-standard vector.
12. Differentiation based on convenience, which is easily copied by competitors.

In the consumer packaged goods industry, two product-positioning strategies (specialised and all-in-one) are commonly used (Chernev, 2007). Chernev describes specialised positioning as a narrower positioning in which a single attribute describes a product. For example, detergent Era is described as a detergent with 'powerful stain removal', Cheer helps protect against fading and Gain offers 'great cleaning power'. All-in-one positioning is a broader positioning in which a product is described by a combination of attributes. For example, Tide combines all of the above features (*ibid.*).

See also: **product development strategies**

Further reading
Chernev, Alexander (2007) 'Jack of All Trades or Master of One? Product Differentiation and Compensatory Reasoning in Consumer Choice', *Journal of Consumer Research*, 33(4), 430–44.
Frambach, R. T., Prabhu, J., Verhallen T. M. M. (2003) 'The Influence of Business Strategy on New Product Activity: The Role of Market Orientation', *International Journal of Research in Marketing*, 20, 377–39.
Hill, Charles W. L. (1988) 'Differentiation Versus Low Cost or Differentiation and Low Cost: A Contingency Framework', *Academy of Management Review*, 13(3), 401–12.
McGrath, Michael E. (2000) *Product Strategy for High Technology Companies*, 2nd edn, New York, McGraw-Hill.
Porter, Michael E. (2008) 'The Five Competitive Forces that Shape Strategy', *Harvard Business Review*, 86(1), 78–93.

D

Diffusion of innovation

Diffusion of innovation is an interdisciplinary field that studies 'the spread of new ideas, opinions, or products throughout a society' (Valente, 1995). Rogers

(2003) defined it as 'the process by which an innovation is communicated through certain channels over time among the members of a social system'. In spite of a large body of research, however, diffusion models of innovation fail to explain innovation failures. The *Bass model* can predict the future adoption rate of an innovation only after its launch. Firms thus use less formal methods to test innovation (such as **focus groups**) and to obtain subjective perceptions about the probability of adoption (Thiriot and Kant, 2008).

Innovations generally diffuse through a society following a five-stage process. As defined by Rogers (*ibid.*) these are:

1. Knowledge or awareness stage.
2. Persuasion or interest stage.
3. Decision or evaluation stage.
4. Implementation or trial stage.
5. Confirmation or adoption stage.

In the awareness stage, individuals get the first exposure to an innovation. In the interest stage, they become interested in it; and they decide whether to adopt or reject it in the evaluation stage. They determine the usefulness of the innovation in the trial stage and finalise their decision to continue using the innovation in the adoption stage.

Diffusion of an innovation in a social system is measured in terms of its rate of adoption. It is represented in the form of an S-shaped adoption curve. The five categories of adopters as defined by Rogers (2003) are innovators, early adopters, early majority, late majority and laggards.

Technology adoption occurs after a firm becomes aware of an innovation through contact with the technology's originators or prior adopters. However, *software process innovations* require certain learning-related obstacles to be overcome as a pre-condition to adoption (Attewell, 1992; Fichman & Kemerer, 1997). Supply-side agents, such as consultants and technology solution providers, play an important role in innovation adoption if adoption requires seamless integration with legacy systems, processes and structures (Ettlie and Reza, 1992). Supply-side agents are relevant in customer-interfacing industries, such as internet banking (Curran and Meuter, 2005).

See also: **product life cycle, technology life cycle**

Further reading
Attewell, P. (1992) 'Technology Diffusion and Organizational Learning: The Case of Business Computing', *Organization Science*, 3, 1–19.
Curran, J. M. and Meuter, M. L. (2005) 'Self-Service Technology Adoption: Comparing Three Technologies', *Journal of Services Marketing*, 19(2), 103–13.
Ettlie, J. E. and Reza, E. M. (1992) 'Organizational Integration and Process Innovation', *Academy of Management Journal*, 35, 795–827.
Fichman, R. G. & Kemerer, C. F. (1997) 'The Assimilation of Software Process Innovations: An Organizational Learning Perspective', *Management Science*, 43, 1345–63.
Rogers, E. M. (2003) *Diffusion of Innovations*, 5th edn, New York, Free Press.
Thiriot, Samuel and Kant, Jean-Daniel (2008) 'Using Associative Networks to Represent Adopters' Beliefs in a Multiagent Model of Innovation Diffusion', *Advances in Complex Systems*, 11(2), 261–72.
Valente, T. W. (1995) *Network Models of the Diffusion of Information*, Cresskill, NJ, Hampton.

Discontinuities

Discontinuities refer to interruptions in the life cycle of a product, process or technology brought about by shifts in technological, market, political and other domains. In periods of discontinuities, incumbent companies must adapt their practices significantly or learn new ones. If incumbents fail to learn proactively, they fail (Phillips *et al.*, 2006). This failure is evident in Christensen's work on disk drives and other technologies. He found that market leaders failed when new markets emerged that required combinations of technologies (Christensen, 1997). Utterback (1996) found that only a quarter of existing competitors either initiated **radical innovation** or adapted quickly to remain among the market leaders once a disruptive, radical innovation occurred. In disruptive environments, market forces move rapidly and new leaders emerge at the expense of incumbents. Andy Grove, CEO of Intel, referred to such disruptions as 'major inflexion points' in the evolution of an industry.

Joseph Schumpeter (1961[1911]) stressed the importance of discontinuities in the innovation process and postulated that product and process life cycles are often interrupted by a new cycle of creativity and discontinuous change. The academic literature views discontinuities as discrete zero/one binary events in the time path of industry evolution (Campo-Rembado and Taylor, 2007). However, innovators are continually attempting to make breakthroughs throughout the life cycle of an industry. The role of **incremental innovations** and the availability of capital may be underestimated in the existing literature (*ibid.*).

See also: **creative destruction, technology discontinuities**

Further reading
Campo-Rembado, Miguela A. and Taylor, Alva H. (2007) 'Relative Technological Discontinuities: Insights from the Identification of Innovation Regime Shifts', *Academy of Management Proceedings*, 1–6.
Christensen, C. (1997) *The Innovator's Dilemma*, Cambridge, MA, Harvard Business School Press.
Phillips, Wendy, Noke, Hannah, Bessant, John and Lamming, Richard (2006) 'Beyond the Steady State: Managing Discontinuous Product and Process Innovation', *International Journal of Innovation Management*, 10(2), 175–96.
Schumpeter, Joseph A. (1961[1911]) *The Theory of Economic Development*, New York, Oxford University Press.
Utterback, James (1996) *Mastering the Dynamics of Innovation*, Boston, MA, Harvard Business School Press.

D

Discontinuous change

See **discontinuities**

Discovery-driven growth (DDG)

Rita McGrath at Columbia University pioneered the concept of discovery-driven growth (DDG). Its goal is to deliver maximum results with minimal risk. DDG uses a road map for creating flexible enterprise architectures. Companies such as Amazon, DuPont and Hewlett-Packard use this approach effectively. McGrath and

MacMillan (2009) report DDG as a step-by-step process involving five different activities:

1. Defining what success looks like.
2. Benchmarking to use the right metrics.
3. Determining operational capabilities needed to complete an initiative.
4. Documenting key assumptions.
5. Planning around key checkpoints.

The cornerstone of DDG is **discovery-driven planning**. This enables organisations to build plans to pursue the future they envision, learn where their real future lies and test assumptions about that future at low cost.

Further reading
McGrath, Rita Gunther and MacMillan, Ian C. (2009) *Discovery Driven Growth,* Boston, MA, Harvard Business School Press.

Discovery-driven planning

Discovery-driven planning is a practical, strategic planning tool for **new ventures**. It systematically converts assumptions at the start of new ventures into knowledge as the venture unfolds. The real potential of the venture becomes clear over time. McGrath and MacMillan (1995) define discovery-driven planning as a disciplined process that makes use of four related evolving documents: a reverse income statement, pro forma operation specs, a key assumption checklist and a milestone planning chart.

While discovery-driven planning addresses uncertainties at low cost, it is a necessary but not sufficient condition to guide emerging and high-uncertainty ventures (Thompson and MacMillan, 2010). Thompson and MacMillan (2010) propose a five-step process that includes scoping the venture, mobilising supporters, discovery-driven planning, planning for disengagement and anticipation of the unanticipated to develop new **business models** for highly uncertain ventures.

Chief technology officers (CTOs) can use discovery-driven planning to select among and **real options analysis** to evaluate and invest in uncertain projects (MacMillan *et al.*, 2006).

Further reading
MacMillan, Ian C., Van Putten, Alexander B., McGrath, Rita Gunther and Thompson, James D. (2006) 'Using Real Options Discipline for Highly Uncertain Technology Investments', *Research Technology Management,* 49(1), 29–37.
McGrath, Rita Gunther and MacMillan, Ian C. (1995) 'Discovery-Driven Planning', *Harvard Business Review,* 73(4), 44–54.
Thompson, James D. and MacMillan, Ian C. (2010) 'Making Social Ventures Work', *Harvard Business Review,* 88(9), 66–73.

Disruptive innovation

Clayton Christensen coined the term disruptive innovation for a process by which a product or service enters a market at the bottom and moves 'up market', eventually displacing established competitors. Christensen *et al.* (2004) define a disruptive innovation as 'innovation that cannot be used by customers in mainstream markets' because it is inferior to existing products. Disruptive innovations serve underserved market segments. They create new growth opportunities by changing the basis for competition. Some disruptive innovations require a different set of trade-offs, offering inferior performance along one dimension in exchange for additional benefits related to simplicity, convenience and low prices (Anthony *et al.*, 2008). They do not entail frontier technologies but make use of low-cost and mature technologies. Canon introduced slower but more affordable tabletop photocopiers to disrupt Xerox. Toyota introduced less sophisticated but cheaper car models and disrupted General Motors; and in turn, Korean, Chinese and Indian automobile manufacturers disrupted Toyota (Hwang and Christensen, 2008). In the short run, low-end disruptions do not distort markets for incumbent players. However, if those incumbents fail to respond to a disruptive innovation in the short run, this may be disastrous in the long run.

More recently Christensen has recognised that there are disruptions not only at the low-price but also in the mid-price range. Disruptors such as Starbucks came into the market with a mid-priced offering, creating an exceptional customer experience and disrupting sit-down restaurants. Christensen recommends that organisations identify a disruptive innovation when they see one; and that they respond to the threat with a business unit outside of the core business.

However, the widespread confusion about the exact meaning of the term 'disruptive innovation' prevents managers from distinguishing **sustaining innovations** from **radical innovations** (Schmidt and Druehl, 2008).

Procter & Gamble (P&G) has a history of successful disruptive innovations. Its Crest Whitestrips allow people to whiten their teeth simply and easily at home, for instance. Swiffer consumable cloths, which make it easy to clean quickly, yielded $1 billion in new revenue.

Further reading
Anthony, Scott D., Altman, Elizabeth J., Johnson, Mark W. and Sinfield, Joseph V. (2008) *The Innovator's Guide to Growth: Putting Disruptive Innovation to Work*, Boston, MA, Harvard Business School Press.
Christensen, C. M., Anthony, S. D. and Roth, E. A. (2004) *Seeing What's Next*, Boston, MA, Harvard Business School Press.
Hwang, Jason and Christensen, Clayton M. (2008) 'Disruptive Innovation In Health Care Delivery: A Framework For Business-Model Innovation', *Health Affairs*, 27(5), 1329–35.
Schmidt, Glen M. and Druehl, Cheryl T. (2008) 'When Is a Disruptive Innovation Disruptive?', *Journal of Product Innovation Management*, 25(4), 347–69.

Divergent thinking

D

Divergent thinking is the ability to generate multiple solutions to an open-ended problem (Guilford, 1967). Divergent thinking techniques are in use to expand thinking processes for generating and recording a large number of new or stimulating ideas. Divergent thinking is valuable early in the idea-generation phase of the **new product development process**. In contrast to **convergent thinking**, which converges the information gathered on the central problem, divergent thinking branches off (diverges) and seeks new perspectives and **creativity** (Rosenau *et al.*, 1996). It focuses on the generation of novel ideas, builds on other ideas and seeks new and attractive combinations of ideas. **Brainstorming** is a divergent thinking technique.

Divergent thinking tests measure individuals' creative potential. Torrance's (1974) *Torrance Test of Creative Thinking* (based on divergent thinking) and Mednick's (1962) *Remote Associates Test* (based on associative hierarchies) are two tests used to measure creative potential. An example of a divergent thinking task is the *Alternative Uses Test* (Guilford, 1967), which asks participants to generate novel uses for everyday objects. These tests can be utilised in staffing for innovation-related projects.

See also: **creativity measurement**

Further reading
Guilford, J. P. (1967) *The Nature of Human Intelligence,* New York, McGraw-Hill.
Mednick, S. (1962) 'The Associative Basis of the Creative Process', *Psychological Review,* 69, 220–32.
Rosenau, Milton D., Jr., Griffin, Abbie, Castellion, George and Anschuetz, Ned (eds) (1996) *The PDMA Handbook of New Product Development,* New York: John Wiley and Sons.
Torrance, E. P. (1974) *Torrance Test of Creative Thinking,* Lexington, MA, Personal Press.

Dominant design

A dominant design is a particular *product design* architecture that defines the specifications for the entire product category (Utterback, 1994). A technological breakthrough can lead to discontinuity, which initiates a period of intense trial and error, technology variation and selection. A period of incremental technical progress follows this time of turbulence (that is, **incremental innovations** until the technology matures) and the emergence of a single standard or dominant design. Dominant designs do not use frontier technologies; they incorporate more mature technologies (Anderson and Tushman, 1990).

Dominant designs are ubiquitous. Diverse product categories such as VCRs, nuclear reactors, automatically controlled machine tools and watches have dominant product designs (Utterback, 1994). Competition before the emergence of a dominant design is between competing designs. Competition after its emergence is within the domain of a dominant design.

Dominant designs emerge differently in different industries. The dominant design for DVD players emerged three years after the product's introduction in 1996. The dominant design for the modern fax machine emerged after 23 years. The first modern fax machine was introduced in 1960, and the dominant design, the GIII, did not emerge until 1983 (Baum *et al.*, 1995). In some other industries, the dominant design may not emerge for a long time or not emerge at all. New firms with competing technologies enter the market for long periods in such industries (Christensen *et al.*, 1998). For example, the camcorder category, introduced in 1984, supported various designs more than 20 years later. These designs include the VHS (home video system), VHS-C (home video system–compact), super VHS, super VHS-C, 8 mm, Hi-8 and digital 8 mm formats (Srinivasan *et al.*, 2006).

Dominant design contributes to **commoditisation**. *Radical product innovation* ends with the emergence of a dominant product design. Christensen *et al.* (1998) suggests that a rapid succession of disruptive innovations occurred in the disk-drive industry after the emergence of the dominant design. These innovations occurred at the level of product components, rather than at the level of the overall design or product architecture.

See also: **technology life cycle**

Further reading

Anderson, Philip and Tushman, Michael L. (1990) Technological Discontinuities and Dominant Designs: A Cyclical Model of Technological Change, *Administrative Science Quarterly*, 35, 604–-33

Baum, Joel A. C., Korn, Helaine J. and Kotha, Suresh (1995) 'Dominant Designs and Population Dynamics in Telecommunications Services: Founding and Failure of Facsimile Transmission Service Organizations, 1965–1992', *Social Science Research*, 24(2), 97–135.

Christensen, C., Suarez, A. and Utterback, J. (1998) 'Strategies for Survival in Fast-Changing Industries', *Management Science*, 44, S207–20.

Srinivasan, Raji, Lilien, Gary L. and Rangaswami, Arvind (2006) 'The Emergence of Dominant Designs', *Journal of Marketing*, 70(2), 1–17.

Suarez, F. and Utterback, J. (1995) 'Dominant Designs and the Survival of Firms', *Strategic Management Journal*, 16, 415–30.

Utterback, James M. (1994) *Mastering the Dynamics of Innovation: How Companies Can Seize Opportunities in the Face of Technological Change*, Boston, MA, Harvard Business School Press.

Due diligence (technology and patents)

Technology due diligence is a process for evaluating alternative technologies and technology services (Andriole, 2007). Technology due diligence is useful for assessing 'evidence of use and the value of the technology (patent portfolio)' (Miller *et al.*, 2009). For due diligence purposes, technologies can be viewed as complex and discrete. *Complex technologies* can only be utilised by employing others' patents. For example, a DNA-chip system includes several components protected by patents (Andriole, 2007). Mutual licensing of complementary technologies is a common practice in the electronic, semiconductor, telecommunication and software industries. Businesses use cross-licensing or private agreements to complement their technologies. *Discrete technologies* dominate when a new product contains one particular technology. 3M's Post-It Note is an example of a discrete technology. The probability of blockage (such as patent litigation) by the competition is greater with complex technologies than discrete technologies. In the licensing of products or process technologies, patent due diligence can identify issues that may adversely affect the use of a product or process, or restrict development or **commercialisation** (Miao, 2007). Formal due diligence allows an organisation to invest in early-stage technologies and technology service models that are most likely to succeed (that is, they cross the *chasm;* Andriole, 2007).

Patent due diligence is an important part of the buying, selling, merger or acquisition of technology-based companies. The goal of patent due diligence for the buyer is to buy a better company at a lower price. The goal for the seller is to close the deal at the highest possible price. The seller needs to establish an investment-grade patent portfolio with the highest possible market evaluation while ensuring that the portfolio will not hinder the deal.

D

Glazier (2003) identified five key due diligence questions that drive technology deals:

1. Are the target's products protected from copying by competitors?
2. Can the target make its product without infringing the patents of competitors?
3. Does the target own its patents and trade secrets?
4. What patent strategies will improve the shareholder value of the target going forward?
5. Do the price and terms of the acquisition need changing due to the answers to the above patent points, or are there other remedies for problems identified in the patent portfolio?

Professional legal advice is necessary for both buyers and sellers when negotiating technology deals.

Further reading
Andriole, Stephen J. (2007) 'Mining for Digital Gold: Technology Due Diligence for CIOs', *Communications of the Association for Information Systems*, 20, Article 24.
Glazier, Stephen (2003) *Technology Deals*, Washington, DC, LBI Law and Business Institute.
Miao, Emily (2007) 'Due Diligence Review in Patent Licensing Transactions', *Licensing Journal*, 27(10), 27–30.
Miller, Jeff, Minhas, Micky, Radovsky, Natasha and Thumm, Michael (2009) 'Managing Patent Transactions: What Makes for Good Patent Due Diligence?', *Computer & Internet Lawyer*, 26(11), 10–12.

Dynamic capabilities

The dynamic capabilities of a firm are its abilities to expand or contract its **core competences**, on an on-going basis, to meet the challenges of a rapidly evolving Schumpeterian world of competition and imitation (Teece *et al.*, 1997). In contrast to the resource-based view of a firm that emphasises resource choice (that is, selecting of appropriate resources), dynamic capabilities emphasise resource development and renewal.

Dynamic capabilities serve as a buffer between a firm's resources and the changing business environment. They support a firm in sustaining its **competitive advantage** by adjusting its resources. Some dynamic capabilities integrate firm resources (e.g. in product development and strategic decision making), while others can involve reconfiguration of resources within a firm (as in knowledge brokering). Some dynamic capabilities can be related to gaining and releasing resources (e.g. knowledge creation routines, alliance, acquisition and exit routines).

The dynamic capabilities orientation of strategy is different from Michael Porter's *five forces framework* and the framework for **strategic management**. The dynamic capabilities framework views **competitive advantage** as coming from high-performance routines operating within the firm. It steers managers in acquiring, shedding, integrating or recombining resources to create new value. This approach creates unique and difficult-to-imitate competitive advantages (Teece and Pisano, 1994). O'Connor (2008) offers a management framework for building *major innovation (MI) dynamic capabilities* (the capabilities for both radical and novel innovations) by drawing on systems theory and recent advances in dynamic capability theory. O'Connor argues that dynamic capabilities are a complex phenomenon and that MI dynamic capabilities are more than operating routines and repeatable processes.

Further reading
O'Connor, Gina Colarelli (2008) 'Major Innovation as a Dynamic Capability: A Systems Approach', *Journal of Product Innovation Management*, 25(4), 313–30.
Teece, D. J., Pisano, G. and Shuen, A. (1997) 'Dynamic Capabilities and Strategic Management', *Strategic Management Journal*, (18)7, 509–33.
Teece, D. J. and Pisano, G. (1994) 'The Dynamic Capabilities of Firms: An Introduction', *Industrial and Corporate Change*, 3(3), 537–56.

Dynamics of innovation

Innovation is a risky activity. It is not possible to know in advance whether a project will be successful, especially if it involves a radical or novel technology (Hill and Utterback, 1980). A successful example of the market introduction of a new system that overcame both technology and use hurdles can be found in the Eastman story (Utterback, 1995). In the summer of 1888, Eastman built a new system to change the picture-taking business. The picture-taking system he built was simple. The amateur photographer had to perform only three basic tasks: 'pull the cord' (to cock the shutter), 'turn the key' (to advance the film) and 'press the button' (*ibid.*). Eastman's system also overcame two restraints of technology in the 1880s: the perishability of photographic material; and the difficulty of picture taking for the average person. Eastman's intuitive understanding of the dynamics of innovation was crucial to his solving of both problems and to his company's early survival and success (*ibid.*).

Abernathy and Utterback (1978) described the dynamics of innovation in a firm by a model that comprises three phases: fluid, transitional and specific). The fluid phase is exploratory. **Innovation** in the fluid phase moves forward with uncertainty about the target market for the product and the technology. In the transitional phase, the firm's attention shifts from the inventor's bench to the factory floor and **product and process innovations** become tightly linked. Factory automation and managerial control begin to influence the culture and product changes become costly. In this stage, a culture of **incremental innovation** and adaptation takes root in the firm. The specific phase is the death phase of the **product life cycle**. **Flexible manufacturing** and **mass customisation** strategies can help a firm renew its products and stay away from this stage.

See also: **diffusion of innovations**

Further reading
Abernathy, William J. and Utterback, James M. (1978) 'Patterns of Innovation in Industry', *Technology Review*, 80(7), June–July, 40–47.
Hill, Christopher T. and Utterback, James M. (1980) 'The Dynamics of Product and Process Innovation', *Management Review*, 69(91), 14–20.
Utterback, James M. (1994) *Mastering the Dynamics of Innovation: How Companies Can Seize Opportunities in the Face of Technological Change*, Boston, MA, Harvard Business School Press.
Utterback, James M. (1995) 'Developing Technologies: The Eastman Kodak Story', *McKinsey Quarterly*, 1, 130–43.

Ee

Early supplier involvement (ESI)

Early supplier involvement (ESI) is a collaborative practice in **new product development (NPD)**. Early refers to supplier involvement in the design specification phase of the **NPD process**. With their early involvement, suppliers commit resources to both NPD and manufacturing of the new product. However, there are also costs in involving suppliers in the development process. This can lead to loss of proprietary knowledge, reduced control over the development process and increased costs related to managing the collaboration (Bruce *et al.*, 1995). There is also the possibility of a supplier turning into a competitor.

The one-size-fits-all approach does not work for coordinating supplier involvement in the NPD process. Using the Tetra Pak project as a case study, Lakemond *et al.* (2006) found that the company's project success can be traced to its recognition that supplier collaboration needs coordination throughout the development process. Tetra Pak also tailored supplier coordination from project to project and supplier to supplier.

The benefits of ESI include an increase in development speed, faster time to market, reductions in development costs, improved supplier innovation and future assurance of the availability of a supplier's production capacity (Bonaccorsi and Lipparini, 1994). Eisenhardt and Tabrizi (1995) found that supplier involvement decreased product development time (that is, decreased cycle time) when the products developed were mature and development goals well defined. According to Clark (1993), much of the Japanese advantage in concept-to-market time came from supplier involvement in the NPD process. Companies using the operational management and **strategic management** approaches to supplier involvement in the product development process capture both short- and long-term benefits (van Echtelt *et al.*, 2008).

However, ESI is not widely practised in NPD. One significant road block is the lack of an effective interface mechanism between product customers and suppliers. This lack restricts ESI in NPD (Huang and Mak, 2000).

Further reading
Bonaccorsi, A. and Lipparini, A. (1994) 'Strategic Partnerships in New Product Development: An Italian Case Study', *Journal of Product Innovation Management*, 11, 134–45.
Bruce, M., Leverick, F., Littler, D. and Wilson, D. (1995) 'Success Factors for Collaborative Product Development: A Study of Suppliers of Information and Communication Technology', *R&D Management*, 25(1), 33–44.
Clark, K. (1993) 'Project Scope and Project Performance: The Effect of Parts Strategy and Supplier Involvement on Product Development', *Management Science*, 35(10), 1247–63.

Eisenhardt, Kathleen M. and Tabrizi, Behnam N. (1995) 'Accelerating Adaptive Processes: Product Innovation in the Global Computer Industry', Administrative Science Quarterly, 40, 84–110.

Huang, G. Q. and Mak, K. L. (2000) 'Modeling the Customer–Supplier Interface over the World-Wide Web to Facilitate Early Supplier Involvement in the New Product Development', Proceedings of the Institution of Mechanical Engineers – Part B – Engineering Manufacture, 214(9), 759–69.

Lakemond, Nicolette, Berggren, Christian and van Weele, Arjan (2006) 'Coordinating Supplier Involvement in Product Development Projects: A Differentiated Coordination Typology', R&D Management, 36(1), 55–66.

van Echtelt, Ferrie E. A., Wynstra, Finn and van Weele, Arjan J. (2008) 'Managing Supplier Involvement in New Product Development: A Multiple-Case Study', Journal of Product Innovation Management, 25(2), 180–201.

Earned value management (EVM)

The use of earned value management (EVM) began in the US Department of Defense. It focuses project management's attention 'to what is to be done and when it is to be done'. The basis for earned value is work performed and not money spent (Pratt, 2006). It helps stakeholders such as project teams, managers and supervisors to determine project progress against plan. The *project manager* (PM) calculates earned value at predetermined intervals based on the project plan. The PM calculates the first earned value when the project is about 20 per cent complete. If the cost efficiency suggests that the project will require more than the budgeted amount of resources for completion, then management has a choice whether to approve or disapprove continuation of the project (*ibid.*). EVM also enables teams to predict where they will be in the future, allowing managers to allocate critical resources to stay on the plan, or modify plans before projects spin out of control (*ibid.*).

Significant numbers of projects experience delays, go over budget or underdeliver results. By imposing a disciplined decision system for managing and controlling projects, EVM improves the overall chances of delivering on agreed deliverables at the time of the *project charter*. However, EVM is not accepted as a project management tool by US defence contractors because the US government pays contractors to develop a given system at a cost-plus basis (Zosh, 2009). Zosh observes that US defence contractors have a conflict of interest: they gain by increasing the scope of work. EVM is thus counter to their corporate interests.

Further reading
Pratt, Mary K. (2006) 'Earned Value Management', Computerworld, 40(14), 48.
Zosh, Daniel A. (2009) 'Advancing EVM and Government Contracting Efficiencies', Defense AT&L, 38(6), 20–23.

E

Ecodesign

The ecodesign principle recognises the environmental impact of a new product during the design specification phase of the **new product development (NPD) process** (Lewis and Gertsakis, 2001; European Commission, 2005). According to the European Commission, ecodesign-based NPD organisations need to consider the resource-circulation approach for components in the product system at the design phase of the NPD process.

Ecodesign also recognises the environmental impact of upgraded products, maintenance of existing products, extensions of the **product life cycle**, product or component reuse and recycling of materials. Environmental requirements are different from customer requirements. Brezet (1997) proposes a four-step model of ecodesign innovation: product improvement (improvement to comply with environmental standards), product redesign (use product life-cycle approach to increase reuse, minimise energy use etc.), function innovation (such as use of electronic communications in place of paper-based communications) and system innovation (such as industry transformation with information technology products).

Ecodesign aims to reduce a product's environmental impact without sacrificing other design criteria, such as cost and functionality. Grote *et al.* (2007) developed a design decision methodology to design complex products by applying ecodesign principles, without economic trade-offs. Complex products have longer design phases compared to simpler products.

In the past, many managers believed that optimisation of environmental performance leads to increases in costs (Hargroves and Smith 2005). However, environmental performance optimisation can also increase economic performance. A new consciousness that cost savings and improved environmental performance can complement each other is taking root among government and industry leaders. Fujimoto *et al.* (2009) presented a future scenario in which advanced information and communication technologies (ICT) enable development of a low-carbon society and compatible, emotionally healthy lifestyles.

Further reading
Brezet, H. (1997) 'Dynamics in Ecodesign Practice', *UNEP Industry and Environment Review*, 20(1–2), 21–4.
European Commission (2005) *Ecodesign: Your Future*, http://ec.europa.eu/enterprise/policies/sustainable-business/ecodesign/index_en.htm (accessed on 30 September 2010).
Fujimoto, Jun, Poland, Dean and Matsumoto, Mitsutaka (2009) 'Low-Carbon Society Scenario: ICT and Ecodesign', *Information Society*, 25(2), 139–51.
Grote, C. A., Jones, R. M., Blount, G. N., Goodyer, J. and Shayler, M. (2007) 'An Approach to the EuP Directive and the Application of the Economic Eco-design for Complex Products', *International Journal of Production Research*, 45(18–19), 4099–117.
Hargroves, K. and Smith, M. H. (2005) *The Natural Advantage of Nations: Business Opportunities, Innovation and Governance in the 21st Century*, London, Earthscan/James & James.
Lewis, H. and Gertsakis, J. (2001) *Design + Environment: A Global Guide To Designing Greener Goods*, Sheffield, Greenleaf

E

Eco-innovation

Coined by Peter James and Fussler Claude in 1996, the term eco-innovation is interchangeably used with *'environmental innovation'*. Eco-innovation is broadly defined as any innovation that benefits the environment. It is the commercial application of knowledge for ecological improvements (Claude and James, 1996). In eco-innovation, new products and processes create customer and business value while minimising environmental impact and improving the quality of life. In the mobile telecommunications industry, Nokia and Samsung implemented ecodesign in mobile phones in support of **sustainable development** (Nasr, 2010).

Samsung Group manufactured a touchscreen mobile phone from recycled plastic, and used a solar panel to charge the battery.

In recent years, eco-innovation has become an important component of **national innovation systems**. The Japanese government promotes eco-innovation by supporting energy efficiency and green information technology projects. *Ecos* (April–May 2008 issue) reported that the World Business Council for Sustainable Development (WBCSD) and IBM partnered to promote open sharing of patents to further eco-innovation. In another example, Nokia offered a patented technology for free to reuse old mobile phones safely by transforming them into new products.

Further reading
Claude, F. and James, P. (1996) *Driving Eco-Innovation: A Breakthrough Discipline for Innovation and Sustainability,* London: Pitman.
Nasr, Nabil (2010) 'Eco-innovation Is Next', *Industrial Engineer,* 42(3), 26.

Ecological technology innovation (ETI)

Ecological technology innovation (ETI) promotes *sustainable development*. The goal of ETI is to reduce incremental increases in the cost of manufactured goods by taking into consideration the limited availability of resources and increasing public demands to protect the environment. Xia *et al.* (2009) reported that 40.5 per cent of the Chinese industrial sectors studied implemented effective ETI. However, the majority of industrial sectors adopting ETI policies were in the non-high-tech industry. China's high-tech manufacturing industries had high total resource input and a low level of actual ETI output (*ibid.*).

Joseph Huber (2004) argues that incremental improvements are insufficient to achieve a sustainable environment. He emphasises the use of radically new technologies, products and practices early in the **product life cycle** to address environmental issues. He proposes the use of *technological environmental innovations (TEI)* to prevent environmental pollution and degradation.

Further reading
Huber, Joseph (2004) *New Technologies and Environmental Innovation,* Cheltenham, Edward Elgar.
Xia, Weili, Chen, Chen and Jijiao, Jiang (2009) 'Measure of the Ecological Technological Innovation Efficiency of China's Industrial Sectors-Based on Super-efficiency DEA Model' in Hu Shuhua and Hamsa Thota (eds), *Proceedings of the 4th International Conference on Product Innovation Management,* Wuhan, China, Hubei People's Press, Part 2: 389–97.

E

Ecosystem

See **innovation ecosystem, social ecosystem**

E-innovation

E-innovation is an emerging, internet-based **platform** for innovation (Lan, 2004). The internet expands the traditional **innovation process** in both its use and delivery. The characteristics of e-innovation platform are their digital form, distributed nature, decoupling of innovation activities, enabling of change and offering a new game with new rules. Amazon, eBay and PayPal have implemented web service platforms to support e-commerce. Skype offers voice-over-internet (VoIP) telephony and Christie's online auctions attract shoppers from the USA, UK, France, Italy and the Netherlands. Online auctions are increasingly accepted as a mainstream method of e-commerce. The shift from hard-copy books to soft-copy e-books delivery is changing the landscape for learning and is a tipping point for libraries (Hellman, 2010). Internet-based technology platforms are delivering legal services (such as complex legal documents) to underserved consumers and small businesses (S. F. W., 2009). E-innovations are transforming human life in multiple fields.

Wu and Hisa (2008) used an electronic commerce (e-commerce) innovation model to study differences in technological knowledge and **business model**. They found that the innovation from internet-enabled commerce (I-commerce) to mobile commerce (M-commerce) is radical, leading to drastic changes in the business model (*ibid.*). However, they found that in innovation from M-commerce to ubiquitous commerce (U-commerce), disruptive changes occurred in both technological and business model dimensions (*ibid.*). M-commerce is all the transactions (such as mobile banking, mobile entertainment, mobile advertising and mobile ticketing) made using mobile transaction devices such as personal digital assistants (PDAs) and mobile phones (Kao, 2009).

Further reading

Hellman, Eric (2010) 'Libraries, Ebooks, and Competition', *Library Journal*, 135(13), 223.
Kao, Danny Tengti (2009) 'The Impact of Transaction Trust on Consumers' Intentions to Adopt M-Commerce: A Cross-Cultural Investigation', *CyberPsychology & Behavior*, 12(2), 225–9.
Lan, P. (2004) 'E-innovation: An Emerging Platform for a Networked Economy', *International Journal of Electronic Business*, 2(1), 93.
S. F. W. (2009) 'Internet Obsessive', *ABA Journal*, 95(9), 46–7.
Wu, Jen-Her, and Hisa, Tzyh-Lih (2008) 'Developing E-Business Dynamic Capabilities: An Analysis of E-Commerce Innovation from I-, M-, to U-Commerce', *Journal of Organizational Computing & Electronic Commerce*, 18(2), 95–111.

Empathic design

Empathic design is a set of techniques for the observation of consumers using products or services in the course of normal, everyday routines (Leonard and Rayport, 1997). Empathic design consists of observation, data capture, reflection and analysis, **brainstorming** for solutions and development of prototypes of possible solutions. In empathic design, developers empathise with customers rather than listen to them as in **lead users**. Empathic design is a participation-observation technique, in which product-in-use is one of the tools used. In one study, 'teams of researchers filmed customers using their cars in various situations, to capture users' problems, improvised solutions, and contextual and mood information. The footage formed a permanent record' (Evans and Burns, 2007).

Empathic design techniques do not replace **market research**. They generate ideas to understand the unarticulated needs that require additional testing (Leonard and Rayport, 1997). Trained observers have extensive knowledge of the company's technical and operational capabilities and can spot solutions to unarticulated needs. They often identify **user innovations** when they see them in the field. They can quickly determine the suitability of observed innovations for duplication or feasibility to develop them further and make them relevant for the broader market. Deep, empathic understanding of users' unarticulated needs challenges industry assumptions and leads to a shift in strategy at the business unit level or the level of the corporation.

In high-tech industry, empathic techniques uncover different designs for electronic products, home appliances, MP3 players and accessories. Youngchan *et al.* (2008) recommend integrating creative consumer focus group interviews, indepth interviews and empathic design to uncover fresh ideas during the product design concept development.

See also: **ethnography research**

Further reading
Evans, S. and Burns, A. D. (2007) 'An Investigation of Customer Delight during Product Evaluation: Implications for the Development of Desirable Products', *Proceedings of the Institution of Mechanical Engineers – Part B – Engineering Manufacture*, 221(11), 625–38.
Leonard, Dorothy and Rayport, Jeffrey F. (1997) 'Spark Innovation through Empathic Design', *Harvard Business Review*, 75(6), 102–13.
Youngchan, Kim, Yongseob, Kim, Hangseop, Lim and Malhotra, Naresh K. (2008) 'Back to the Drawing Board',
Marketing Research, 20(3), 36–42.

Endogenous growth paradigm

See **growth theory**

Enterprise Europe Network (EEN)

Enterprise Europe Network (EEN) is an international collaboration platform for business cooperation, technology cooperation and research partnerships for *small and medium-sized enterprises (SMEs)*. EEN supports SMEs' need for integrated services, which combine information, contacts, commercial prospects and financial support to exploit opportunities. Industrial sectors served by EEN include aerospace, defence, ICT and mobile technology, auto and motors, automation, agro-food, food and packaging, maritime, solar technology, environment, biotechnology, engineering and materials, security, renewable energies and energy saving, nanotechnology and technological textiles. The network partners provide information and help SMEs to access *European Union (EU)* policies, programmes and funding opportunities.

Collaboration with countries beyond the EU allows European SMEs to reach distant markets and meet demands not always present in Europe. For example, the network's activities supported a more than doubling of *Chilean exports* to and imports from the EU in 2002–2008 (Broechler, 2009).

Further reading
Broechler, Raimund (2009) personal communication, Intrasoft International, http://www.enterprise-europe-network.ec.europa.eu/index_en.htm (accessed on December 8, 2009).

E

Entrepreneur

An entrepreneur initiates, organises and operates his or her own new business venture. In this definition, **entrepreneurship** is synonymous with **new venture** creation (Timmons *et al.*, 1985). An entrepreneur takes risks. He or she also profits from the new venture.

There are other definitions of entrepreneurship. Livesay (1982) suggests that entrepreneurship is purposeful. It is a successful initiative to establish, maintain and generate a profit-oriented business. Carland *et al.* (1984) distinguish entrepreneurs from small business owners. They argue that innovative strategy oriented to profitability and long-term growth is a key characteristics of entrepreneurs. Landes *et al.* (2010) distinguish between *replicative entrepreneurs*, who follow other successful businesses and establish more new businesses, and truly innovative entrepreneurs, 'who create new products and services and change the very nature of the market'.

Joseph Schumpeter (1961[1911]) described the entrepreneur as the agent of innovation. He made a distinction between invention and entrepreneurial innovation. *Entrepreneurial innovation* includes the commercialisation of inventions. It also includes the introduction of new production methods and new organisational forms to obtain business benefits. According to Schumpeter, both invention and entrepreneurial innovation require skill and risk taking.

Further reading
Carland, J. W., Hoy, F., Boulton, W. R., and Carland, J. A. C. (1984) 'Differentiating Entrepreneurs from Small Business Owners: A Conceptualization', *Academy of Management Review*, 9, 354–9.
Landes, David S., Mokyr, Joel and Baumol, William J. (eds) (2010) *The invention of enterprise: Entrepreneurship From Ancient Mesopotamia to Modern Times*, Princeton, NJ, Princeton University Press.
Livesay, H. C. (1982) 'Entrepreneurial History', in C. A. Kent, D. L. Sexton and K. H. Veper (eds), *Encyclopedia of Entrepreneurship*, Englewood Cliffs, NJ, Prentice-Hall.
Schumpeter, Joseph A. (1961 [1911]) *The Theory of Economic Development*, New York, Oxford University Press.
Timmons, J. A., Smollen, L. E. and Dungee, A. (1985) *New Venture Creation*, 2nd edn, Homewood, IL, Richard D. Irwin.

Entrepreneurial model of new ventures

The entrepreneurial model of new ventures can be viewed from financial and venture charter perspectives. Financing of new ventures is a multistage, evolutionary process (Eckhardt *et al.*, 2006). Freeman and Engle (2007) view the development of entrepreneurial ventures in four stages. According to them, the inventor-entrepreneur leads the **innovation process** in the first stage. This is the period before income flows. The inventors have the most knowledge and can effectively lead innovation teams. In the second stage, the company generates revenues from sales. Customer demands become clearer. The founder has to reconcile differences between his or her views of the business versus the views of customers about the nature of the business. In addition, as it grows the enterprise formalises organisational routines to maintain discipline and control. In the third stage, the venture continues to build up its capabilities and achieves alignment of interest among stakeholders (*ibid.*). The venture's success also draws attention from established market players. Some established players may enter the market in the third phase. Increased competition triggers need to be scaled up. The need for additional resources to increase the scale of operations often leads to an initial public offering of securities. The company enters the fourth stage with its public offering. The ability of a venture to obtain external funding depends on objective criteria such as the completion of organising activities, marketing activities and the sales volume (Eckhardt *et al.*, 2006).

Ambos and Birkinshaw (2007) studied the evolution of *technology ventures* through the lens of a *venture charter*. This contains three key elements: 'the product markets targeted, the capabilities held by the venture and the aspired scope of the venture' (*ibid.*). Charters of ventures evolve during the life cycle from a capability-driven charter to an aspiration-driven charter, back to a capability-driven charter. For example, Stem Cell's charter transitioned from a heavy emphasis on technology in its first charter to a heavy emphasis on aspiration and growth in its second charter and then back to a focus on its technology in its third charter (*ibid.*). The evolution model of technology ventures underscores the dramatic changes that successful new ventures endure to obtain the resources needed to sustain growth.

Further reading
Ambos, Tina C. and Birkinshaw, Julian (2007) 'How Do New Ventures Evolve? The Process of Charter Change in Technology Ventures', *Academy of Management Proceedings*, 1–6.
Eckhardt, Jonathan T., Shane, Scott and Delmar, Frédéric (2006) 'Multistage Selection and the Financing of New Ventures', *Management Science*, 52(2), 220–32.
Freeman, John and Engel, Jerome S. (2007) 'Models of Innovation: Startups and Mature Corporations', *California Management Review*, 50(1), 94–119.

Entrepreneurial opportunity

In **entrepreneurship** theory, the concept of opportunity has two viewpoints. The first is the opportunity discovery perspective, where opportunities are something waiting to be discovered by the **entrepreneur**. The second is the opportunity enactment perspective, where the entrepreneur enacts the opportunity.

Shane and Venkataraman (2000) emphasise three steps in the **entrepreneurial process**: the identification of opportunities; the discovery, evaluation and utilisation of opportunities; and entrepreneurs who discover, evaluate and exploit them. In this perspective, an opportunity is most likely to be a product or service. It also includes new organisational methods and the creation or discovery of new raw materials. The entrepreneur here is the one pursuing the opportunity and his focus on opportunities extends entrepreneurship into the market.

Block and Wagner (2010) define an entrepreneurial opportunity as 'a situation in which new goods or services can be introduced and sold at the higher price than their cost of production'. In a study of opportunity identification and exploitation to create *social value*, Corner and Ho (2007) found that opportunities in social enterprises were 'neither purely created nor purely discovered'. They advanced the idea of a '*collective entrepreneur*'. According to them, the collective or group-level action pattern may more accurately reflect **social entrepreneurship** than traditional entrepreneurship theory.

See also: **opportunity identification**

Further reading
Block, Joern H. and Wagner, Marcus (2010) 'Necessity and Opportunity Entrepreneurs in Germany: Characteristics and Earnings Differentials', *Schmalenbach Business Review*, 62(2), 15474.
Corner, Patricia Doyle and Ho, Marcus (2010) 'How Opportunities Develop in Social Entrepreneurship', *Entrepreneurship: Theory & Practice*, 34(4), 635–59.
Shane, S. and Venkataraman, S. (2000) 'The Promise of Entrepreneurship as a Field of Research', *Academy of Management Research,* 25(1), 217–27.

Entrepreneurial orientation (EO)

Entrepreneurial orientation (EO) reflects a firm's capacity to identify and exploit untapped opportunities as an organising principle (Lumpkin and Dess, 1996). It is an environmental management capacity by which firms initiate actions to alter the competitive landscape to their advantage (Atuahene-Gima and Ko, 2001). EO is also a strategic construct that reflects the extent to which firms are innovative, proactive and risk taking in their behaviour and management philosophies (Anderson, 2009). For high-EO firms, strategic planning is a disciplinary device. The strategic planning process enables a high-EO firm to focus and concentrate its entrepreneurial initiatives while avoiding reckless innovation that may put performance at risk. **Market orientation** complements EO to improve the quality and quantity of the firm's innovations (Baker and Sinkula, 2009).

Further reading
Anderson, Brian S. (2009) 'Understanding the Relationship between Entrepreneurial Orientation and Strategic Learning', *Academy of Management Proceedings*, 1–6.
Atuahene-Gima, K. and Ko, A. (2001) 'An Empirical Investigation of the Effects of Market Orientation and Entrepreneurship Orientation Alignment on Product Innovation', *Organization Science,* 12(1), 54–74.
Baker, William E. and Sinkula, James M. (2009) 'The Complementary Effects of Market Orientation and Entrepreneurial Orientation on Profitability in Small Businesses', *Journal of Small Business Management*, 47(4), 443–64.
Lumpkin, G. T. and Dess, G. G. (1996) 'Clarifying the Entrepreneurial Orientation Construct and Linking it to Performance', *Academy of Management Review*, 21, 135–72.

E

Entrepreneurial process

At the level of the firm, the entrepreneurial process consists of three phases (Evald *et al.*, 2006). The first is the emergence of the firm. In this phase, the focus is on what happens before the firm is legally established. It begins when an **entrepreneur** makes a decision to establish a new business and ends with the establishment of the **new venture**. The second phase begins with the first day of the newly established firm. The focus is on activities that occur immediately after the firm is legally established. It ends when the firm is well established. The third phase is the mature firm and begins after the firm is well established.

Further reading
Evald, M. R. G., Klyver, Kim and Svendsen, S. G. (2006) 'The Changing Importance of the Strength of Ties throughout the Entrepreneurial Process', *Journal of Enterprising Culture*, 14(1), 1–26.

Entrepreneurship

The definition of entrepreneurship is different in business and academia. In business, entrepreneurship describes new venture creation and the management of small businesses. Schumpeter (1961[1911]) described entrepreneurship as the ability to make an invention into an innovation through the combination of knowledge, capabilities, skills and resources. In Shailer's view, entrepreneurship is a process and refers to an owner-managed firm (Shailer, 1993). In addition to the creation of a new business, Paul Westhead (Westhead and Wright, 2000) sees entrepreneurship as imagination, creativity, innovativeness, calculated risk taking, opportunity recognition, pursuit and exploitation.

In academia, entrepreneurship is a heterogeneous phenomenon. It is complex and multidimensional, involving elements of rationality, intuition and improvisation. There are up to 25 different schools of thought on entrepreneurship. Fayolle (2002) identified the significant features of six entrepreneurial schools of thought:

- The classic school focuses on **innovation** and **creativity** and **opportunity identification**.
- The great person school focuses on the inborn characteristics of entrepreneurs and their success stories.
- The intrapreneurship school focuses on entrepreneurship behaviour in existing organisations.
- The management school focuses on the pursuit of business opportunities and the use of management tools to achieve them.
- The leadership school focuses on the leadership qualities of entrepreneurs.
- The psychological characteristics school focuses on the specific and unique psychological traits of entrepreneurs.

There are two conflicting *theories of entrepreneurship*: discovery and creation (Alvarez and Barney, 2007). In the discovery theory, entrepreneurs collect and analyse data to understand the possible outcomes; they discover and exploit opportunities, which are a product of the environment. In the creation theory, entrepreneurs' actions are the source of opportunities and the actions taken by the entrepreneurs uncover those opportunities (*ibid.*). In this view, opportunities depend on entrepreneurs' subjective perceptions and are created through social interactions and learning processes. Edelman and Yli-Renko (2010) conducted studies to bridge the gap between the discovery and creation theories of entrepreneurship. They found that entrepreneurs' opportunity perceptions mediate between the environment and efforts to start a new venture (*ibid.*).

E

Further reading
Alvarez, S. and Barney, J. (2007) 'Discovery and Creation: Alternative Theories of Entrepreneurial Action', *Strategic Entrepreneurship Journal*, 1(1–2), 11–26.

Edelman, Linda and Yli-Renko, Helena (2010) 'The Impact of Environment and Entrepreneurial Perceptions on Venture-Creation Efforts: Bridging the Discovery and Creation Views of Entrepreneurship', *Entrepreneurship: Theory & Practice*, 34(5), 833–56.

Fayolle, Alain (2002) 'Insights to Research on the Entrepreneurial Process from a Study on Perceptions of Entrepreneurship and Entrepreneurs, *Journal of Enterprising Culture*, 10(4), 257.

Schumpeter, Joseph A. (1961 [1911]) *The Theory of Economic Development*, New York, Oxford University Press.

Shailer, G. E. P. (1993) *The Art and Science of Entrepreneurship*, Cambridge, MA, Ballinger.

Westhead, P. and Wright, M. (2000) *Advances in Entrepreneurship*, London, Edward Elgar.

Environmental innovation

See **eco-innovation**

Ethnography research

Ethnography is about understanding how people live their lives (Anderson, 2009). It is a branch of anthropology. Traditional market researchers ask specific and highly practical questions when they talk to consumers. Anthropological researchers make visits to consumers' homes or offices. They observe them and listen to their conversations in a non-intrusive way. While this observational method appears inefficient, it gives insights about the context in which customers use a product. This observation also sheds light on the meaning that product holds in the daily lives of consumers.

People cannot articulate what they need from products or services. By understanding how people interact and use their products, ethnographers discover hidden trends that can shape the company's future strategies. An excellent example is ethnography research with a smartphone. With smartphones anthropologists can contrast the technology perspectives of teenagers, who grew up with mobile phones, with those of their parents and grandparents, who embraced them only after becoming proficient with personal computers (PCs).

Ethnography has proved valuable at Intel. The company employs two dozen anthropologists and other trained ethnographers (*ibid.*). However, there is ongoing ethical concern that research into 'public' behaviour may be morally problematic if privacy rights are not safeguarded and the moral significance of 'insight inflicted' is not appropriately considered (Schrag, 2009). Ethical issues in electronic communities can be overcome by using codes of conduct (*deontology*) and following the principle of the greatest good for the greatest number (*teleontology*) (Hair and Clark, 2007).

See also: **market research**

Further reading

Anderson, Ken (2009) 'Ethnographic Research: A Key to Strategy', *Harvard Business Review*, 87(3), 24.

Hair, Neil and Clark, Moira (2007) 'The Ethical Dilemmas and Challenges of Ethnographic Research in Electronic Communities', *International Journal of Market Research*, 49(6), 781–800.

Schrag, Brian (2009) 'Piercing the Veil: Ethical Issues in Ethnographic Research', *Science & Engineering Ethics*, 15(2), 135–60.

European Innovation Scoreboard (EIS)

See **innovation scoreboard**

Evolutionary innovation

The implications of evolutionary innovation are different in biology and economy. In *evolutionary biology*, there are three classes of evolutionary processes: adaptation, speciation and innovation (Shpak and Wagner, 2000). Shpak and Wagner (2000) define adaptation as the modification of organisms by random variation and selection, leading to an increased fit of organisms to their living conditions; and speciation as the origin of a population that does not exchange genes with its parental population. Speciation is not necessarily an adaptive process and adaptation does not require speciation. Innovation refers to any major transition in evolution. Some examples of key transitions are the origin of multicellular organisms, the emergence of new metazoan body plans or a significant modification of an existing body plan such as the origin of birds from dinosaurs (*ibid.*). There is disagreement among biologists as to whether innovation has any meaning that is qualitatively different from adaptation. Biologists also examine whether a formal theory of evolutionary innovation needs to be embedded in a more general mathematical framework.

According to Joseph Schumpeter's (1950, 1961[1911]) model of the evolutionary process, ideas, innovations and technologies compete for resources in a market environment. The technology of products and processes most suited to the existing market conditions thrives. Utterback suggested that a wide range of innovations ultimately leads to a dominant product design (Utterback and Abernathy, 1975). **Dominant design** leads to standard products, narrowing diversity. However, the requirement to meet **customer needs** leads to product **differentiation.** In differentiation firms face the choice of either accepting new technologies and replacing existing manufacturing technologies with new ones or exiting the game completely. This is 'do or die' technology adoption. Maital *et al.* (1994) found that economies of scale and scope require high-tech firms to adopt new, complex technologies and dispose of old ones; whereas older, simpler technologies continue to exist among firms that seek 'niche' market strategies. Economic evolution occurs in part when 'firms watch other firms and try to learn from their experience' (Nelson, 1990).

In the health-care industry, Consoli and Mina (2009) utilised an evolutionary approach to gain a better understanding of the health-care service–industrial complex. Distribution of *medical innovation* occurs across time, scientific knowledge, medical manufacturing technologies and the provision of health-care services (*ibid.*). *Innovation in services* is part of the productivity revolution. Information and communications technologies (ICT) enable it by coordinating the production and delivery of services. Potts and Mandeville (2007) propose a theory of innovation in services based on the development of new markets that exploit the potential of ICT.

Further reading
Consoli, Davide and Mina, Andrea (2009) 'An Evolutionary Perspective on Health Innovation Systems', *Journal of Evolutionary Economics*, 19(2), 297–319.

E

Maital, S., Grupp, H., Frenkel, A. and Koschatzky, K. (1994) 'The Relation between the Average Complexity of High-Tech Products and Their Diversity: An Empirical Test of Evolutionary Models', *Journal of Evolutionary Economics*, 4, 273–88.

Nelson (1990) reference to come.

Potts, Jason and Mandeville, Tom (2007) 'Toward an Evolutionary Theory of Innovation and Growth in the Service Economy', *Prometheus*, 25(2), 147–59.

Schumpeter, J. (1950) *Capitalism, Socialism and Democracy*, 3rd edn, New York, Harper.

Schumpeter, J. (1961[1911]) *The Theory of Economic Development*, New York, Oxford University Press.

Shpak, Max and Wagner, Günter P. (2000) 'Asymmetry of Configuration Space Induced by Unequal Crossover: Implications for a Mathematical Theory of Evolutionary Innovation', *Artificial Life*, 6(1), 25–43.

Utterback, J. M. and Abernathy, W. J. (1975) 'A Dynamic Model of Product and Process Innovation', *Omega*, 3, 639–56.

Expected commercial value (ECV)

Expected commercial value (ECV) determines the commercial value of projects. It is a variation of the decision tree methodology. ECV recognises that new product projects are investments made in increments (Kahn, 2006). It discounts future revenues by the probability of commercial success, probability of technical success and the costs of development and commercialisation. According to Scott Edgett, 'ECV starts with *net present value (NPV)* calculations. It is multiplied by the probability of commercial success minus the commercialization costs. The resulting value is then multiplied by the probability of technical success minus the development cost' (Donnelly, 2000).

In new product **portfolio management**, the commercial worth of individual projects can be calculated using ECV. Each ECV is then used in making the decision to either fund or not fund the next stage of development of a **new product development (NPD)** project. Inaccurate estimation of probabilities of technical and commercial success is a serious problem when ECV determines the trade value of **radical innovation** projects. It penalises uncertainty in technology or commercial success even when the returns are high. This bias can result in funding of an imbalanced portfolio of NPD projects. Imbalance in resource allocation among NPD projects will contribute to the long-term demise of a company. Built-in bias in ECV can be eliminated using **Delphi forecasting** methodology.

See also: **earned value management**

Further reading
Donnelly, George (2000) 'A P&L for R&D', *CFO*, 16(2), 44–50.
Kahn, Kenneth B. (2006) *New Product Forecasting: An Applied Approach*, New York, M.E. Sharp.

E

Experience design

In his book *Experience Design,* Nathan Shedroff (2001) defines experience design as 'the deliberate, careful creation of a total experience for an audience'. It is a way to make an emotional connection with customers through the construction of physical and subtle service elements. For instance, 90-year-old greeting card company Hallmark Cards helps people develop emotional connections (Robinette,

2001). Service providers gain **competitive advantage** by gaining customer loyalty. They continuously innovate and create new customer experiences. Experience design has gained popularity in hospitality and retail businesses. Some examples of customer service experience concepts are boutique hotels (such as Starwood's W hotels), 'try and buy' retail concepts (such as American Girl Stores) and full-experience portfolios (such as Lego International) (Pullman and Gross, 2004).

In the internet-based e-business environment, creation of digital experiences has become a necessity. User experience tools and techniques can be used to create websites and software products to help customers find information quickly, make purchases or participate in social media. Microsoft Windows 7 is an example of a contemporary product designed for creating an effective digital experience.

Further reading
Pullman, Madeleine E. and Gross, Michael A. (2004) 'Ability of Experience Design Elements to Elicit Emotions and Loyalty Behaviors', *Decision Sciences*, 35(3), 551–78.
Robinette, Scott (2001) 'Get Emotional', *Harvard Business Review*, 79(5), 24–5.
Shedroff, Nathan (2001) *Experience Design*, Indianapolis, IN, New Riders.

Exploitation

See **organisational learning**

Exploration

See **organisational learning**

Exploratory learning

See **organisational learning**

E

Failure mode and effects analysis (FMEA)

Failure mode and effects analysis (FEMA) is a risk-quantification method for assessing the risk of failure in a system. According to the Society for Risk Analysis, FEMA 'systematically analyses all contributing component failure modes and identifies the effects on the system'. FEMA determines the relationship between element failures and system failures, malfunctions, performance degradation, operational constraints and degradation of integrity. It identifies priorities for process controls inspection tests during manufacture and installation, qualification, approval and acceptance and also for start-up tests.

In the food industry, FEMA can be used to classify risk ruled by factors of severity (S), probability of occurrence (O) and probability of detection (D) of raw materials at risk (Arvanitoyannis and Varzakas, 2007). FEMA, cause-and-effect analysis and *Pareto diagrams* are complementary tools used in conjunction with *hazard analysis and critical control points (HACCP)* programmes in the food industry. Examples of FEMA use include ready-to-eat vegetables (Varzakas and Arvanitoyannis, 2009), potato chips manufacturing (Arvanitoyannis and Varzakas, 2007) and genetically modified organism (GMO) detection (Arvanitoyannis and Savelides, 2007).

FEMA is also used in the reliable design of medical products, including the medical device, its components, component function failure modes, cause of failure, severity level, probability of occurrence, risk level and design control (Fries, 2006). In the reliable design of medical devices, *robust design FMEA (DFMEA)* is an enhancement to the standard FMEA that anticipates the safety and reliability of failure modes through the use of *parameter diagram (P-diagram)*. *Robust DFMEA* incorporates inputs from a cross-functional team with expertise in design, human factors, manufacturing, testing, service, quality, reliability, clinical, regulatory, supplier or other fields, as appropriate (*ibid.*).

DFMEA is an essential tool of the *design for Six Sigma process (DFSS)*.

See also: **risk-quantification techniques, Six Sigma for product development**

Further reading
Arvanitoyannis, Ioannis S. and Varzakas, Theodoros H. (2007) 'Application of Failure Mode and Effect Analysis (FMEA), Cause and Effect Analysis and Pareto Diagram in Conjunction with HACCP to a Potato Chips Manufacturing Plant', *International Journal of Food Science & Technology*, 42(12), 1424–42.
Arvanitoyannis, Ioannis S. and Savelides, Socrates C. (2007) 'Application of Failure Mode and Effect Analysis and Cause and Effect Analysis and Pareto Diagram in Conjunction with HACCP to a Chocolate-Producing Industry: A Case Study of Tentative GMO Detection at Pilot Plant Scale', *International Journal of Food Science & Technology*, 42 (11), 1265–89.

Fries, Richard C. (2006) *Reliable Design of Medical Devices,* 2nd edn, Boca Raton, FL, Taylor & Francis.

Society for Risk Analysis (SRA), http://www.sra.org/resources_glossary_p-r.php (accessed on October 26, 2009).

Varzakas, Theodoros H. and Arvanitoyannis, Ioannis S. (2009) 'Application of Failure Mode and Effect Analysis and Cause and Effect Analysis on Processing of Ready to Eat Vegetables – Part II', *International Journal of Food Science & Technology,* 44(5), 932–9.

Fast-follower strategy

A fast follower takes advantage of new technology or a new market created by a pioneering company (**first-to-market strategy**). Sometimes a fast follower may obtain more value from the new technology or a new market than the pioneering company. For example, Sony was the first to market with the Betamax format home video recording technology and fast followers RCA and Matsushita introduced the VHS format. Sony positioned Betamax on excellent quality home recording while RCA and Matsushita positioned VHS for longer play. Both formats battled for market dominance. Ultimately the VHS format became the dominant technology and Betamax declined.

According to McGrath (2000), two main fast-follower strategies are to wait until a new market is clear; and to reverse engineer successful, competitive products. First-to-market Litton positioned the microwave as a substitute for traditional cooking; it was a failure. Japanese and Korean microwave manufacturers benefited from Litton's failure by repositioning the microwave as a secondary method of cooking.

Reverse engineering is the process of designing a product by copying the functionality of a successful, competitive product. It reduces development costs as well as market uncertainty. PC clone manufacturers successfully reverse engineered the IBM PC. Fast-follower companies with a superior **new product development process** can take advantage of faster **time to market (TTM)** or faster product development **cycle times**. Fast followers can also supply **target markets** at significantly lower investment cost than first-to-market players. Fast followers can buy second-generation manufacturing equipment instead of the first-generation equipment purchased by first-to-market players. Second-generation equipment maintains higher throughput and gets better yield and utilisation rates for the same investment (Gray, 2006). Toyota Motor Corporation exemplifies a fast-follower automobile manufacturer which transitioned from fast follower to innovation leader in hybrid passenger vehicles by demonstrating leadership in both product and process technologies (Teresko, 2004).

Further reading

Gray, Bruce (2006) 'Maximizing Capital Productivity Is Key to the 'Fast Follower' Strategy', *Solid State Technology,* 49(12), 76–7.

McGrath, Michael E. (2000) *Product Strategy for Technology Companies,* 2nd edn, New York, McGraw-Hill.

Teresko, John (2004) 'Asia Yesterday's Fast Followers, Today's Global Leaders', *Industry Week,* 253(2), 22–9.

Field testing

See **product use testing (PUT)**

Financial innovation

'Innovation to gain competitive advantage' has been a well-accepted adage in the financial products and services industry since 1990s. According to Federal Reserve Board Chairman Alan Greenspan, financial innovations helped the US economy to withstand shocks in 2002 (*Toronto Star*, November 20, 2002) in the aftermath of the terrorist attacks in New York and the collapse of a number of high-profile corporations. US lenders became more diversified and borrowers became less dependent on a single market for financing. Complex financial instruments (such as derivatives) and securities were key contributors to growth.

While financial innovation can help businesses and investors lower the cost of capital and manage risks more efficiently, 'much of the innovation activity has appeared more devoted to circumventing regulation than to reducing businesses cost of capital', according to Lord Turner, chairman of the Financial Services Authority in the UK (*Times*, January 23, 2009). Lax governance contributed largely to the 2008 global financial crisis and the global recession that followed it. Regulatory oversight of financial institutions was also inadequate to discourage the excessive risk-taking behaviour associated with complex financial instruments.

Frame and White (2009) define financial innovation as 'something new that reduces costs, reduces risks, or provides an improved product/service/instrument that better satisfies financial system participants' demands'. Frame and White (2009) classify financial innovations into 'new products (i.e. subprime mortgages) or services (i.e. Internet banking); new production processes (i.e. credit scoring); or new organizational forms (i.e. Internet-only banks)'. When done well, financial innovations can enrich **customer experiences**. Kiwibank of New Zealand won 2007 'Best Direct Channels Initiative at the Financial Innovation Awards' for its mobile banking facility. It enables its customers to view and manage their financial transactions from their mobile phones.

Further reading
'Financial innovations help economy weather shocks', *Toronto Star*, November 20, 2002.
'Financial Innovation and its Costs', *Times*, January 23, 2009.
Frame, W. Scott and White, Lawrence J. (2009) 'Technological Change, Financial Innovation, and Diffusion in Banking', Working Paper Series, Federal Reserve Bank of Atlanta, 10, 1–31.

Firm growth

In 1959, Penrose published *The Theory of the Growth of the Firm*. Firm growth is a multidimensional phenomenon that looks into how a firm grows over time. The characteristics of a firm as well as its environment influence its resource levels and its growth. Eisenhardt and Schoonhoven (1990) reported that, besides organisational environment, three key factors influencing firm growth are market stage, strategy and management team:

F

- *Market stage:* In growth markets, **incremental innovations** (such as price/performance improvements) trigger market turbulence and create large markets and rapid growth. Morone and Testa (2008) report that Italian SMEs' sales growth was positively associated with firm size, **process innovation**, **product innovation** and organisational changes. Managers of successful and unsuccessful SMEs differ in the innovations they launch into the market. In successful SMEs, founders and/or top managers identify incremental opportunities. They enrich and exploit market opportunities with incremental rather than **radical innovations**.
- *Strategy:* One key distinction can be made between organic growth firms and acquisition growth firms. Organic growth is associated with job creation, whereas growth through acquisition is considered a change of job from one firm to another. McKelvie and Wiklund (2010) encourage a shift of focus in academic research from 'growth change' to 'growth mode' (such as organic, acquisition or hybrid growth) in order to provide satisfactory answers to the question of how growth occurs.
- *Management team:* Small firms have little or no in-house R&D resources. The top management of small firms obtain their knowledge through cooperation with large firms, universities and other research institutions. They also gain knowledge through selective hiring of new employees.

See also: **breakthrough growth, discovery-driven growth**

Further reading

Eisenhardt, Kathleen M. and Schoonhoven, Claudia Bird (1990) 'Organizational Growth: Linking Founding Team, Strategy, Environment and Growth among US Semiconductor Ventures', *Administrative Science Quarterly*, 35(3), 504–29.

McKelvie, Alexander and Wiklund, Johan (2010) 'Advancing Firm Growth Research: A Focus on Growth Mode instead of Growth Rate', *Entrepreneurship Theory & Practice*, 34(2), 261–88.Morone, Piergiuseppe and Testa, Giuseppina (2008) 'Firms Growth, Size and Innovation: An Investigation into the Italian Manufacturing Sector', *Economics of Innovation & New Technology*, 17(4), 311–29.

Penrose, E. (1959) *The Theory of the Growth of the Firm*, New York, Oxford University Press.

First-mover advantage

Tellis and Golder (2001) state that pioneering and first-to-market companies achieve long-term market leadership and enduring market share and become successful. The German company Neumann is a leading manufacturer of high-end microphones for studios and sustains its competitive advantage 'by being first with a new technology' (Kopel and Löffler, 2008). The Austrian Airlines Group gained first-mover advantage in the eastern European market (*ibid.*). In e-retailing, Amazon retains the dominant share of internet retailing. It had a compelling initial **value proposition**, offering the ability to search through and choose from millions of book titles. No brick-and-mortar store could offer this capability. Amazon also disrupted the pattern of a 'solitary shopping experience' by creating a 'collaborative shopping experience' where opinions seemingly matter as much as the dollars. It cemented its first-mover advantage with dependability of service,

F

which made it one of the most trusted retailers. Amazon easily beat late-entrant, brick-and-mortar retailers such as Wal-Mart (Laing *et al.*, 2009).

First movers can lose their advantage if their choice is imperfectly observed or if there are substantial observation costs (Vardy, 2004). Kopel and Löffler (2008) found a unique equilibrium in which 'the follower can overcome the first-mover advantage' of the quantity leader and obtain a higher profit than the leader. This occurs under two conditions: both first-mover and second-mover firms invest in process R&D; and only the follower delegates quantity decisions to managers. Under these two conditions the follower can overcome the first-mover advantage of the quantity leader and obtain a higher profit than the leader (*ibid.*).

See also: **fast-follower strategy**

Further reading
Kopel, Michael and Löffler, Clemens (2008) 'Commitment, First-Mover, and Second-Mover Advantage', *Journal of Economics*, 94(2), 143–66.
Liang, T. P., Czaplewski, Andrew J., Klein, Gary and Jiang, James J. (2009) 'Leveraging First-Mover Advantages in Internet-based Consumer Services', *Communications of the ACM*, 52(6), 146–8.
Suarez, Fernando and Lanzolla, Gianvito (2005) 'The Half-Truth of First-Mover Advantage', *Harvard Business Review*, 83(4), 121–7.
Tellis, G. J. and Golder, P. N. (2001) *Will and Vision: How Latecomers Grow to Dominate Markets*, New York, McGraw-Hill.
Vardy, F. (2004) 'The Value of Commitment in Stackelberg Games with Observation Costs', *Games and Economic Behavior*, 49, 374–400.

Five generations of innovation models

Rothwell (1991, 1992) suggests that innovation models evolved in five successive generations from the 1950s to the 1990s. Each generation had successive waves of **technological innovation** with a corresponding evolution in corporate strategy (Hobday, 2005). In the first generation (technology push; Rothwell, 1986), the **innovation process** was linear and sequential. The emphasis was on R&D push and the market was the recipient of the results of the R&D. In the second generation (market pull; Rothwell, 1986) the market drove the innovation process. It was a linear and sequential process and the emphasis was on marketing. The market was the source of ideas and provided guidance to R&D. In the third generation (coupling models; Rothwell, 1992), the innovation process was sequential, with feedback loops from later to earlier stages. It involved push or pull–push combinations. The emphasis was on integration at the R&D–marketing interface. In the fourth generation (integrated model; Rothwell, 1992), the innovation process consisted of parallel development with integrated development teams. It linked upstream suppliers and partners and was coupled with **lead users**. The emphasis was on the integration between R&D and manufacturing (that is, **design for manufacture**). Horizontal collaboration included joint ventures and strategic partnerships. In the fifth generation (systems integration and networking model; Rothwell, 1992), the innovation process was fully integrated, and advanced information technology supported parallel development. R&D used expert systems and **simulation.** The **fuzzy front end** involved customers in the innovation process. Strategic integration included co-development of new products and **computer-aided design (CAD)** system linkages with key suppliers.

Horizontal linkages included joint ventures, collaborative research and marketing arrangements. The emphasis was on corporate agility and speed of development. There was also an emphasis on quality and other non-price factors.

The five generations of innovation models are not appropriate for *catch-up innovation* in developing countries due to the heterogeneity and unpredictability of innovation processes (Hobday, 2005).

Further reading

Hobday, Michael (2005) 'Firm-Level Innovation Models: Perspectives on Research in Developed and Developing Countries, *Technology Analysis & Strategic Management*, 17(2), 121–46.

Rothwell, Roy (1986) 'Innovation and Re-Innovation: A Role for the User', *Journal of Marketing Management*, 2(2), 109–23.

Rothwell, R. (1991) 'Successful Industrial Innovation: Critical Factors for the 1990s', extended version of a paper presented at the Science Policy Research Unit's 25th Anniversary Conference: 'SPRU at 25: Perspectives on the Future of Science and Technology Policy', SPRU, University of Sussex, 3–4 July.

Rothwell, R. (1992) 'Developments towards the Fifth Generation Model of Innovation', *Technology Analysis and Strategic Management*, 4(1), 73–5.

Flexibility

Flexibility was originally defined in the operations management field as a company's ability to meet market needs without excessive costs, time, organisational disruption or loss of performance (Aggarwal, 1997). In the innovation field, flexibility includes the ability to develop and introduce new products with flexibilities in *product design* and **product innovation** (MacCormack *et al.*, 2001). Clark and Baldwin (1997) found that the flexibility of product design and its ability to embed innovations throughout the **product life cycle** come from implementing modularity and platform-based approaches.

Flexibility in product innovation refers to using an iterative product development process that emphasises learning and adaptation rather than planning and implementation. **Service life-cycle flexibility** refers to the ability to launch innovations during the service life cycle at low costs and in a short time by internet-based service companies (Verganti, 2005).

Service life-cycle flexibility in internet-based services is essential, and it must be pursued. Buganza and Verganti (2006) identified three life-cycle flexibility (LCF) dimensions to turn development process flexibility into LCF capability: frequency of adaptation, rapidity of adaptation and quality of adaptation. Managerial practices to increase LCF include managing back-end technological competences; sharing front-end technological competences with external suppliers; utilising open and established technologies; having low formalisation of new service development (NSD) procedures; and maintaining high formalisation of NSD organisation (Buganza and Verganti, 2006).

F

Further reading

Aggarwal, S. (1997) 'Flexibility Management: The Ultimate Strategy', *Industrial Management*, January–February, 26–30.

Buganza, Tommaso and Verganti, Roberto (2006) 'Life-Cycle Flexibility: How to Measure and Improve the Innovative Capability in Turbulent Environments', *Journal of Product Innovation Management*, 23(5), 393–407.

Clark, K. and Baldwin, C. (1997) 'Managing in an Age of Modularity', *Harvard Business Review*, 75(5), 84–93.

MacCormack, A., Verganti, R. and Iansiti, M. (2001) 'Developing Products on Internet Time: The Anatomy of a Flexible Development Process', *Management Science*, 47(1), 133–50.

Verganti, Roberto and Buganza, Tommaso (2005) 'Design Inertia: Designing for Life-Cycle Flexibility in Internet-Based Services', *Journal of Product Innovation Management*, 22, 223–37.

Focus group

A focus group is a qualitative research method. It refers to a group experience made possible by careful selection of a small group of consumers (the preferred size is between seven and eight) brought together for their knowledge of the product or service under research. A trained facilitator leads the focus group session, investigates what the participants currently use and seeks new information about what they would like to see to improve the product. The **new product development (NPD) team** (marketing function) sponsoring the focus group session can view the discussion from behind a one-way mirror. The company that employs the facilitator provides a video tape and transcript to the NPD team after the session is complete. Focus groups can be used to gain exploratory information during new product development, to generate options and to explore new markets. Typical applications of focus groups include to identify customer needs, to identify how products are used, to test new products and to understand consumer buying habits and the buying process (Hague *et al.*, 2004).

A focus group can be used as a stand-alone method; as supplementary to a survey; and as part of a multimethod design. Focus groups can be offline (face to face) or online.

Characteristics of offline focus groups are the people participating in them, who have a commonality of experience and interest, the depth of information sought and the topic for discussion. Consumer packaged goods companies use offline focus groups widely.

In online focus groups, respondents can talk to each other through a net-phone arrangement. They can also write their answers on a blackboard. Special online focus group software gives the researcher the opportunity to guide the group discussion. Online focus group research is fast and inexpensive. However, it cannot provide insightful answers about motivations or backgrounds. Online focus groups do not offer face-to-face interaction between respondents. On a positive note, they are flexible and researchers can raise additional questions or request feedback. Typical applications of online focus groups include a quick scan of visual examples of designs, marketing and sales offers or promotions, or in the pre- or ex-post research phases, brand-identity research and positioning research. Online focus groups may be better if respondents in the target group are busy, as long as respondents have a high-speed internet connection (Brüggen and Willems, 2009).

See also: **market research**

Further reading
Brüggen, Elisabeth and Willems, Pieter (2009) 'A Critical Comparison of Offline Focus Groups, Online Focus Groups and e-Delphi', *International Journal of Market Research*, 51(3), 363–81.

Hague, Paul, Hague, Nick and Morgan, Carol-Ann (2004) *Market Research in Practice: A Guide to the Basics,* London, Kogan Page.

Front end of innovation (fuzzy front end)

Front end of innovation (FEI) is a divergent process. It comprises all activities that come before the systematic and well-structured **new product and process development** (NPPD) process. FEI activities include opportunity identification and analysis, idea generation, idea evaluation and selection, concept definition and development, and in some cases strategic business planning and programme planning (Cooper, 2001; Koen *et al.*, 2001; Cagan and Vogel, 2002). To create **customer value**, both parts of the **innovation process** (that is, NPPD and FEI) must be aligned with business strategy.

The Industrial Research Institute (IRI) has developed a common language and consistent definition of key elements of the front end in its New Concept Development (NCD) model. This contains three-parts (Koen *et al.*, 2001):

1. Five front-end NCD elements comprising **opportunity identification**; opportunity analysis; idea genesis (also called **idea generation and enrichment**); idea selection; and concept and technology development.
2. The engine that powers these elements, which includes leadership and organizational culture.
3. External factors influencing organisational capabilities (such as business strategy, the enabling science and distribution channels, customers and competitors).

Sustained and successful product development occurs when FEI activities are within the **core strategic vision (CSV)** of an organisation. By using information technology (IT) tools, companies become proficient collaborators. Successful companies utilise IT tools in FEI to exchange ideas and solutions; for ideation (such as **brainstorming)**; forming innovation networks with partners and customers; gathering competitive intelligence; organising and accessing organisational information and learning; **data mining** for simulation, optimisation and modelling; and **visualisation** (Gordon *et al.*, 2008).

The front end of innovation can also be viewed through the lens of self-organisational behaviour. Self-organisation means that new structures, patterns or properties emerge spontaneously without being externally imposed on the system (Koch and Leitner, 2008). Koch and Leitner (2008) found self-organisation in the Austrian semiconductor industry, where employees intrinsically looked for opportunities to innovate and pursued their own ideas. They built coalitions with other employees to gain attention and support for their idea, and built prototypes. They kept their idea secret until it was mature enough to seek formal approval. After a management decision to accept the idea, it became an official project in the new product development portfolio.

F

See also: **idea management, network innovation**

Further reading
Cagan, J. and Vogel, C. (2002) *Creating Breakthrough Products: Innovation from Product Planning to Program Approval,* Englewood Cliffs, NJ, Financial Times/Prentice Hall.

Cooper, R. (2001) *Winning at New Products – Accelerating the Process from Idea to Launch*, 3rd edn, Cambridge, MA, Perseus.

Gordon, Steven, Tarafdar, Monideepa, Cook, Robert, Maksimoski, Richard and Rogowitz, Bernice (2008) 'Improving the Front End of Innovation with Information Technology', *Research Technology Management*, 51(3), 50–58.

Koch, Rudolph and Leitner, Karl-Heinz (2008) 'The Dynamics and Functions of Self-Organization in the Fuzzy Front End: Empirical Evidence from the Austrian Semiconductor Industry,' *Creativity & Innovation Management*, 17(3), 216–26.

Koen, P., Ajamian, G., Burkart, R., Clamen, A., Davidson, J., D'Amore, R., Elkins, C., Herald, K., Incorvia, M., Johnson, A., Karol, R., Seibert, R., Slavejkov, A. and Wagner, K. (2001) 'Providing Clarity and a Common Language to the "Fuzzy Front End"', *Research Technology Management*, 44(2), 46–55.

F

Game-changing innovation

What is a 'game-changer'? Game-changers break the rules (e.g. Apple; Bernard, 2009) or change the rules (e.g. Intuit; Norman, 2003). Apple broke the rules with the iPod, by ensuring that songs purchased from the Apple iTunes music store would play only on its iPod, and by tightly managing its suppliers (Bernard, 2009). Intuit changed the rules by launching personal finance tools for individuals rather than for professional accountants.

Game-changing companies transform a market by introducing product, technology or business model innovations. Scott Cook launched his **product innovation**, Quicken personal finance software, in 1984 after talking to people and learning about how they pay their bills, the problems they have in doing so and what solutions they wished they had (Norman, 2003). According to Cook, eBay is also an example of a game-changing innovation achieved by listening to customers (*ibid.*). Amazon exemplifies game-changing innovation that capitalises on technology for customisation (*ibid.*). Its software tracks a customer's purchasing history, including what products each person browsed, to make new suggestions about similar offerings (*ibid.*).

According to Michael Cader, publisher of *Publishers Lunch*, Apple is promoting a new and radical **business model** for e-books (ereads.com). Apple favours the agency model over the traditional wholesale model popular with physical book and e-book sales. Under the agency model, the publisher is the owner of the e-book files and the publisher pays a commission to its authorised selling partner such as Apple, which becomes a fixed-price distributor (ereads.com).

See also: **radical innovation**

Further reading
'Apple Promoting a New (and Radical!) Business Model for Selling E-Books?', edreads.com, January 19, 2010, http://ereads.com/2010/01/apple-promoting-new-and-radical.html (accessed on December 31, 2010).
Bernard, Stan (2009) 'Think Different', *Pharmaceutical Executive*, 29(10), 26–8.
Norman, Jan (2003) 'Intuit Founder Tells Accountants' Group That Success Comes by Changing Rules', *Orange County Register*, June 10, 2003.

Game theory

Von Neumann and Morgenstern (1944) founded game theory with the publication of their book *The Theory of Games and Economic Behavior*. Game theory studies the interaction behaviour of cooperating or competing individuals (Roughgarden, 2010).

There are two types of games in game theory: rules based and freewheeling (Kleindl, 1999). In *rules-based games*, players have rules of engagement. For example, contract negotiations limit player behaviour (*ibid.*). In *freewheeling games*, players interact without any external constraints such as contracts (*ibid.*). In a rules-based game, it is possible to figure out all possible moves and all possible reactions. This leads to an if–then scenario building (*ibid.*). When firms evaluate strategic alternatives, they would like to avoid choices that put themselves at a competitive disadvantage (*ibid.*). So they can use game theory to formulate **strategy**, predict the outcome of strategic situations and develop the best play in the marketplace (*ibid.*). Firms can also use game theory to choose 'what not to do'. For instance, Bell Atlantic uses game theory to manage its business process (MIT Open Courseware).

Game theory is applicable to marketing and **entrepreneurship**. Game theory research in the marketing literature covers competitive entry, competitive decision making and pricing and advertising (Kleindl, 1999). Game theory studies in industry include investigation of **innovation** in the music industry, **product and process innovation**, obtaining of patents and relationships between **entrepreneurs** and *venture capitalists* (*ibid.*).

Algorithmic game theory (AGT) is the interface of theoretical computer science and game theory (Roughgarden, 2010). It is different from game theory in its application domain, such as networks, and its quantitative engineering approach. AGT helps in understanding of how large networks form, how web-based network users behave and how to ensure proper network performance. Examples of AGT applications are single auctions at eBay and Amazon, and sponsored search auctions at Google (Roughgarden, 2010).

Further reading
Kleindl, Brad (1999) 'A Game Theoretic Perspective on Market Oriented versus Innovative Strategic Choice', *Journal of Strategic Marketing*, 7, 265–74.
MIT Open Courseware, Game Theory for Managers, Lecture Notes, http://ocw.mit.edu/NR/rdonlyres/Sloan-School-of-Management/15-040Spring2004/DC1CE8BA-FF0B-4242-84AE-02F8ABF834C1/0/lec1.pdf (accessed on December 7, 2009).
Roughgarden, Tim (2010) 'Algorithmic Game Theory', *Communications of the ACM*, 53(7), 78–86.
Von Neumann, J. and Morgenstern, O. (1944) *The Theory of Games and Economic Behavior*, Princeton, NJ, Princeton University Press.

Gamma testing

G

A gamma test is a product use test. Product developers use gamma testing to assess the extent to which the new product meets the needs of target customers, the extent to which it becomes a solution for the targeted customer's problem and the level of customer satisfaction it achieves. Gamma testing is the third stage of software product testing and is sometimes performed after **beta testing** but before commercial launch (Douglas and Covington, 2009). It is used in the pharmaceutical industry to evaluate new drug formulations (Kahn, 2006).

See also: **alpha testing**

Further reading
Downing, Downing and Covington, Melody Maudlin (2009) *Dictionary of Computer and Internet Terms*, Hauppauge, NY, Barron's Educational Series.

Kahn, Kenneth B. (2006) *New Product Forecasting: An Applied Approach*, Armonk, NY, M.E. Sharpe.

Gatekeepers

Gates function as stop/go and prioritisation points in the **new product development (NPD) process**, where decisions for the future of the project are made (Grönlund *et al.*, 2010). Gatekeepers are a cross-functional group of senior managers who evaluate projects on the basis of quality of execution, business rationale and the quality of the action plan, and make go/no go decisions (*ibid.*). The role of gatekeepers becomes critical in **new product development (NPD)** projects when tasks are defined locally while the technology deployed is changing rapidly (Tushman and Katz, 1980). Gatekeepers have the knowledge and expertise to link NPD teams with external sources of technology and act as interpreters of change. The *designer* is a key interpreter who facilitates access to the ongoing discussion about design languages with manufacturing clients. Key interpreters bring bits of relevant knowledge and help their clients understand the design discourse as well as positioning themselves appropriately in this discourse.

The designer Mendini has been a key gatekeeper at the Italian design firm Alessi, providing it with access to design discourse. Design firm IDEO acts as a *technology broker* with access to as many as 40 different industries (Hargadon, 2003). It leverages its network to develop solutions across industries, generating creative sparks in the process (Hargadon and Sutton, 1997). However, investigations have questioned the value of gatekeepers and technology brokers, as their use might contribute to a loss of firms' ability to deal with long-term changes.

Further reading
Grönlund, Johan, Sjödin, David Rönnberg and Frishammar, Johan (2010) 'Open Innovation and the Stage-Gate Process: A Revised Model for New Product Development', *California Management Review*, 52(3), 106–31.
Hargadon, Andrew (2003) 'Retooling R&D: Technology Brokering and the Pursuit of Innovation', *Ivey Business Journal*, 68(2), 1–7.
Hargadon, Andrew and Sutton, Robert I. (1997) 'Technology Brokering and Innovation in a Product Development Firm', *Administrative Science Quarterly*, 42(4), 716–49.
Tushman, Michael L. and Katz, Ralph (1980) 'External Communication and Project Performance: An Investigation into the Role of Gatekeepers', *Management Science*, 26(11), 1071–85.

Generative learning

See **organisational learning**

GIM product design procedure

GIM (GRAI Integrated Methodology) is widely practised in French, British and European companies. GIM models design and specify advanced manufacturing systems. The product design process and product design activities are also modelled with GIM.

The GIM methodology for product development consists of three levels: concept, structure and realisation (Huang, 1996). The concept level defines the requirements and final objectives of design. A functional analysis establishes and formalises the functions satisfying the product requirements. The structure level corresponds to product architecture. A technical analysis gives a structure according to the functions defined at a concept level. The realisation level clarifies the answers to the functional requirements. Each structure is defined as giving a technological solution.

The steps in GIM product design consist of looking for solutions, eliminating solutions, listing needs, searching for candidates and choosing a solution.

Further reading
Huang, George Q. (1996) 'GIM: GRAI Integrated Methodology for Product Development' in George Q. Huang (ed.), *Design fox X: Concurrent Engineering Imperatives*, London, Chapman & Hall, pp. 153–72.

Global product launch process

The process of a global product launch consists of more than a company's ability to gain access to a market. It requires a company to understand the key design issues in its global regions and respond with sensitivity to the differences. Each targeted country will have notable differences across market channel, demand parameters, customs, language and colloquialisms, and technology infrastructure (Bruce *et al.*, 2007). If there are greater differences across these design categories, more **customisation** is required (*ibid.*).

Channel owners (that is, brand owners, retailers and distributors) can make or break a new product. They have a significant role in the timing of a new product launch. They also can guide the new product's design in attributes such as colour, style or shape. Collaboration with channel owners can secure preference status for a new product versus a competitive product developed without such collaboration.

Harvey and Griffith (2007) recommend *global virtual teams* for coordination. Effective logistics ensure that demand can be met across all global regions (Bruce *et al.*, 2007). In the therapeutic drugs industry, each country is a 'different market with a different need' and launch windows are narrow across countries (MacCarthy and Gascoigne, 2007). With sensitivity to countries' cultures and the languages used on the internet, a global launch manager can ensure a successful launch of a new product by properly positioning and marketing it.

Further reading
Bruce, Margaret, Daly, Lucy and Kahn, Kenneth B. (2007) 'Delineating Design Factors that Influence the Global Product Launch Process', *Journal of Product Innovation Management*, 24(5), 456–70.
Harvey, Michael G. and Griffith, David A. (2007) 'The Role of Globalization, Time Acceleration, and Virtual Global Teams in Fostering Successful Global Product Launches', *Journal of Product Innovation Management*, 24(5), 486–50.
MacCarthy, John and Gascoigne, David (2007) 'Decoding the DNA of Global Product Launch: What Defines True Launch Excellence?', *Pharmaceutical Executive*, 27(7), Special section, 1–8.

Global strategy

The goal of global strategy is to manage large differences that occur at the borders of global markets. In his book *Revisiting Global Strategy*, Ghemawat argues that an industry will cross borders successfully if it is a good manager of differences in cultural, administrative, geographical and economic dimensions (CAGE factors) (Ghemawat, 2007b). He is a proponent of semi-globalisation and a critic of Thomas Friedman's *'world is flat'* argument.

Ghemawat *(*2007a) presents the triple challenges of globalisation as *adaptation, aggregation* and a*rbitrage*. Heavy advertisement to develop the local market characterises adaptation, in which companies maximise local relevance. Aggregation is characterised by extensive leverage of fixed costs such as R&D. With aggregation companies deliver economies of scale from fixed investments in regional and global operations. Arbitrage is characterised by exploiting differences in cost structures in national or regional markets, so that companies establish different parts of the supply chain in different places (call centers in India, factories in China and retail shops in western Europe, for instance).

Ghemawat (2007a) proposed the *AAA triangle* as a **strategy map** for global managers. In the AAA Triangle, the percentages of sales spent on adaptation, aggregation and arbitrage are plotted on the sides of the triangle. Scores above the median draw managers' attention because they deserve a strategic focus.

Further reading
Ghemawat, Pankaj (2007a). 'Managing Differences: The Central Challenge of Global Strategy', *Harvard Business Review*, 85(3), 58–68.
Ghemawat, Pankaj (2007b) *Redefining Global Strategy: Crossing Borders in a World Where Differences Still Matter*, Boston, MA, Harvard Business School Press.

Globalisation of innovation

Globalisation refers to the intensification of worldwide social relations. Events occurring many miles away influence local economies because economic activities are integrated globally (Archibugi and Iammarino, 2002).

Globalisation underscores the international scope of the creation and dissemination of technologies. According to Archibugi and Iammarino (2002) there are three components to the globalisation of innovation. The first is the international exploitation of nationally generated innovations in science-based sectors. The second is the global generation of innovations by *multinational enterprises* (MNEs). In Canada, the Netherlands and the UK, the research & development performed by foreign subsidiaries accounts for more than 20 per cent of total R&D in manufacturing. The third component is global techno-scientific collaboration. An example is collaboration in the biotechnology sector, which is growing rapidly between Europe and the US.

Globalisation of innovation is not uniform across developed and developing nations. Advanced nations play a prominent role in shaping its nature. However, all innovations are local. Regional development fuels the global innovation leadership of the USA, where universities play a key role in regional development (Drabenstott, 2009).

See also: **technology diffusion**

Further reading
Archibugi, Daniele and Iammarino, Simona (2002) 'The Globalization of Technological

Innovation: Definition and Evidence', *Review of International Political Economy*, 9(1), 98–122.
Drabenstott, Mark (2009) 'Universities, Innovation and Regional Development: A View from the United States', *Higher Education Management & Policy*, 20(2), 43–55.

Green design

Green design is also called **ecodesign** (*ecological design*) or *environmental design* (*design for environment*). Green design is the choice for **sustainable development**. Victor Papanek (2005) elaborated a human-centred approach to design with a particularly strong emphasis on social and ecological consciousness.

Green design fully considers the life-cycle impact of products on resources and the environment and incorporates those considerations into the design process. To **industrial design** green design is the 3Rs, namely '*reduce, recycle and reuse*'. Green design reduces material and energy consumption, lowers emissions of harmful substances, and designs products and components to facilitate separation recovery, recycling or reuse.

In 2006, the Chinese government implemented a series of policies in support of green design in the automotive industry, and the development of energy-saving and environmental protection technologies (Zhou, 2009).

See also: **ecological technology innovation (ETI)**

Further reading
Papanek, Victor J. (2005) *Design for the Real World: Human Ecology and Social Change*, 2nd edn, Chicago, IL, Academy Chicago Publishers.
Zhou, Pei (2009) 'Green Design Technology of Chinese Automobile Industry during Independence Innovation' in Shuhua, Hu and Thota, Hamsa (eds), *Proceedings of the 4th International Conference on Product Innovation Management*, Wuhan, China, Hubei People's Press, Part 1, 211–14.

Green engineering

Shapiro (2008) describes green engineering as the design and development of products and processes to derive environmental and economic benefits. It is same as any other engineering innovation: first evaluate; then improve. Green should be inherently better, not just more environmentally sustainable (Pokrandt, 2010). Green engineering applications range from monitoring the health of forests to upgrading inefficient production facilities and equipment to make them more energy efficient. Sometimes green engineering projects look like conventional process optimisation projects. For example, when Nucor Steel acquired the Marion Steel Co. in 2005, it automated systems throughout the mini-mill, which also conserved energy. Direct connections to AC power lines are a green engineering approach in the power-distribution field.

Cradle-to-cradle design complements green engineering. This incorporates material flows in regenerative cycles powered by solar energy (Hogan, 2007). McDonough *et al.* (2003) show several examples of companies redesigning both products and processes so that at the end of life they become technical or biological nutrients (Hogan, 2007).

G

Further reading
Hogan, Patricia A. (2007) 'Engineers Are Thinking Green Too! Green Engineering and Cradle-to-Cradle Design', *NEACT Journal*, 25(2), 15–18.
McDonough, W., Braungart, M., Anastas, P. and Zimmerman, J. B. (2003) 'Applying the Principles of Green Engineering to Cradle-to-Cradle Design', *Journal of Environmental Science and Technology*, 37, 434A–441A.
Pokrandt, Rachel (2010) 'What Is Green?', *Technology Teacher*, 69(6), 5–10.
Shapiro, Joel (2008) 'What Is Green Engineering?', *Machine Design*, 80(12), 42.

Green innovation

Chen *et al.* (2006) define green innovation as 'hardware or software innovation that is related to green products or processes'. They extend the definition to innovation in energy saving, pollution prevention, waste recycling and green product design technologies.

Patel *et al.* (2010) found that companies such as Zipcar, Toyota and Greenbox take different approaches to green innovation. Zipcar makes green innovations attractive to the market and profitable while driving significant change in consumer behaviour. It offers drivers an alternative to owning or renting a car via a subscription model, which also provides the company with a steady cash flow (*ibid.*). Toyota funds *sustainability innovations* (such as hybrid technologies) requiring new-to-the-world technologies. It ensures that hybrid innovations are financially and economically feasible and scalable. An example is the mass-produced hybrid vehicle Prius. Greenbox, which makes plastic moving boxes from recycled materials, pursues fully recyclable and up-cycled products.

Further reading
Chen, Yu-Shan, Lai, Shyh-Bao and Wen, Chao-Tung (2006) 'The Influence of Green Innovation Performance on Corporate Advantage in Taiwan', *Journal of Business Ethics*, 67(4), 331–9.
Patel, Hitendra, McNally, Tyler and Jonash, Ronald (2010) 'Innovation's Green Frontier', *Retail Merchandiser*, 50(5), 18–21.

Green process innovation

The purpose of green process innovation is to increase the effectiveness of environmental management practices while protecting the environment. Chen *et al.* (2006) define the performance of green process innovation in terms of energy saving, pollution prevention, waste recycling or no toxic discharge.

Chen *et al.* (2006) identified four parameters for evaluating the performance of green process innovation in manufacturing companies: the manufacturing process effectively reduces hazardous emissions or waste; recycles and reuses waste; reduces consumption of water, electricity, coal or oil; or reduces the use of raw materials.

In Taiwan, the electronics component and semiconductor industries were the two top performers in green process innovation (*ibid.*). In the *active pharmaceutical ingredients (API) manufacturing* industry, some green manufacturing programmes focus on waste management. For example, AstraZeneca reduced its hazardous waste by 6 per cent from 2001 to 2004 (Chiarello, 2005). Other

G

companies such as Bristol-Myers Squibb and Pfizer have developed more environmentally friendly processes. Bristol-Myers Squibb used a bio-renewable source and plant cell fermentation to make its anticancer drug Taxol and Pfizer redesigned the commercial manufacturing process for its antidepressant drug Zoloft (ibid.).

Further reading
Chen, Yu-Shan, Lai, Shyh-Bao and Wen, Chao-Tung (2006) 'The Influence of Green Innovation Performance on Corporate Advantage in Taiwan', Journal of Business Ethics, 67(4), 331–9.
Chiarello, Kaylynn (2005) 'Industry Takes Steps Toward Greener API Manufacturing', Pharmaceutical Technology, 29(11), 44–5.

Green product innovation

Green product innovation begins during product or design development. According to Chen *et al.* (2006), a green company chooses materials that cause the least amount of pollution and consume the least amount of energy and resources. It uses the lower amount of materials in production and evaluates whether the product is easy to recycle, reuse and decompose.

Many firms are generating revenues with green product innovations. In its ethical performance online report, Royal Philips Electronics reported a 33 per cent increase in sales of energy-efficient light bulbs and energy-saving devices in 2007. Many companies in Ireland use waste as the raw material to generate higher revenues and employ more people. Examples of innovation are turning plastic bottles into clothes hangers and making pots from waste cardboard and paper (Kelly, 2010). In Taiwan, the consumer electronics industry was the best in green product innovation among the six information and electronics industries (Chen *et al.*, 2006). Electronic billing products are also green product innovations.

Further reading
Chen, Yu-Shan, Lai, Shyh-Bao and Wen, Chao-Tung (2006) 'The Influence of Green Innovation Performance on Corporate Advantage in Taiwan', *Journal of Business Ethics*, 67(4), 331–9.
Kelly, Olivia (2010) 'Drive to "Upcycle" Products at Green Forum', *Irish Times*, 4 May.
Philips online Ethical Performance Report 2007, https://www.ethicalperformance.com/reports/reportdetail.php?reportid=44&other=299 (accessed on October 14, 2010).

Growth theory

There are three growth theories for policy makers: the *AK paradigm*, the *product variety paradigm* and the *Schumpeterian framework* (Aghion and Howitt, 1998). According to the AK paradigm, high-growth rates are sustained by saving a large share of gross domestic product (GDP). It anticipates that some of the savings will flow to finance a higher rate of technological progress, which in turn will lead to faster growth. In this paradigm, saving and capital accumulation are the keys, not **creativity** and **innovation**.

The product-variety paradigm is driven by innovation, as in the Schumpeterian paradigm. However, innovations in the product-variety paradigm result in new products of the same kind. This model cannot generate context-dependent growth, so it is of little use to policy makers.

The Schumpeterian framework consists of two innovation-based growth models: the *Romer model* and the *Aghion and Howitt model*. In the Romer (1990) model, productivity growth in the product-variety paradigm occurs with new, but not necessarily improved, varieties of products. The Aghion and Howitt (1998) model is a 'Schumpeterian' growth theory. It focuses on innovations that make old products obsolete, so it involves Schumpeter's *creative destruction* forces. Aghion and Howitt (2006) conclude that the Schumpeterian paradigm delivers systematic and yet context-dependent policy prescriptions better than the product-variety or 'one-size-fits-all' AK paradigms.

Further reading
Aghion, Philippe and Howitt, Peter (1998) *Endogenous Growth Theory*, Boston, MA, MIT Press.
Aghion, Philippe and Howitt, Peter (2006) 'Appropriate Growth Policy: A Unifying Framework', *Journal of the European Economic Association,* 4(2/3), 269–314.
Romer, Paul (1990) 'Endogenous Technical Change', *Journal of Political Economy,* 98, 71–102.

G

Harmony

Harmony is an ancient concept from China. China's President Hu Jintao defines his concept of a harmonious society as 'a society with socialist democracy and rule of law, fairness and justice, integrity and friendly love, fullness of vitality, social stability and orderliness, and harmonious interaction between human and nature' (Shen, 2008). *Confucianism* achieves *optimal harmony* by extending each person's virtuous life. *Daoism* achieves harmony by way of coordinating opposition in strife (*ibid*.). For Confucianism, 'the dimension of meaningfulness in human existence is to be understood within the context of ethical relations among human beings, nature and heaven, in a pattern of life imbued with a sense of beauty in an orderly ensemble that is harmony' (*ibid*.). How could these ancient Chinese philosophies be applied to relieve the enormous social and ecological costs that China has paid on its way to becoming a global economic power?

Thota (2009) proposes a new innovation paradigm for measuring *innovation performance* to achieve a balanced gain of economic, social and ecological value. He defines this balanced gain as 'innovation harmony'.

> Innovation Performance (Gain) = Balanced gain of (economic + social + ecological) value.

In this definition, Thota extends Confucian and Daoist principles to innovation in a global economy. Individuals realise their contribution to their organisation's growth, their country's growth and the overall impact of their organisational effort. In Thota's view, with increased awareness of the greater benefit to their community and one's role as ancestor to the coming generations, both the prosperity of the future and environmentally conscious use of resources in the present can be balanced.

The evolution of the concept of harmony can be seen from a different perspective and for entirely different reasons in a *post-scientific society*. From the perspective of the *theory of comparative advantage*, the US held scientific leadership in the post-Second World War scientific society. In the early twenty-first century, a post-scientific society is emerging and the US is moving from scientific leadership to leadership in the integration of knowledge and harvesting capabilities to 'serve the needs of individuals, families, companies, communities, and society as a whole' (Hill, 2007). In a post-scientific society, China also has an opportunity to reengineer its **national innovation system** to achieve a balanced gain of economic, social and ecological value.

Further reading
Hill, Christopher T. (2007) 'The Post-Scientific Society', *Issues in Science & Technology*, 24(1), 78–84.
Shen, Vincent (2008) 'Optimal Harmony, Mutual Enrichment and Strangification', *Diogenes*, 220, 108–21.
Thota, Hamsa (2009) 'Science & Technology for Global Manufacturing Harmony: Towards a new Paradigm', presented at the Fifth International Conference on Global Manufacturing and China, Zhejiang University, Hangzhou, China, September 5, www.innovationbd.com.

Horizons of growth (McKinsey framework)

Horizons for Growth principles provide a framework for prospecting and a means for categorising a firm's portfolio of products (that is, **portfolio management**) in three time horizons. Introduced in the book *Alchemy of Growth* (1999) by Baghai *et al.*, the *three horizons model* balances the competing demands of focusing on the present while investing for the future. It is also known as the *McKinsey framework*. Horizon 1 represents 'business as usual'. It is about improving short-term performance and extending present operations – doing **incremental innovations**. Horizon 2 is about bringing the next generation of high-growth opportunities in the pipeline to market. **Innovation** is needed to find new ways of doing things. Horizon 3 is where the future is imagined, researched and developed. This requires seeding options 'in the present' for the future – in the form of research, pilot projects and possibly investment in start-ups. Horizon 3 investments sustain businesses into the future.

The time horizons are measured over a period of years, and each includes different sets of strategies to address the development needs of the organisation. This framework helps managers deal with uncertainty by balancing investments between present and future source of revenues and growth. Baghai *et al.* (1999) suggest that companies must be adept at managing across the three horizons concurrently to sustain growth over time. According to Geoffrey Moore (2007), large enterprises excel in Horizon 1 operations and also perform effectively in Horizon 3. They fail in Horizon 2 strategies because they lie between the budget and the long-term plan. Moore (2007) calls this 'no man's land' and urges managers to focus on Horizon 2 to succeed in the long run.

Further reading
Baghai, M., Coley, S. and White, D. (1999) *Alchemy of Growth,* New York, Orion.
Moore, Geoffrey A. (2007) 'To Succeed in the Long Term, Focus on the Middle Term', *Harvard Business Review*, 85 7/8), 84–90.

H

Horizontal innovation

See **product development strategies**

Human-centred design

See **user-centred design**

Idea generation and enrichment

Idea generation and enrichment is a divergent process in new product development. It is also an iterative process. Idea generation is an important activity in engineering design. The generation of a large number of ideas or solutions increases the chances of finding better ideas to satisfy functional requirements. Perttula and Sipilä (2007) reported that idea exposure or exchange prior to idea generation could increase or impede designers' idea-generation performance.

Idea enrichment is building on other people's ideas. This is one of the rules of **brainstorming** (Osborn, 1953). Some brainstorming groups may succumb to an illusion of productivity. Such groups do not fully utilise the resources and ideas of their members and perform worse than or no better than individuals on their own. This potential drawback can be overcome by *brainwriting*, a technique in which group members write down their ideas on paper and share them with the group (Paulus and Yang, 2000). Heslin (2009) argues that, under certain contextual boundary conditions, brainwriting is better than brainstorming for idea generation. The group's creative performance in idea generation and enrichment can also be increased by the use of facilitator who encourages the group to develop initial ideas (Jones and Kelly, 2009).

See also: **concept, concept definition, concept testing**

Further reading
Heslin, Peter A. (2009) 'Better than Brainstorming? Potential Contextual Boundary Conditions to Brainwriting for Idea Generation in Organizations', *Journal of Occupational & Organizational Psychology*, 82(1), 129–45.
Jones and Kelly (2009) Reference to come.
Osborn, A.F. (1953) *Applied Imagination: Principles and Procedures of Creative Thinking*, New York, Charles Scribner's Sons.
Paulus, P. B. and Yang, H. C. (2000) 'Idea Generation in Groups: A Basis for Creativity in Organizations', *Organizational Behavior and Human Decision Processes*, 82, 76–87.
Perttula, Matti and Sipilä, Pekka (2007) 'The Idea Exposure Paradigm in Design Idea Generation', *Journal of Engineering Design*, 18(1), 93–102.
VanGundy, A. B. (1983) 'Brainwriting for New Product Ideas: An Alternative to Brainstorming', *Journal of Consumer Marketing*, 1, 67–74.

Idea management

Idea management is the process of 'searching for, generating, and implementing ideas that determines innovativeness or lack of it in organizations' (Saatcioglu, 2002). As an organisational process, idea management structures members' acting and thinking towards stability and change. It is also a system. McGrath

(2004) defines idea management as the process of gathering ideas from a variety of internal and external sources, organising them and prioritising them to guide business strategy (*ibid.*). There are two types of ideas, product ideas and development ideas (Karanjikar, 2007). *Product ideas* are solicited in the ideation stage of new product development. *Development ideas* relate to the quality of execution and reduction in cycle time. Development ideas can also influence product idea screening at the **front end of innovation** (*ibid.*).

*Saatcioglu (*2002) viewed **innovation** as part of the idea management process, which is subject to influence by managerial behaviour. Vandenbosch *et al.* (2006) proposed five idea management archetypes: incrementalists, consensus builders, debaters, searchers and assessors. *Incrementalists* prefer to maintain the status quo. *Consensus builders* delegate key managerial and organisational issues and tasks to others. *Debaters* welcome different viewpoints and seek out opportunities. *Searchers* solicit ideas and invite conflicting points of view. *Assessors* are renaissance people. Each archetype exerts a different influence on the development of ideas and the quality of execution of the **innovation process**.

In an *idea management system*, all ideas are categorised and available for review and comparison. They can also be enriched with supporting documents and links in a web-based system. If **product strategy** is the front end of product development, then idea management is the front end of product strategy.

When idea management is fully integrated into *development chain management (DCM) systems*, ideas can be transformed into products or specifications for *product enhancement*.

Further reading

Karanjikar, Mukund R. (2007) 'Funnel-Reverse-Funnel: The Future Model of Idea Management in New Product Development', *Futures Research Quarterly*, 23(3), 21–6.
McGrath, Michael E. (2004) *Next Generation Product Development*, New York, McGraw-Hill.
Saatcioglu, Argun (2002) 'Using Grounded Inquiry to Explore Idea Management for Innovativeness', *Academy of Management Proceedings & Membership Directory*, C1–C6.
Vandenbosch, Betty, Saatcioglu, Argun and Fay, Sharon (2006) 'Idea Management: A Systemic View', *Journal of Management Studies*, 43(2), 259–88.

IMP³rove®

Supported by the European Commission under the *Europe INNOVA* initiative, IMP³rove (www.improve-innovation.eu) is an online benchmarking tool for assessing innovation management performance in small and medium-sized enterprises (SMEs). Based on *A.T. Kearney's house of innovation* framework, IMP³rove takes a holistic approach to improving innovation success in an SME by covering four dimensions of innovation management including innovation strategy, innovation organization and culture, management processes and enabling factors.

The European standardisation agency CEN's Standardization of Innovation Capability Rating for SMEs (2008) lists the IMP³rove methodology as a procedure for measuring innovation management performance. This is a first step to IMP³rove being established as a common European standard for innovation management assessment.

Further reading
CEN (2008) CWA 15899:2008, *Standardization of an Innovation Capability Rating for SMEs*, 1 December.

Incremental innovation

Incremental innovations modify existing technology, focus on cost reduction or feature improvements in existing processes, products or functional areas, and employ low resources as compared to the overall activities of the firm (Manimala *et al.*, 2005). Incremental innovations can be viewed as competence-enhancing innovations (Tushman and Anderson, 1986).

Incremental innovations provide low potential risk for product upgrades. Companies that recognise the value of incremental innovations find efficiencies within their work processes. They implement continuous quality improvement programmes such as *Six Sigma* and *kaizen* (Pizysuski, 2008). While they might not generate long-term value on their own, material and operational innovations play an essential, supportive role in **breakthrough innovations**.

Some researchers have suggested that a '*curse of innovation*' can afflict firms that focus innovation efforts solely on *breakthrough products*. Breakthrough products often fail in the marketplace because managers consistently overvalue their benefits relative to existing products, whereas consumers systematically undervalue them or forgo them altogether in favour of more familiar products. Other researchers argue for **ambidextrous organisational** capabilities. Together, incremental and breakthrough innovations sustain firms' competitiveness (Kanter, 2010).

Further reading
Kanter, Rosabeth Moss (2010) 'Block-by-Blockbuster Innovation', *Harvard Business Review*, 88(5): 38.
Manimala, Mathew J., Jose, P. D. and Thomas, K. Raju (2005) 'Organizational Design for Enhancing the Impact of Incremental Innovations: A Qualitative Analysis of Innovative Cases in the Context of a Developing Economy', *Creativity & Innovation Management*, 14(4), 413–24.
Pizysuski, Martin (2008) 'Recognizing and Managing Incremental Innovation: A Strategic Management Perspective for Corporate Managers', *Corporate Business Taxation Monthly*, 9(4), 15–39.
Tushman, M. L. and Anderson, P. (1986) 'Technological Discontinuities and Organizational Environments', *Administrative Science Quarterly*, 31, 439–65.

Independent innovation

Independent innovation is a term used to explain China's aspirations to become an **innovation-oriented country.** Chen Jin was the first Chinese scholar to use the term in 1994 (Zuo and Zhou, 2009). Since then, different Chinese scholars have attributed different meanings to independent innovation. For Shi Peigong, independent innovation in enterprises means that 'enterprises gain valuable research and development results through their own efforts while overcoming technical difficulties based on their own resources, and they get profitable business results'. Liu Yulin defined independent innovation as 'to create innovation of its own intellectual property rights'. For Fu Jiaji and other scholars, independent innovation means that 'through their own efforts and exploration, enterprises generate technological breakthroughs, and they overcome technical difficulties. On the basis of their own resources, they encourage innovation's follow-up link; complete the commercialization of technological achievements; and deliver business profits.' Thus the definition of independent innovation is evolving in China.

Further reading
Zuo, Jihong and Zhou, Liangjin (2009) 'Research on Index System of High-tech Zone's Independent Innovation Capability' in Hu Shuhua and Hamsa Thota (eds) *Proceedings of the 4th International Conference on Product Innovation Management*, Wuhan, China, Hubei People's Press, Part 3, 641–6.

Indigenous innovation

Lazonick (2004) describes indigenous innovation as an innovation that evolves locally and utilises local knowledge and materials, skills and resources to solve local problems. Indigenous innovations are necessary for developing nations. Munir and Hu (2006) suggest that developing countries can build up their national indigenous innovation capacity. This depends on a country's **absorptive capacity**, its policy infrastructure and its entrepreneurial ambitions and **creativity**. Environment that is conducive to indigenous innovation has to be built on a nation's core competitiveness and core technologies, and provides a learning environment based on internal R&D and external **alliances.**

Further reading
Lazonick, W. (2004) 'Indigenous Innovation and Economic Development: Lessons from China's Leap into the Information Age', *Industry and Innovation,* 11(4), 273.
Munir, Zunaira and Hu, Shuhua (2006). 'Indigenous Innovation Framework for Raising China's National Innovation Capacity', *Proceedings of IABE Annual Conference 2006*, Las Vegas, October.

Information technology (IT) innovation

Information technology (IT) is innovative if the organisation adopting it perceives it as a new technology (Rogers, 2003). IT becomes an enabling tool if organisations use it to innovate business processes. **Business process innovation** with IT is a journey (Swanson and Ramiller, 2004) involving four core processes: comprehension, adoption, implementation and assimilation. Organisations learn from their environments about how IT supports business process innovation. This comprehension guides their rationale and decision on whether to adopt the IT innovation. Adoption begins with the installation of hardware and software, followed by changes in business processes. Users are trained and, in due course, the IT innovation is assimilated into the organisational routines. Wal-Mart has successfully innovated with IT over time. Harrah's, Equifax, RR Donnelley and Harley-Davidson are other successful implementers of IT innovations (Stratopoulos and Lim, 2010).

Some organisations adopt IT innovation without adequate comprehension of what it is and how it may improve their business processes. Failure then occurs because they find it extremely difficult to implement the IT they have adopted. They experience delays in its deployment or abandon efforts to integrate it altogether.

Further reading
Rogers, E. M. (2003) *Diffusion of Innovations,* New York, Free Press.
Stratopoulos, Theophanis C. and Lim, Jee-Hae (2010) 'IT Innovation Persistence: An Oxymoron?', *Communications of the ACM,* 53(5), 142–6.
Swanson, E. B. and Ramiller, Neil C. (2004) 'Innovating Mindfully with Information Technology', *MIS Quarterly,* 28(4), 553–83.

Innovation

Innovation can be both technological and non-technological. According to Dosi (1988), technological innovation involves solving problems. For example, the transformation of heat into movement and the shaping of materials with certain processes, or producing compounds with biological, chemical or pharmaceutical value while meeting cost and marketability requirements, are technological innovations (*ibid.*). The technology-specific and sector-specific knowledge required for such solutions implies different degrees of tacitness of knowledge underlying innovation success. While *technological innovations* dominate the academic literature, non-technological innovations in **business models,** as well as innovations creating value in social, economic or ecological dimensions, are essential to sustain firms' competitiveness.

Chesbrough and Teece (1996) distinguish between **autonomous innovations** and **systemic innovations** because each requires a different organisational design to support innovation activity. Autonomous innovations can be pursued independently of other innovations. For example, a high-performance engine can be designed without the complete redesign of an engine or automobile (that is, autonomous innovation), while *lean manufacturing* is a systemic innovation because it requires changes in product design, supplier management, information technology and so on (*ibid.*). A number of researchers have introduced concepts for understanding innovation and technical change. They include radical vs incremental (Ettlie *et al.,* 1984; Dewar and Dutton, 1986), competence-enhancing vs competence-destroying (Anderson and Tushman, 1990), architectural and generational (Henderson and Clark, 1990), disruptive (Christensen and Rosenbloom, 1995), core/peripheral (Tushman and Murmann, 1998) and modular (Baldwin and Clark, 2000).

The following is a partial list of definitions of innovation used in academic and business settings at the level of the individual, firm, society and nation.

Joseph Schumpeter
- Schumpeter interpreted innovation in the broader sense of *new 'combinations'* of producers and means of production, which includes new products, new methods of production, opening up of new markets, utilisation of new raw materials or even the reorganisation of a sector of the economy.

Peter Drucker
- Innovation is not invention. It is a term in economics rather than technology. Innovation may result in a lower price, a new and better product, a new convenience or the definition of a new want. Innovation may be finding new uses for old products.

- The most productive innovation is a different product or service creating a new potential for satisfaction, rather than an improvement. Typically, this new and different product may cost more, but overall its effect is to make the economy more productive.
- Non-technological innovation – social or economic innovation – is at least as important as technical innovation.
- Managers must convert society's needs into opportunities for profitable business. That too is a definition of innovation.

Ed Roberts

- Ed Roberts stresses technology-driven innovation by defining innovation as:

  Innovation = Invention + Commercial exploitation

Everett Rogers

- An innovation is an idea, practice or project that is perceived as new by an individual or other unit of adoption.

Curt Carlson

- Innovation is the creation and delivery of new **customer value** in the marketplace with sufficient value for those who are producing it.

PDMA

- The Product Development and Management Association (PDMA) glossary of new product development defines innovation as 'A new idea, method, or device. The act of creating a new product or process. The act includes **invention** as well as the work required to bring an idea or concept into final form.'

US National Innovation Initiative Report

- The US National innovation Initiative Report (2004) published by the US Council on Competitiveness defines innovation as 'The intersection of invention and insight, leading to the creation of social and economic value.'

APQC

- American Productivity and Quality Center (APQC) define innovation as 'The embodiment, combination and/or synthesis of knowledge into new and unique combinations. Examples include new or modified processes or products, techniques, managerial tools, organizational approaches, patents, licenses, and new **business models**. These new knowledge-based products, processes, or services increase the performance or competitiveness of an organization.'

OECD and Eurostat

- An innovation is the implementation of a new or significantly improved product (good or service) or process, a new marketing method or a new organisational method in business practices, workplace organisation or external relations (OECD/Eurostat, 2005).
- The 1997 Oslo Manual defines technological **product and process innovation** as 'implemented technologically new products and processes and

significant technological improvements in products and processes' (OECD/Eurostat, 1997).

Further reading

Anderson, P. and Tushman, M. (1990) 'Technological Discontinuities and Dominant Designs: A Cyclical Model of Technological Change, *Administrative Science Quarterly*, 35(4), 604–33.

Baldwin, C. and Clark, K. (2000) *Design Rules: The Power of Modularity,* Cambridge, MA, MIT Press.

Chesbrough, Henry W. and Teece, David J. (1996) 'When Is Virtual Virtuous? Organizing for Innovation', *Harvard Business Review*, 74(1), 65–73.

Christensen, C. and Rosenbloom, R. (1995) 'Explaining the Attacker's Advantage: Technological Paradigms, Organizational Dynamics and the Value Network', *Research Policy*, 24, 233–57.

Dewar, R. D. and Dutton, J. E. (1986) 'The Adoption of Radical and Incremental Innovations: An Empirical Analysis', *Management Science*, 32(11), 1422–33.

Dosi, Giovanni (1988) 'Sources, Procedures, and Microeconomic Effects of Innovation', *Journal of Economic Literature*, 26(3), 1120–71.

Ettlie, J., Bridges, W. and O'Keefe, R. (1984) 'Organization Strategy and Structural Differences for Radical versus Incremental Innovation', *Management Science*, 30, 682–95.

Henderson, R. and Clark, K. (1990) 'Architectural Innovation: The Reconfiguration of Existing Product Technologies and the Failure of Existing Firms', *Administration Science Quarterly*, 35, 9–30.

OECD/Eurostat (1997) *Oslo Manual: Proposed Guidelines for Collecting and Interpreting Technological Innovation Data*, 2nd edn, Paris, OECD.

OECD/Eurostat (2005) *Oslo Manual: Guidelines for Collecting and Interpreting Innovation Data*, 3rd edn, Paris, OECD.

Rogers, E. M. (2003) *Diffusion of Innovations*, 5th edn, New York, Free Press.

Tushman, M., and Murmann, J. P. (1998) 'Dominant Designs, Technology Cycles, and Organizational Outcomes', Research in Organizational Behavior, 20, 231–66.

Innovation capitalist

Innovation capitalists are firms that search and appraise intellectual property (that is, ideas and technologies) from the external environment (that is, innovators, inventors, research parks, universities and so on). Innovation capitalists often have extensive knowledge of their industry. They understand market opportunities for ideas and further develop and refine them until their market potential is validated. They then sell the potential business opportunities to large client firms. An innovation capitalist reduces a client company's acquisition costs and early-stage risks, and also profits from innovation (Nambisan and Sawhney, 2007).

Nambisan and Sawhney (2007) expect innovation capitalists to play a crucial role in the early twenty-first century, much as *venture capitalists* did during the technology boom of the 1990s. Unlike venture capitalists who add value through capital investments, innovation capitalists add value through their substantial market knowledge, leadership at **the front end of innovation** and large networks.

Further reading

Nambisan, Satish and Sawhney, Mohan (2007) 'Meet the Innovation Capitalist', Harvard Business Review, 85(3), 24.

Innovation culture

In innovation teams culture emerges during the interactions of team members with each other, directed at reducing the anxiety and uncertainty associated with innovation projects. Team members executing innovation projects share a common view of the workplace and operate under common belief and value systems (Detert *et al.*, 2000; Jassawalla and Sashittal, 2002). This is *organisational culture*. While new product development (NPD) team members bring with them their personal culture and learning behaviours, the dominant shaping force is organisational culture. It can be supportive or destructive when carrying out innovation projects and it shapes **innovation processes** (Jassawalla and Sashittal, 2002). The shaping is reflected in a consistent pattern of behaviours among team members. Personal culture is the sum of the spiritual and material values, symbols, traditions, rules and behaviours of individual team members. Learning behaviour is moulded by national, regional and ethnic culture (and that learning could be voluntary, mandatory, rewarded or punished). Each NPD team member has a choice of embracing personal culture, professional culture or organisational culture.

Jassawalla and Sashittal (2002) identify three distinctive elements of highly innovation-supportive cultures as the guiding values, beliefs and assumptions of participants; behaviours; and related new product outcomes.

- *Guiding values, beliefs and assumptions of participants*: In highly supportive innovation cultures, cross-functional team members take the initiative and exhibit creativity and risk-taking behaviours. Mutual trust is valuable to them. Each member is an equal stakeholder in the co-creative effort. Extended team members including leading customers and key suppliers, and members of other functional groups are involved early in the product development process (*ibid.*). Team members embrace change because change is energising and refreshing.
- *Behaviours:* In highly supportive innovation cultures team members have autonomy and control over how they complete a **product innovation** project. They exhibit high levels of co-creative, collaborative behaviour. They willingly receive feedback from other team members.
- *Related new product outcomes*: In high-supportive innovation cultures, new products from new technologies are developed on time and within cost budgets and achieve market success. In low-supportive innovation cultures, adoption of new product processes is low because of the status quo in core cultural values and beliefs (*ibid.*). Low-supportive innovation cultures adopt **incremental products and process innovations**.

Google is a company with a high-innovation culture. It built **innovation** into its organisational design by budgeting innovation into job descriptions. Technical employees spend 20 per cent of their time on technical projects of their own choosing and managers spend 10 per cent of their time on entirely new businesses and products (Iyer and Davenport, 2008).

Further reading
Agin, Erika and Gibson, Tracy (2010) 'Developing an Innovation Culture', *T+D*, 64(7), 52–5.

Detert, J. R., Schroeder, R. G. and Mauriel, J. J. (2000) 'A Framework for Linking Culture and Improvement Initiatives in Organizations', *Academy of Management Review*, 25(4), 850–63.

Iyer, Bala and Davenport, Thomas H. (2008) 'Reverse Engineering Google's Innovation Machine', *Harvard Business Review*, 86(4), 58–68.

Jassawalla, Avan R. and Sashittal, Hemant C. (2002) 'Cultures that Support Product-Innovation Processes', *Academy of Management Executive,* 16(3), 42–54.

Innovation diamond

The innovation diamond, also called the *initiatives diamond*, was developed by Procter & Gamble (P&G) management. It evolved from Cooper-Klein Schmidt's triangle framework of process, resources and strategy (Cooper and Mills, 2005).

P&G's initiatives diamond includes four major drivers for achieving **new product development** (NPD) results: **product innovation** and **technology strategy,** effective and efficient idea-to-launch process, **portfolio management** and people.

See also: **launch, launch plan, launch strategy, new product development team**

Further reading
Cooper, Robert G. and Mills, Michael S. (2005) 'Succeeding at New Product Development the P&G Way. A Key Element Is Using the Innovation Diamond', *Visions*, 29(4), 9.

Innovation economics (economics of innovation)

The economics of innovation provides an economic explanation for technological change. Antonelli (2009) identified four heuristic frameworks that contributed to the development of the economics of innovation: 'classical legacies of Adam Smith and Karl Marx, the Schumpeterian legacy, the Arrovian legacy, and the Marshallian legacy'. The classical legacy elaborates a theory of economic growth based on the intentional introduction of **technological innovations**. The Schumpeterian legacy appreciates the role of **entrepreneurship** in the introduction of new technologies. The Arrovian legacy analyses the economic characteristics of knowledge. The Marshallian legacy has led to the development of new *complexity theory* based on the interplay between heterogeneity, complementarity and competition. This interplay characterises the **innovation process** (*ibid.*). Economists understand that **product innovation** builds on **invention**, and **process innovation** is a reconfiguration of the production process (Li *et al.*, 2007). In economic terms, technological progress yields increased technical efficiency (productivity).

Economic growth has led to a loss oif biodiversity. This may be mitigated by *end-use innovation* that increases technical efficiency (Czek, 2008). For example, innovation economists view the reduction of carbon dioxide and other greenhouse gas emissions as essentially an innovation challenge. Proponents of *innovation economics in energy* argue that the government must spend more money on research and development to develop cost-effective non-carbon or low-carbon energy alternatives (Atkinson, 2009).

See also: **harmony**

Further reading
Antonelli, Cristiano (2009) 'The Economics of Innovation: From the Classical Legacies to the Economics of Complexity', *Economics of Innovation & New Technology*, 18(7), 611–46.
Atkinson, Rob (2009) 'Innovation Economics Can Fight Global Warming', *Business Week Online*, http://www.businessweek.com (accessed on June 18, 2009).
Czek, Brian (2008) 'Prospects for Reconciling the Conflict between Economic Growth and Biodiversity Conservation with Technological Progress Conservation', *Biology*, 22(6), 1389–98.
Li, Y., Liu, Y., and Ren, F. (2007) 'Product Innovation and Process in Innovation in SOEs: Evidence from the Chinese Transition', *Journal of Technology Transfer*, 32, 63–85.

Innovation ecosystem

The innovation ecosystem enables *collaborative innovation*, in which firms with complementary resources combine them to provide a coherent, customer-facing solution (Adner, 2006). The collaborative solution creates value that no one company can produce on its own. Along with new opportunities, the innovation ecosystem also gives rise to new risks, one of which is that a company's success depends on the fulfilment of the commitments of others, such as collaborating partners who are part of the ecosystem. In some cases, the additional risks may also arise from upstream players in the value chain. These players may not behave as members of the innovation ecosystem.

This was the case with the initial failure of high-definition television (HDTV) set manufacturers (*ibid.*). Philips, Sony and Thomson are leading manufacturers of television consoles. They made significant R&D investments in developing high-definition television technology. Their HDTV sets had outstanding picture quality and were ready for the mass market. However, their breakthroughs in television console technology were not matched by the producers of television shows. The studio production equipment was not adopted concurrently to capture images in high definition, so the HDTV category became a failure.

In mobile, broadband communications, Nokia and Nortel have focused on developing their ecosystems. In 2008, Nokia collaborated to promote innovation across speech, predictive text and mobile interface innovations; and Nortel encouraged the early adoption of 4G LTE wireless broadband. In the energy sector, the US Department of Energy is funding projects to bolster the innovation ecosystem by promoting technology transfer from university laboratories into the market in energy efficiency and renewable energy fields (Agency Group 05, 2010).

See also: **innovation strategy**

Further reading
Adner, Ron (2006) 'Match Your Innovation Strategy to Your Innovation Ecosystem', *Harvard Business Review*, 84(4), 98.
Agency Group 05 (2010) 'DOE Awards $5.3 Million to Support the Development of University-based Technology Commercialization', FDCH Regulatory Intelligence Database, 15 September.

Innovation engine

See **front end of innovation**

Innovation governance

Governance of innovation is a significant challenge. In response, governance structures in the USA and Europe have been evolving since the mid-1990s (Shapira and Kuhlmann, 2003). There are many types of innovations and one governance structure may not fit all. The Spanish **Community Innovation Survey (CIS)** defines product, process, marketing and organisational innovations differently (Fernández-Ribas, 2009). It defines **product innovation** as the market introduction of a new (or significantly improved) good or service. **Process innovation** is the use of new or significantly improved methods for the production or supply of goods and services. According to the CIS, both **product and process innovations** must be new to the firm. However, they do not have to be new to the company's market. *Organisational innovation* consists in new or significant changes in firm structure and knowledge management methods; and **marketing innovation** in new or significant changes in design, packaging or distribution methods (*ibid.*).

Tylecote and Conesa (1999) identified three innovation governance challenges for national *financial and corporate governance (F&CG) systems*: the degree of novelty of innovation, degree of visibility of innovation and appropriability of innovation (*ibid.*).

- *Degree of novelty of innovation*: Industries characterised by a high degree of novelty need shareholders/stakeholders with high levels of industry-specific expertise. This allows outsiders with industry-specific expertise to assess the quality of the firm's plans. Such governance is available in outsider-dominated F&CG systems.
- *Degree of visibility of innovation*: Industries where the **innovation process** has a low degree of visibility benefit from an insider-dominated F&CG system. Such a governance structure encourages a high level of firm-specific perceptive. In industries where the innovation process is relatively visible, outsider-dominated corporate governance can be quite effective.
- *Appropriability of innovation*: High appropriability favours shareholder supremacy as against stakeholder inclusion.

Ramirez and Tylecote (2004) found that the pharmaceutical industry is high in novelty of innovation, degree of visibility of innovation and appropriability of innovation. According to conventional logic this should favour the outsider-dominated corporate governance structure of the UK versus the insider-dominated Swedish system for the UK/Swedish pharmaceutical company AstraZeneca. After the merger of Astra AB and Zeneca Group plc, AstraZeneca became an UK/Swedish/US hybrid. The US member (the outsider) made a crucial difference in the group's governance. The hybrid governance structure protected the company from short-term pressures within the UK stock market, and allowed it to continue investments in long-term innovation (*ibid.*). In the US, the process of *stem cell innovation* in which the company is involved is accompanied by the evolution of multidimensional governance through both public and private governance (Salter and Salter, 2010).

Further reading
Fernández-Ribas, Andrea (2009) 'Public Support to Private Innovation in Multi-Level Governance Systems: An Empirical Investigation', *Science & Public Policy*, 36(6), 457–67.
Moore, Mark and Hartley, Jean (2008) 'Innovations in Governance', *Public Management Review*, 10(1), 3–20.
Ramirez, Paulina and Tylecote, Andrew (2004) 'Hybrid Corporate Governance and its Effects on Innovation: A Case Study of AstraZeneca', *Technology Analysis & Strategic Management*, 16(1), 97–119.
Salter, Brian and Salter, Charlotte (2010) 'Governing Innovation in the Biomedicine Knowledge Economy: Stem Cell Science in the USA', *Science & Public Policy*, 37(2), 87–100.
Shapira, P. and Kuhlman, S. (eds) (2003) *Learning from Science and Technology Policy Evaluation: Experiences in the United States and Europe*, Cheltenham, Edward Elgar.
Tylecote, A. and Conesa, E. (1999) 'Corporate Governance, Innovation Systems and Industrial Performance', *Industry and Innovation*, 6(1), 25–50.

Innovation harmony

See **harmony**

Innovation killers

Innovation occurs at the intersection of invention and insight (NII Report, 2005). Joyce Wycoff (2003) recognised that the context for innovation emanates from interactions between the firm and its operating environment and that organisational culture nurtures and develops innovative ideas (Koulopoulus, 2010). Different cultures shape innovation models and structures differently (Pohlmann, 2005). While **ambidextrous organisations** master culture to support **incremental and radical innovations**, success factors for nurturing innovation are not always clear; nevertheless, failure factors that stifle innovation are well known (*ibid.*). Context factors that kill innovation include lack of the necessary time, effort and/or money. However, time, effort and resources will not guarantee the successful development of innovation. Isolated work without a socially networked atmosphere is an innovation killer, since innovation is based on cooperation and creativity in a socially well-embedded environment. *Tayloristic organisations* with specialisation of tasks and a *manufacturing mindset* are not likely to develop innovative technologies and products. In addition, incentives are not necessarily the motivator of innovation; sometimes more incentives discourage innovative behaviour and too many incentives lead to incentive reaching. Furthermore, bureaucratic behaviours destroy the incentive to innovate.

Corporate 'regime shifts' (changes in administrations that increase or facilitate access to resources) can both help and retard innovations (Feldman, 2007). The 3M Company lost its innovative edge under the leadership of James McNerney during 2001–05. He emphasised execution, streamlined operations and implemented Six Sigma programmes. However, efficiencies came at a price (Gunther *et* *al.*, 2010). 3M has since revived its innovation capabilities with George Buckley as chairman and CEO (ibid.*.).

Christensen *et al.* (2008) identified that paying too much attention to the company's most profitable customers, creating new products that do not help customers do the jobs they want to do and misguided application of financial analysis tools as killers of innovation. The discounted cash flow (DCF) and net present value (NPV) methods underestimate the real returns and benefits of more innovative products. Fixed and sunk cost considerations also disadvantage innovations in established firms. The emphasis on earnings per share as the primary driver of share price diverts resources towards **incremental innovations** at the expense of longer-horizon innovation investments (*ibid.*).

Further reading
Balsano, Thomas J., Goodrich, Nina E., Lee, Richard K., Miley, John W., Morse, Terri F. and Roberts, David A. (2008) 'Identify Your Innovation Enablers and Inhibitors', *Research Technology Management*, 51(6), 23–33.
Braganza, Ashley, Awazu, Yukika and Desouza, Kevin C. (2009) 'Sustaining Innovation Is Challenging to Incumbents', *Research Technology Management*, 52(4), 46–56.
Christensen, Clayton M., Kaufman, Stephen P. and Shih, Willy C. (2008) 'Innovation Killers', *Harvard Business Review*, 86(1), 98–105.
Feldman, Jonathan Michael (2007) 'The Managerial Equation and Innovation Platforms: The Case of Linköping and Berzelius Science Park', *European Planning Studies*, 15(8), 1027–45.
Gunther, Marc, Adamo, Marilyn and Feldman, Betsy (2010) '3M's Innovation Revival', *Fortune*, 162(5), 73–6.
Koulopoulus, Thomas (2010) 'How to Kill an Idea in 10 Easy Steps', *Journal for Quality & Participation*, 32(4), 8–11.
National Innovation Initiative Report: Innovate America, May 2005, http://www.compete.org/publications/detail/202/innovate-america (accessed on November 8, 2010).
Pohlmann, Markus (2005) 'The Evolution of Innovation: Cultural Backgrounds and the Use of Innovation Models', *Technology Analysis & Strategic Management*, 17(1), 9–19.
Wycoff, Joyce (2003) 'The "Big 10" Innovation Killers: How to Keep Your Innovation System Alive and Well', *Journal for Quality & Participation*, 26(2), 17–48.

Innovation management

Dankbaar (2003) defines innovation management 'as the creation of preconditions to promote human creativity, including strategic commitment and context management'. Innovation management is also 'seen as a process to foster the application of knowledge' (Igartua *et al.*, 2010). Brown (1997) described the management of innovation as the management of 'people, culture, communication, and organization of business processes, as well as technology'. Senior management has a significant role in ensuring the success of innovation programmes. It is management's role to articulate and communicate clear goals, foster and support innovation (open communication, share knowledge, build trust), engage more at the **front end of product development** and less at the back end, initiate NPD teams with a **product innovation charter**, establish strategy and implement processes to execute innovation (e.g. **product strategy, portfolio management** and the **new product development process**).

One hallmark of innovation leadership in the USA is a pioneering spirit (Lechleiter, 2010). Pioneering leaders of American innovation include Thomas Edison, Henry Ford, the Wright Brothers, the Mercks, Upjohns, Pfizers, Lillys, Thomas Watson, Steve Jobs and Bill Gates (*ibid.*). Henry Ford innovated in pursuit of his vision to build a reliable vehicle for the masses. Ford 'democratized the automobile'. With the Microsoft Windows operating system, Bill Gates brought modern computing to the masses.

As a process, innovation management is complex. It ranges from technical invention to final **commercialisation**. According to Roberts (2007), the overall management of **technological innovation** includes 'the organization and direction of human and capital resources toward effectively creating new knowledge, generation of technical ideas for new and enhanced products, manufacturing processes and services, developing those ideas into working prototypes and transferring them into manufacturing, distribution and sale. They may be based on a single dominant technology or multiple technologies.'

The outcomes of technological innovation may be incremental or radical in degree. They may be embodied in modifications of existing or entirely new products, processes or services, and may be oriented towards consumer, industrial or governmental use.

In his book *Innovation and Entrepreneurship: Practice and Principles*, Peter Drucker (1985) identified seven sources for innovative opportunity. They are the unexpected, incongruity, process need, changes in industry or market structure, demographics, changes in perception, mood and meaning, and new scientific and non-scientific knowledge.

Capturing value from innovation is a complex problem for managers of innovation. Returns on innovation can be diminished by imitators, customers, suppliers and other providers of complementary products and services (Teece, 1986). Innovation managers must understand market outcomes when they commercialise new technologies (Chesbrough *et al.*, 2006). The Profiting from Innovation (PFI) framework developed by Teece (1986) highlights the importance to commercial success of owning critical complementary technologies and/or controlling the bottleneck asset(s) in the value chain (Pisano and Teece, 2007). Sometimes it may be beneficial for technology managers to push technology into the public domain rather than keeping it proprietary and at other times it may be beneficial to promote modularity (*ibid.*).

The Product Development and Management Association (PDMA) Research Foundation identified seven pillars for successful innovation management. The seven pillars are measuring what matters, setting up for product development, using emerging NPD tools/methodologies, managing the development portfolio, identifying valid strategic opportunities, understanding customer needs and influencing others (Castellion, 2010).

Goffin and Mitchell (2005) introduced the *Innovation Pentathlon Framework* as an integral framework for management of innovation. This identified five elements of innovation management, the first three of which represent the development funnel: managing creativity and knowledge; prioritisation, which includes selecting and managing the portfolio; and implementation of new products, processes and services.

See also: **innovation process, innovative climate, technology-push innovation**

Further reading
Bessant, J. R. and Tidd, J. (2007) *Innovation and Entrepreneurship*, Hoboken, NJ, John Wiley & Sons.
Brown, D. (1997) 'Innovation Management Tools: A Review of Selected Methodologies', *EIMS Studies*, 30, Luxembourg, European Commission Directorate-General XIII Telecommunications, Information Market and Exploitation of Research.
Castellion, George (2010) personal communication.
Chesbrough, H., Vanhaverbeke, W. and West, J. (2006) *Open Innovation: Researching a New Paradigm*, Oxford, Oxford University Press.
Dankbaar, B. (2003) *Innovation Management in the Knowledge Economy*, Series on technology management, 7, London, Imperial College Press.
Drucker, Peter (1985) *Innovation and Entrepreneurship: Practice and Principles*, New York, Harper & Row.
Goffin, Keith and Rick, Mitchell (2005) *Innovation Management: Strategy and Implementation using the Pentathlon Framework*, New York, Palgrave Macmillan.
Igartua, Juan Ignacio, Garrigós, Jose Albors and Hervas-Oliver, Jose Luis (2010) 'How Innovation Management Techniques Support an Open Innovation Strategy', *Research Technology Management*, 53(3), 41–52.
Lechleiter, John C. (2010) 'The Frontier Within: Advancing American Medical Innovation', *Vital Speeches of the Day*, 76(9), 392–7.
New Product Institute of the PDMA Research Foundation (2009) 'Develop Your Own Best Practices in Innovation.' Workshop conducted at the PDMA Annual International Conference, Anaheim, California, November 2, http://newproductinstitute.org (accessed on December 2, 2009).
Pisano, Gary P. and Teece, David J. (2007) 'How to Capture Value from Innovation: Shaping Intellectual Property and Industry Architecture', *California Management Review*, 50(1), 278–96.
Roberts, Edward B. (2007) 'Managing Invention and Innovation', *Research Technology Management*, 50(1), 35–54.
Teece, D. J. (1986) 'Profiting from Technological Innovation', *Research Policy*, 15(6), 285–305.
Tidd, J., Bessant, J. R. and Pavitt, K. (2005) *Managing Innovation: Integrating Technological, Market and Organization Change,* 3rd edn, Chichester: John Wiley & Sons.

Innovation mindset

Thomas Kuczmarski (1998) described an innovation mindset as an attitude that stimulates and motivates individual employees, as well as cross-functional innovation teams, to adopt a belief in creating something new. People with an innovation mindset view failure as an intrinsic part of innovation. Companies practising an innovation mindset have a **new product strategy** and place innovation within a strategic and business context. Characteristics of an innovation mindset include:

1. Considering innovation as **creative problem solving**, not **blue sky ideas** and **brainstorming**.
2. Having a well-defined **new product development (NPD) process**.
3. Beginning the NPD process with front-end 'problem identification' and 'need intensity' research.

4. Using multifunctional teams with dedicated team members.
5. Developing compensation incentives to simulate an entrepreneurial environment.

According to Kuczmarski, top management commitment is the cornerstone of successful innovation.

Many of Kuczmarski's insights into the innovation mindset are substantiated as the practices of the best-performing companies in the **comparative performance assessment studies (CPAS)** of the Product Development and Management Association (PDMA Research Foundation).

Further reading
Kuczmarski, Thomas (1998) 'The Ten Traits of an Innovation Mindset', *Journal for Quality & Participation*, 21(6), 44.

Innovation-oriented country

China is an example of a country investing to become innovation oriented. Chinese President Hu Jintao outlined plans for building China into such a country at a national conference on science and technology held in Beijing in 2006 (chinanet.com). He outlined a new role for companies that would play a prominent part in innovation. The Chinese government is improving innovation-related laws and regulations and creating a favourable business environment for companies to invest in innovation. In addition, the plans call for funding cutting-edge technologies and basic research. For example, China is investing in material sciences and nanotechnology research with a long-term perspective. This is to help its universities and research institutions assume a leading role in **technological innovation**. Chinese innovation strategy consists of leapfrogging advanced countries with technological developments in key areas, funding breakthrough innovations in key technologies and utilising core technologies to achieve sustained and coordinated economic and social development. Goals for scientific and technological development include the development of high-tech manufacturing and information industries; raising agricultural production capacity; achieving breakthroughs in energy conservation, exploration and clean energy technologies; and the optimisation of the energy infrastructure.

In 2009 Chinese vice-premier Zhang Dejiang announced the *big plane project* to develop a global airliner. The goals for this project include increased fuel efficiency (the aeroplane will use between 12 per cent and 15 per cent less fuel) and a reduction in carbon emissions. The funding of the project is an example of how China is investing to become an innovation-oriented country. While China has put a great deal of focus on building technological innovation capabilities in key industries, it could be argued that it falls short on building its workforce for the twenty-first century, when there is a need for scientists, engineers and technologists with creativity, autonomy, problem-solving skills, industriousness and a passion for knowledge to not only create but realise the value from innovations.

Further reading
'China Sets to Build Innovation-Oriented Country', April 16, 2006, http://www.chinanet.com.cn/cn/bwdt/details.jsp?id=1495 (accessed on November 22, 2009).

Innovation outsourcing

Outsourcing is a strategic decision to substitute external sourcing for internal activity or to use externally provided activities to extend a firm's capabilities (Love and Roper, 2001). Innovation outsourcing is a common practice in R&D management. It is a new model of innovation involving global networks of partners. A global network might include 'U.S. chipmakers, Taiwanese engineers, Indian software developers, and Chinese factories' (Engardio *et al.*, 2005).

Both mature and novel technologies are targets for outsourcing. Loch *et al.* (2009) found five common drivers for successful outsourcing of technology. Two critical success drivers across both mature and novel technologies are a project-specific partner and in-house competence. Success drivers for each innovation source include managing expectations (customers tend to overestimate their capacity to sell the technology) and protecting one's intellectual property while collaborating with competitors and **new ventures**. The success drivers shift from project issues to manufacturing and distribution issues such as producibility and system compatibility when a mature technology is outsourced.

Love and Roper (2001) reported that European manufacturers utilised external technology outsourcing as a substitute for in-house development. For instance, Motorola buys complete designs for mature technologies used in low-end phones while retaining in-house innovation for high-end phones like the Razr (Engardio *et al.*, 2005). The outsourcing benefit is doubtful when a company outsources crucial technology or an integral item for innovation. If so, the outsourcing partners would have gained more knowledge than the customer. This enables outsourcing partners to overcome the customer in product design and innovation. This was the case between Motorola and its outsourced partner, Taiwan's BenQ Corp. Engardio *et al.* (2005) report that Motorola hired BenQ to design and manufacture its high-volume mobile phones. Afterwards, BenQ began selling phones in the Chinese market under its own brand. The probability of innovation outsourcing failure increases where the customer depends on partners for knowledge of critical innovation elements.

Further reading
Engardio, Pete, Einhorn, Bruce, Kripalani, Manjeet, Reinhardt, Andy, Nussbaum, Bruce and Burrows, Peter (2005) 'Outsourcing Innovation', *BusinessWeek*, 21 March, 84–94.
Loch, Christoph, Zhijian, Cui, Grossmann, Bernd and Ru, He (2009) 'Outsourcing Innovation: A Comparison of External Providers at Siemens', *Research Technology Management*, INSEAD Faculty Research & News, September/October, http://www.insead.edu/facultyresearch/research/documents/RD0909_001.pdf (accessed on December 7, 2009).
Love, James H. and Roper, Stephen (2001) 'Outsourcing in the Innovation Process: Locational and Strategic Determinants', *Papers in Regional Science*, 80(3), 317–36.
'The Customer Perspective: Sharing One's Knowledge May Be Required', *Bangkok Post*, February 17.

Innovation platforms

The term 'innovation platform' describes a given set of organisations or networks that incubate and sustain innovative teams tied to a given sector (Feldman, 2007). Consoli and Patrucco (2008) consider innovation platforms as a particular case of *technology infrastructure* that supports collective innovation. *Collective innovation* requires the coordination of complex, distributed innovation processes and the orchestration of supplier networks, outsourcing and user–producer relations (Crémér *et al.*, 2007).

For inter-organisational coordination, innovation platforms define the levels of engagement of each new player, the unidirectional or bidirectional flow of information and knowledge, and the extent of exchange across organisations. Innovation platforms are purposefully open to the entry of new players and new competences. The extent of contribution by each new player depends on the relative value of the internal competences measured against the collective goal. Both technology and innovation platforms provide directed and coordinated organisation versus the 'spontaneous' organisation found in market processes. However, innovation platforms are different from *technology platforms,* which are the hubs of technology industries. Some examples of technology platforms are Microsoft Windows (PC operating system), Intel processors (PC hardware) and the Sony PlayStation (game console) (Economides and Katsamakas, 2006).

Information technology (IT) plays a critical role in a firm's value chain. Successful IT innovators gain value by *creative adaptation* of existing functionality to enable entirely new applications (Kohli and Melville, 2009). Kohli and Melville (2009) identify customers; people and creativity; and processes as the three faces of adaptation. Successful companies combine these to form an *IT innovation platform* to create new sources of growth. Parcel service UPS reduced the number of days in the supply chain of customer goods from 30 to 17 and overall costs to customer by 40 per cent. TAL Apparel provides superior customer service using IT as an enabler and adapting systems to fit new purposes in a continuous process (*ibid.*).

See also: **IT innovation**

Further reading

Consoli, Davide and Patrucco, Pier Paolo (2008) 'Innovation Platforms and the Governance of Knowledge: Evidence from Italy and the UK', *Economics of Innovation & New Technology*, 17(7/8), 701–18.

Crémér, J., Garicano, L. and Prat, A. (2007) 'Language and the Theory of the Firm', *Quarterly Journal of Economics*, 122(1), 373–407.

Economides, Nicholas and Katsamakas, Evangelos (2006) 'Two-Sided Competition of Proprietary vs. Open Source Technology Platforms and the Implications for the Software Industry', *Management Science*, 52(7), 1057–71.

Feldman, Jonathan Michael (2007) 'The Managerial Equation and Innovation Platforms: The Case of Linköping and Berzelius Science Park', *European Planning Studies*, 15(8), 1027–45.

Kohli, Rajiv and Melville, Nigel P. (2009) 'Learning to Build an IT Innovation Platform', *Communications of the ACM*, 52(8), 122–6.

Innovation process

In the traditional view, innovation is a closed process controlled by the organisation. It is a structured process followed in the development of practice-based innovations. *Practice-based innovation* is a combination of product, process and organisational structure. The process consists of actions required to introduce a new or improved product (such as a new production process); new technologies;

new **business models** and practices; and new organisational structures. The innovation process is most commonly used to commercialise **technological innovations**. It begins at the **front end of innovation** and continues into **product development, commercialisation** and post-launch audit.

There are many models of the innovation process with a variety of phases and gates. Myers and Marquis (1969) have proposed a *three-step innovation process* for *industrial innovation*: 'idea development (the generation of a design concept), problem solving (technical efforts and problem solving in developing the proposed idea), and implementation (pilot production, inter area coordination)' (Tushman, 1977). They view the innovation process as consisting of three subprocesses:

- **Front end of innovation (FEI) process**
- **New product development (NPD) process**
- **Commercialisation process**

For van de Ven (1986), innovation is a developmental process in an interactive, social context. Van de Ven claimed that the central task in managing the innovation process was to understand the context in which social interactions occur among participants, and the resulting ideas that are developed. For Chesbrough (2003), the innovation process can also mean being open to acquiring new knowledge and exploiting it (that is, **open innovation**). Chesbrough and Teece (2002) argue that the open exchange of knowledge that fuels the innovation process may also give rise to conflicts between participants. Coordination and management are necessary if all participants in the *open innovation process* are to benefit from new products, services, new knowledge or through intellectual property rights (Shapiro and Varian, 1999). Grönlund *et al.* (2010) found that an effective innovation process contains collaborative, open knowledge creation activities and somewhat more closed, internal knowledge-exploitation activities. Laursen and Salter (2006) found that too much openness in the search for new ideas might have a negative impact on performance.

See also: **development funnel**

Further reading
Chesbrough, H. W. (2003) 'The Logic of Open Innovation: Managing Intellectual Property',
California Management Review, 45(3), 33–58.
Chesbrough, H. W. and Teece, D. J. (2002) 'Organizing for Innovation: When Is Virtual Virtuous?'
Harvard Business Review, 80(8), 127–35.
Grönlund, Johan, Sjödin, David Rönnberg and Frishammar, Johan (2010) 'Open Innovation and the Stage-Gate Process: A Revised Model for New Product Development', *California Management Review*, 52(3), 106–31.
Knickel, Karlheinz, Brunori, Gianluca, Rand, Sigrid and Proost, Jet (2009) 'Towards a Better Conceptual Framework for Innovation Processes in Agriculture and Rural Development: From Linear Models to Systemic Approaches', *Journal of Agricultural Education & Extension*, 15(2), 131–46.
Laursen, K. and Salter, A. (2006) 'Open for Innovation: The Role of Openness in Explaining Innovation Performance among UK Manufacturing Firms', *Strategic Management Journal*, 27(2), 131–50.

Myers, Summer and Marquis, Donald (1969) *Successful Industrial Innovation*, Washington, DC, National Science Foundation.

Shapiro, C. and Varian, H. R. (1999) *Information Rules: A Strategic Guide to the Network Economy*, Boston, MA, Harvard Business School Press.

Tushman, Michael L. (1977) 'Special Boundary Roles in the Innovation Process', *Administrative Science Quarterly*, 22(4), 587–605.

van de Ven, A. H. (1986) 'Central Problems in the Management of Innovation', *Management Science*, 32(5), 590–607.

Innovation scoreboard

Innovation scoreboards are instruments used for benchmarking *national innovation performance* on an annual basis. The *Lisbon strategy* forms the basis for the *European Innovation Scoreboard (EIS), Global Innovation Scoreboard (GIS)* and *Regional Innovation Scoreboard (RIS)*.

The European Innovation Scoreboard (EIS) is an annual publication. It is the most widely watched measure of competitiveness in the study of European technology policy (Schibany and Streicher, 2008). For analysis, the EIS uses a set of indicators covering different aspects of innovation performance and draws on statistics from a variety of sources. According to the EIS 2007 report, Sweden is the most innovative country, although it is less efficient in transforming strong innovation inputs into innovation outputs. The USA continued to hold its lead in 11 of the 15 indicators measured and expanded its lead in public R&D expenditures and high-technology exports. The *Summary Innovation Index (SII)* is a composite indicator of average innovation performance. Choice of indicators and methodology for calculating the SII have changed over time.

The Global Innovation Scoreboard (GIS) compares the innovation performance of the EU25 member states to that of the other significant R&D-performing countries in the world, such as Australia, Brazil, Canada, China, India, Israel, Japan, the Russian Federation and the USA. The GIS uses a more limited set of indicators than the EIS.

The Regional Innovation Scoreboard (RIS), again based on a limited number of indicators, compares innovation performance across 15 European countries.

There is an emerging concern in Europe that technology policy has a limited influence on the variables measured by the EIS.

Further reading

Ashmore, C. (2006) 'Viewpoint: Is Innovation Dead in the EU?', *Engineering Management*, 16(2), 8.

Innovation Nation. American: A Magazine of Ideas, May/June 2008, 2(3), 10.

Schibany, Andreas and Streicher, Gerhard (2008) 'The European Innovation Scoreboard: Drowning by Numbers?', *Science and Public Policy*, 35(10), 717–32.

Innovation strategy

An innovation strategy is a **competitive strategy** through which a company seeks **competitive advantage** in its ecosystem by developing new products and services before its competitors. Miller and Floricel (2004) found that *time to prototype* and *time to market* are two key *drivers of innovation strategy*. Industries devel-

oping new scientific knowledge have a long time to prototype, industries developing new engineering practice have a medium time to prototype, and industries
applying known science and engineering practice have short time to prototype
(*ibid.*). Time to market is long where regulatory compliance is burdensome and
short where there are no regulations and customers recognise the benefits of the
prototype (Germeraad, 2010).

Bowonder *et al.* (2010) outlined three *dimensions of innovation strategy* as
customer excitement, *competitive leadership* and *portfolio enrichment*. Firms can
generate customer excitement with *platform offerings* (e.g. Apple iPod, Nokia
mobile phones and Toyota INNOVA car platform), **co-creation** (e.g. 3M, P&G and
Boeing), *cycle time reduction* (e.g. Boeing, Gillette, Honda, Nissan and Toyota) and
brand value enhancement (e.g. P&G Oil of Olay, and Motorola RAZR). Adobe,
Canon, DuPont, P&G, Pixar, Tetra Pak and Toyota are examples of companies that
dominate competitors. Portfolio enrichment is guided by *innovation mutation* (e.g.
Syngenta micro-encapsulation technology, Apple iPod and iPhone and 3M adhe-
sive tapes); *creative destruction* (e.g. Canon electro-optical system cameras and
Microsoft Vista operating system for Windows); *market segmentation* (e.g.
Southwest Airlines and IKEA); and acquisition (e.g. BASF and Unilever) (*ibid.*).

Strategy making in an **innovation ecosystem** is an iterative process (Adner,
2006). Managers develop a **vision** of what market they want to enter, what
product they want to offer for sale and what constitutes successful performance.
This is followed by assessing risks with that plan, including interdependence risks,
initiative risks and integration risks. The **risk-analysis** process often forces
managers to reconsider their performance expectations and rethink their initial
plan.

Mapping the innovation ecosystem is the best way to determine whether
managers have set realistic performance expectations for a chosen innovation
strategy. Companies such as Intel, Nokia, SAP and Cisco create value in the
ecosystem that a single company outside the system could not create (*ibid.*).

Further reading
Adner, Ron Harvard (2006). 'Match Your Innovation Strategy to Your Innovation
  Ecosystem', *Harvard Business Review*, 84(4), 98–107.
Bowonder, B., Dambal, Anirudha, Kumar, Shambhu and Shirodkar, Abhay (2010)
  'Innovation Strategies for Creating Competitive Advantage', *Research Technology
  Management*, 53(3), 19–32.
Cooper, Robert G. and Edgett, Scott J. (2010) 'Developing a Product Innovation and
  Technology Strategy for your Business', *Research Technology Management*, 53(3), 33–
  40.
Germeraad, Paul (2010) 'Integration of Intellectual Property Strategy with Innovation
  Strategy', *Research Technology Management*, 53(3), 10–18.
Miller. R. and Floricel, S. (2004) 'Value Creation and Games of Innovation', *Research
  Technology Management*, 47(6), 25–37.

Innovation structures

Structuring innovation is a difficult task. *Innovation activity* is influenced by interaction between technology, organisational structure and financing. Eurostat
defines innovation activity as the 'introduction of a new or significantly improved

product (goods or service) or process and engagement in innovation projects not yet complete or abandoned. It excludes expenditure in areas linked to innovation activities' (Robson and Kenchatt, 2010). Innovation activity can be located in a large firm, a small firm or a joint venture between a large firm and a small firm, as in the case of large pharma and early-stage bio-tech firms (Gilson, 2010). Established firms benefit from economies of scale or scope in extensions of exiting products or markets. They are best suited **to incremental innovation**, while start-ups or small companies are better at **disruptive innovation**. A disruptive technology reflects a sharp break with existing products and the market for the disruptive technology is initially small. Established firms generally ignore such innovations because returns on an investment would not justify the **business case** (*ibid.*). Successful companies such as General Electric and Shell create new innovation structures to overcome the structural bias for incremental innovations built into their organisational structures.

Laurie and Scheer (2006) identified training units, funding/oversight mechanisms, incubators and autonomous groups as innovation structures used by successful companies. Training units stimulate innovation by teaching the skills and habits required for **disruptive innovation**, for example the Learning & Development unit at Syngenta designs and conducts training courses to promote innovative and leadership qualities. General Electric uses funding/oversight mechanisms to guide innovation. CEO Jeff Immelt created a senior management team, called the Commercial Council. This council prioritises innovation proposals and growth strategies put forward by GE's business leaders. Shell Oil Company uses incubators to accelerate ideas and created the GameChanger programme to foster innovative ideas and prioritise them. The GameChanger unit develops new businesses that are outside and between the company's existing lines of business. It operates outside the constraints and priorities of Shell's regular business operations. Dow Chemical Company uses autonomous growth groups to start new businesses and identify and develop non-core businesses. They explore concepts outside the comfort zone of the core businesses. A modest budget allows the autonomous group to iterate solutions until they achieve success. The group has the option to transfer ideas to the core business or to seek funding for launching new businesses.

See also: **innovation governance**

Further reading
Anthony, Scott D., Johnson, Mark W. and Sinfield, Joseph V. (2008) 'Institutionalizing Innovating', *MIT Sloan Management Review*, 49(2), 49.
Gilson, Ronald J. (2010) 'Locating Innovation: the Endogeneity of Technology, Organizational Structure, and Financial Contracting', *Columbia Law Review*, 110(3), 885–917.
Henderson, R. and Clark, K. (1990) 'Architectural Innovation: The Reconfiguration of Existing Product Technologies and the Failure of Established Firms', *Administrative Science Quarterly*, 35, 9–30.
Laurie, D., Doz, Y. and Scheer, C. (2006) 'Creating New Growth Platforms', *Harvard Business Review*, 84, 80–90.
Robson, Stephanie and Kenchatt, Martin (2010) 'First Findings from the UK Innovation Survey 2009', *Economic & Labour Market Review*, 4(3), 28–35.

Innovation surveys

Innovation surveys measure innovation in the European Union (EU), Organization for Economic Cooperation and Development (OECD) countries and a growing number of non-OECD countries. Innovation surveys can either focus on a specific innovation (object approach) or firm-level innovation activity (subject approach). The Oslo Manual focuses on the subject approach.

The international guidelines established in the OECD/Eurostat Oslo Manual (OECD, 1992; OECD/Eurostat, 1997, 2005) had a significant impact on survey type and content. The third edition of the Oslo Manual (OECD/Eurostat, 2005) has undergone a number of significant changes to keep *innovation measurement* current. The most significant changes in the third edition include a broadened **definition of innovation** and an increased recognition of innovation in services and low-tech manufacturing. A counter-criticism of the third edition is that *innovation measurement* is now focused on activities and processes, not on output measures, which were the original goal of innovation surveys. The focus on outputs was evident in the first edition of the Oslo Manual.

The UK has participated in the innovation surveys since 2001. The 2001 survey reported wider innovation activities in the UK, with higher levels of **product and process innovation** reported in 2005. The 2007 survey found a higher share of firms with innovation activity in the UK, with increased levels of product and process innovation reported in the 2009 survey (Robson and Kenchatt, 2010). The UK IS in 2009 improved the question routing by changing the layout, which improved data quality.

See also: **comparative performance assessment study**

Further reading
Bloch, Carter (2007) 'Assessing Recent Developments in Innovation Measurement: The Third Edition of the Oslo Manual', *Science & Public Policy*, 34(1), 23–34.
Camacho, J. A. and Rodriguez, M. (2008) 'Patterns of Innovation in the Services Sector: Some insights from the Spanish Innovation Survey', *Economics of Innovation & New Technology*, 17(5), 459–71.
OECD/ Eurostat (2005) *Guidelines for Collecting and Interpreting Innovation Data – The Oslo Manual*, 3rd edn, Paris: OECD.
Robson, Stephanie and Kenchatt, Martin (2010) 'First Findings from the UK Innovation Survey 2009', *Economic & Labour Market Review*, 4(3), 28–35.

Innovation value chain

Porter (1980) defined the *value chain* as the sequential set of primary and support activities that a firm performs to turn inputs into value-added outputs for its external customers. The innovation value chain is a framework for measuring *innovation performance*. Hansen and Birkinshaw (2007) view this as a sequential, three-phase process that involves **idea generation**, idea development and the diffusion of developed concepts. The key activities performed in all three phases are a search for ideas within the organisation, in other business units or externally; selecting, enriching and funding them; and promoting and diffusing ideas throughout the company (*ibid.*). Companies typically face one of three broad 'weakest-link' scenarios: they may be idea poor, conversion poor or diffusion poor (.). Best-practice innovation companies such as P&G, Sara Lee, Shell and Siemens modify their innovation practices to keep pace with dynamic changes in the marketplace. They systematically correct flaws found in idea sourcing, idea conversion and diffusion to sustain innovation performance.

The introduction of radical technologies creates unique challenges to the innovation value chain when the introduction draws the attention of non-innovation value-chain actors. For example, non-innovation value-chain actors such as advocacy groups may oppose the introduction of a radical technology such as genetically modified (GMO) seeds in agriculture. *GMO seeds* pose potentially uncertain health and safety side effects. This was the case with Monsanto when it introduced Round-up ready bio-engineered seeds. It faced the difficult task of balancing its 'economic risk (protect intellectual property investments and receive a financial return), with environmental risk (attempt to prevent the spread of unwanted GMOs) and social risk (farmers in the Third World becoming increasingly dependent upon developed nations)' (Hall and Martin, 2005).

See also: **harmony, innovation process**

Further reading
Hall, Jeremy K. and Martin, Michael J. C. (2005) 'Disruptive Technologies, Stakeholders and the Innovation Value-Added Chain: A Framework for Evaluating Radical Technology Development', *R&D Management*, 35(3), 273–84.
Hansen, Morten T. and Birkinshaw, Julian (2007) 'Innovation Value Chain', *Harvard Business Review*, 85(6), 121.
Porter, M.E. (1980) *Competitive Strategy: Techniques for Analyzing Industries and Competitors*, New York: Free Press.

Innovator

In his book *They Made America,* Harold Evans (2004) argues that the terms 'innovation' and 'innovators' should be reserved for real change. He expresses his dismay that there is a widespread acceptance of Joseph Schumpeter's definition of innovation as **entrepreneurship.** Evans makes a distinction between **entrepreneurship** and **innovation**. He argues that entrepreneurship is the assumption of risk, and that it may not be innovative. An **entrepreneur** assumes the risk if he (or she) opens a new car dealership, but this does not mean that he is innovative unless he is the first person to start a car dealership.

For Evans, innovation is not invention. It is inventiveness put to use; that is, a *practical innovation*. Molecular biologist Herbert Boyer was not satisfied with splicing a gene; he started a company to mass-produce hormones. Boyer is an innovator, according to Evans.

An entrepreneur may become the enemy of innovation. David Sarnoff, who headed RCA, was a classic entrepreneur, but he opposed the introduction of FM radio. He invested heavily in AM radios and AM broadcasting through RCA's National Broadcasting Corporation (NBC). He sabotaged the efforts of Edwin Howard Armstrong, the inventor of FM, even though NBC had the right of first refusal for Armstrong's invention. Armstrong the inventor became an innovator and started his own company.

Further reading
Evans, Harold (2004) *They Made America: From the Steam Engine to the Search Engine: Two Centuries of Innovators*, New York, Little, Brown.

Innovator's dilemma

Great companies can do everything right and still lose to **disruptive innovations** or disappear altogether (Christensen, 2000). In *The Innovator's Dilemma*, Christensen discusses the way in which industry dynamics are driven by competition between two innovation regimes: a dominating regime characterised by sustaining (incremental) technologies and a 'disruptive' regime that introduces new and simpler technologies (*ibid.*). According to him, a successful company with established products will be overcome by new entrants unless its managers know when to quit the established businesses. When to quit is the dilemma. The Honda Supercub, Intel's 8088 processor and the hydraulic excavator became successful new products from new entrants while market leaders hung on to established businesses too long. Managers facing the innovator's dilemma need to realise that their current capabilities, cultures and practices have limitations (*ibid.*).

Christensen (2000) urged firms to become proficient in balancing the needs of current customers as well as the anticipated needs of prospective customers. This balance requires both **exploitation and exploration** capabilities and skills in managing **incremental and radical innovations**. These key capabilities and skills are difficult to integrate (Corso and Pellegrini, 2007), yet their mastery is essential to success. Quantum, Control Data, IBM's PC Division, Allen Bradley and Hewlett-Packard's desk-jet initiative succeeded because these companies created organisations to commercialise both incremental and disruptive technologies successfully.

A popular understanding of the innovator's dilemma is that senior managers who fail to invest in disruptive innovations are irrational and are blinded by current customers and high margins. However, in *The Innovator's Solution* (Christensen and Raynor, 2003), Christensen shifted his view that managers who do not respond to disruptive innovation are acting in their organisation's best interest (Henderson, 2006). The innovator's solution distinguishes between low-end disruptions and new market disruptions. *Low-end disrupters* introduce cheaper or lower-quality products or services than existing products with no other performance improvements. *New market disrupters* provide better performance on dimensions the current customers do not materially value (*ibid.*). Christensen's work highlights the role of market-facing competence in shaping a firm's response to disruptive innovation (*ibid.*).

See also: **ambidextrous organisation**

Further reading
Christensen, Clayton M. (2000) *The Innovator's Dilemma*, New York, HarperCollins.
Christensen, Clayton and Raynor, Michael. (2003) *The Innovator's Solution*, Boston, MA, Harvard Business School Press.
Corso, Mariano and Pellegrini, Luisa (2007) 'Continuous and Discontinuous innovation: Overcoming the Innovator's Dilemma', *Creativity & Innovation Management*, 16(4), 333–47.
Henderson, Rebecca (2006) 'The Innovator's Dilemma as a Problem of Organizational Competence', *Journal of Product Innovation Management*, 23(1), 5–11.

Integrated product and process development (IPPD)

The United States Department of Defense (DOD) document DOD 5000.2-R defines integrated product and process development (IPPD) as 'a management technique that simultaneously integrates all essential acquisition activities through the use of multidisciplinary teams to optimize the design, manufacturing, and supportability processes. IPPD facilitates meeting cost and performance objectives from product concept through production, including field support' (Parry, 1999). IPPD gained increased prominence with the publication of *DoD Handbook* in 1996 (Parry, 1999). According to the DoD handbook (1996), the IPPD process involves three key components: tools, teams, and processes (Schaeffer, 1997).

- *Tools* include the management, organisational, analytical, measurement and design tools needed for development. Those highlighted include information technology and decision support aids, integrated master schedules, **design for manufacture**, **rapid prototyping**, virtual prototyping, modelling and **simulation, CAD/CAE/CAM** and **earned-value management**.
- *Teams* consisting of all stakeholders in the product are essential to the integrated development of products and related processes. Development activities are those that collectively lead to the definition of the end product and its associated processes such as manufacturing and support.
- *Processes* such as the application of **systems engineering (SE)** are necessary to IPPD success.

By applying the IPPD process, companies 'design it right the first time' and cut scrap, rework and engineering change proposals. They get reductions in cycle time and costs, while improving quality (*ibid.*).

Further reading
DoD Guide to Integrated Product and Process Development (Version 1.0) (1996) https://www.acquisition.gov/sevensteps/library/dod-guide-to-integrated.pdf (accessed on January 1, 2011).
Parry, Thomas J. (1999) 'IPPD Gains Increased Emphasis through Publication of New DoD Handbook', *Program Manager*, 28(4), 66–70.
Schaeffer, Mark D. (1997) 'IPPD – One Year After', *Program Manager*, 26(1).

Intellectual capital

Intellectual capital is knowledge applied to create economic value. Human capital, instructional capital and individual capital are examples of intellectual capital that enable **innovation**.

OECD (1999) defines intellectual capital as the economic value of a company's *organisational ('structural') capital* and *human capital*. Information and communication technologies (ICT) and software systems, distribution networks and supply chains are examples of structural capital. Human capital includes employees, customers and suppliers. Often, intellectual capital is used interchangeably with *intangible assets* or *knowledge assets*. OECD considers intellectual capital as a subset of knowledge assets.

Further reading
OECD (1999) 'OECD Symposium on Measuring and Reporting of Intellectual Capital, Amsterdam', Paris, OECD.

Intellectual property (IP)

According to the World Intellectual Property Organization, intellectual property refers 'broadly to the creations of the human mind such as inventions, literary and artistic works, symbols, names, images, and designs used in commerce' (http://www.wipo.int/about-ip/en/). Intellectual property rights protect the interests of creators and encourage **innovation** by giving them property rights over their creations.

Intellectual property is divided into two categories: *industrial property* and *copyright*. Industrial property includes patents, trademarks and industrial designs. Copyright includes literary and artistic works.

In the USA, the Patent and Trademark Office issues patents, which are granted for 20 years. Patent rights exclude others from using the patented invention for their own commercial purposes. Utility model patents provide protection for 6–10 years.

Article 15(1) of the Agreement between the World Intellectual Property Organization and the World Trade Organization (TRIPS, 1995) provides a now internationally accepted definition of a *trademark*. It stipulates that 'any sign, or any combination of signs, capable of distinguishing the goods or services of one undertaking from those of other undertakings, shall be capable of constituting a trademark'. A trademark thus indicates the source of goods (Lemley and McKenna, 2010). It distinguishes trademarked goods from goods made by others and trademark rights prevent others from selling goods with a similar mark. However, trademarks do not preclude others from selling the same goods under a different mark. A *servicemark* is similar to a trademark: it distinguishes the source of service.

Copyright is a protection provided to the authors of original works of authorship. Copyright is granted for both published and unpublished intellectual works and gives the owner the exclusive right to sell his or her literary or artistic works or architectural designs.

It is commonly accepted that *IP rights* are necessary to justify innovative activity. However, intellectual property rights may be a hindrance for innovation and general social welfare. Some researchers argue that the monopoly profits generated by intellectual property play a minor role in driving economic growth (Boldrin and Levine, 2004).

Further reading
Agreement between the World Intellectual Property Organization and the World Trade Organization (TRIPS), December 22, 1995, http://www.wipo.int/treaties/en/agreement/trtdocs_wo030.html (accessed on January 1, 2011).
Boldrin, Michelle and Levine, David K. (2004) '2003 Lawrence R. Klein Lecture: The Case Against Intellectual Monopoly', *International Economic Review*, 45(2), 327–50.
Lemley, Mark A. and McKenna, Mark P. (2010) 'Owning Markets', *Michigan Law Review*, 109(2), 137–89.
World Intellectual Property Organization, http://www.wipo.int/about-ip/en/ (accessed on December 10, 2009).

Intelligent enterprise (IE)

An intelligent enterprise is characterised as agile, adaptive, self-aware, with fuzzy borders and the ability to morph into new and better forms (Feng and Wang, 2009). According to Feng and Wang (2009):

- *Agile* means that the enterprise can execute a production order quickly after closing a web-based purchase.
- *Adaptive*, self-regulating and self-optimising mean that it can adjust revenue, profit and cost in the short term, in step with a changing business climate.
- *Self-awareness* means that it learns from them and adapts to the markets.
- Fuzzy borders and a mesh-like structure mean that it can restructure and scale the organisation as needed.

IE can morph into new and better forms over the long term. Morphing includes adjustments of the business and operational parameters as well as significant structural changes (*ibid.*).

Further reading
Feng, Dexiong and Wang, Xiang (2009) 'A Study on a New Pattern of Enterprise-Intelligent Enterprise', in Hu Shuhua and Hamsa Thota (eds) *Proceedings of the 4th International Conference on Product Innovation Management*, Wuhan, China, Hubei People's Press, Part 3, 856–60.

Interaction analysis

Interaction analysis is an analytical method employed in **design anthropology**. It augments traditional anthropology's emphasis on societal and political issues with a detailed analytical orientation to the organisation of users' practices (Button and Sharrock, 1998). Interaction analysis offers a critical understanding of how users actually use a product versus the flawed assumptions underlying the product design. Suchman's (1987) classic study about task completion with a photocopier showed that the assumptions about users' interpretive behaviour that were embodied in the photocopier interface were fundamentally flawed (Burr and Mathews, 2008). Flawed assumptions were responsible for people's inability to carry out certain routine photocopying tasks. Suchman's research outlined alternative conceptions of users that offered much more promise as a basis for the design of systems (*ibid.*).

Further reading
Burr, Jacob and Matthews, Ben (2008) 'Participatory Innovation', *International Journal of Innovation Management*, 12(3), 255–73.
Button, G. and Sharrock, W. W. (1998) 'The Organizational Accountability of Technological Work', *Social Studies of Science*, 28(1), 73–102.
Suchman, L. A. (1987) *Plans and Situated Actions: The Problem of Human-Machine Communication*, Cambridge, Cambridge University Press.

Interactive value creation

See **crowdsourcing**

Intrapreneur

An intrapreneur is an innovator who is willing to take risks to turn an idea into a successful, finished product in large corporations. *Intrapreneurship* is 'developing a **new venture** within an existing organization to exploit a new opportunity and create economic value' (Pinchot and Pellman, 1985). **Entrepreneurship** is developing a new venture outside an existing organisation (Parker, 2009). Intrapreneurship renews and revitalises existing businesses and improves their overall performance. However, *intrapreneuring* can be frustrating and difficult.

The organisational need for stability and internal politics are two killers of intrapreneurship, while a supportive culture is an enabler (Teltumbde, 2006). Effective intrapreneurship requires a management environment in which innovation can flourish and an organisational culture that transforms ordinary employees into successful intra-corporate entrepreneurs (Pryor and Shays, 1993).

Further reading
Parker, Simon C. (2009) 'Intrapreneurship versus Entrepreneurship: An Analysis of PSED II Data', *Academy of Management Proceedings*, 1–6.
Pinchot, Gifford and Pellman, Ron (1985) *Intrapreneuring: Why You Don't Have to Leave the Corporation to Become an Entrepreneur*, Scranton, PA, Harper Collins.
Pryor, Austin K. and Shays, Michael E. (1993) 'Growing the Business with Intrapreneurs', *Business Quarterly*, 57(3), 42–8.
Teltumbde, Anand (2006) 'Entrepreneurs and Intrapreneurs in Corporations', *Vikalpa: The Journal for Decision Makers*, 31(1), 129–32.

Invention

The United States Patent Law defines an invention as 'a new, useful process, machine, improvement, etc. that did not exist before'. Invention is the result of unique intuition or genius. The *invention process* covers all efforts to create new ideas and develop them into new products and processes (Roberts, 2007). Inventions can be solutions to common design problems, minor improvements to an existing system, a fundamental improvement to an existing system, a new generation of a system, or a rare scientific discovery or pioneering invention. **Inventors** bridge science and practical problem-solving skills.

Further reading
Kurzweil, Ray (2004) 'Kurzweil's Rules of Invention', *Technology Review*, 107(4), 46–8.
Roberts, Edward B. (2007) 'Managing Invention and Innovation', *Research Technology Management*, 50(1), 35–54.

Knowledge-driven innovation

Knowledge-driven innovation is the use of knowledge or knowledge assets for innovation. It is a management process. The key issue in knowledge-driven innovation is to utilise organisational knowledge to develop commercially viable new products and services while also building new organisational competences (Sohel-uz-Zaman *et al.*, 2006).

Sohel-uz-Zaman *et al.* (2006) claim that the first step in the *knowledge-driven innovation process* is to prepare a product or service specification from data gathered from customers. The second step is an analysis of organisational competences, skill level and technological capability to deliver the selected product/service (ibid*.)*. If internal capability is sufficient, new knowledge is not required. An organisation can further develop its new product/service idea or implement a product/service improvement plan. If internal capability is not sufficient, organisation needs to acquire new knowledge or create it internally in order to build the new product/service (*ibid.*).

A change in the external environment triggers a need for additional internal competence development and new knowledge. Knowledge-driven innovation can be viewed as a responsive mechanism in the interface between the external environment and the organisation (*ibid.*).

Further reading
Sohel-uz-Zaman, Abu Saleh Md, Kefan, Xie, Munir, Zunaira and Yun, Chen (2006) 'Innovation in Knowledge Reality – A Search for New Dimensions', *Proceedings of International Conference on Innovation & NPD*, Chennai, India, December.

Lateral thinking

What is thinking? 'Thinking is really a functioning knowledge' (Ilyenkov, 2007). Lateral thinking, a term created by Edward de Bono (1971), 'seeks new ways of looking at a problem rather than proceeding by logical steps' (Sloane, 2006). In contrast to *vertical thinking*, which uses the traditional processes of logic, lateral thinking involves a deliberate search for alternatives by focusing on elements that would normally be ignored by logical thinking. Challenging labels, searching for alternatives, fractionation, exposure to other fields and reversal are some of the techniques used in lateral thinking. Sloane (2006) distinguishes between creative thinking, which involves introducing something new or a new idea, and lateral thinking, which involves introducing a new approach.

Further reading
De Bono, E. (1971) *The Use of Lateral Thinking: A Textbook of Creativity,* Harmondsworth, Penguin.
Ilyenkov, E. V. (2007) 'Knowledge and Thinking', *Journal of Russian & East European Psychology*, 45(4), 75–80.
Sloane, Paul (2006) *The Leader's Guide to Lateral Thinking Skills: Unlocking the Creativity and Innovation in You and Your Team*, 2nd edn, London, Kogan Page.

Launch (product launch)

New product introductions involve two significant activities: **product development** and product launch. Product launch is the most expensive phase of the **new product development (NPD) process**. Launch costs often exceed all the combined costs in all earlier development stages of NPD (Beard and Easingwood, 1996). The role of the launch is to maximise the chances of profitability by achieving acceptance of new product in the target market. Four strategic issues in new product launch are what to launch, where to launch, when to launch and how to launch (Song and Montoya-Weiss, 1998).

The launch phase in the NPD process begins when senior management makes a commitment to build the new product and accepts the risk of the high cost of production and marketing. This can involve a 'go' decision for a technical or marketing launch. Technical launch activities include building regular production capacity, initial production runs, review of initial product and process, production scale-up and availability of product for market testing. Marketing launch activities include completion of augmented product requirements, finalisation of marketing plan, creation of launch plan, developing and testing all implementation components, and creation of post-launch marketing plan. The product is also market tested and the product and process, and/or marketing programmes, are revised as necessary. All operations are fine tuned as required for a successful product launch.

The ability to launch new products successfully is critical to the continued growth and profitability of product development firms. However, new product introductions experience a high rate of failure (Montoya-Weiss and Calantone 1994). Hultnik *et al.* (2000) found a number of heuristics about how products are launched that actually lead to failure rather than success.

Further reading
Beard, C. and Easingwood, C. (1996) 'New Product Launch: Marketing Action and Launch Tactics for High-Technology Products', *Industrial Marketing Management*, 25, 87–103.
Crawford, Merle and Di Benedetto, Anthony (2008) *New Products Management*, 9th edn, New York, McGraw-Hill.
Hultnik, Erik Jan, Hart, Susan, Robben, Henry S. J. and Griffin, Abbie (2000) 'Launch Decisions and New Product Success: An Empirical Comparison of Consumer and Industrial Products', *Journal of Product Innovation Management*, 17(1), 5–23.
Montoya-Weiss, Mitzi and Calantone, R. (1994) 'Determinants of New Product Performance: A Review and Meta-Analysis', *Journal of Product Innovation Management*, 11(5), 397–417.
Song, X. M. and Montoya-Weiss, M. M (1998) 'Critical Development Activities for Really New versus Incremental Products', *Journal of Product Innovation Management*, 15, 124–35.

Launch decision

The launch decision is also known as the *product commercialisation decision*. It consists of two distinctly different decisions: strategic and tactical. *Strategic launch decisions* are made before the physical development of the product (Nyström, 1985). These decisions include the firm's overall orientation towards its NPD efforts (firm strategy), the nature of the new product to be developed (product strategy) and the nature of the market and the competitive position of the new product (market strategy) (Hultink *et al.*, 1999). Strategic launch decisions become a subset of the **product innovation charter** as defined by Merle Crawford (Crawford, 1980). *Tactical launch decisions* include the amount of marketing investment, the number of product versions launched, how and where to distribute and promote the new product and its price.

There are differences in launch decisions between consumer and industrial products. Strategic launch decisions in consumer products companies seem to be defensive, whereas they seem to be offensive for industrial product companies (Hultink *et al.*, 2000).

See also: **launch strategy**

Further reading
Crawford, C. Merle (1980) 'Defining the Charter for Product Innovation', *Sloan Management Review*, Fall, 3–12.
Hultink, Erik Jan, Hart, Susan J., Robben, Henry S. J. and Griffin, Abbie J. (1999) 'New Consumer Product Launch: Strategies and Performance, *Journal of Strategic Marketing*, 7(3), 153–74.

Hultink, Erik Jan, Hart, Susan, Robben, Henry S. J. and Griffin, Abbie (2000) 'Launch Decisions and New Product Success: An Empirical Comparison of Consumer and Industrial Products', *Journal of Product Innovation Management*, 17(1), 5–23.

Nyström, H. (1985) 'Product Development Strategy: An Integration of Technology and Marketing', *Journal of Product Innovation Management*, 2, 25–33.

Launch plan

Launch is generally the least well-managed part of the **new product development (NPD) process**. Guiltinan (1999) defined the launch plan as ;a coordinated set of strategies and tactics for introducing a product to a target market'. Calantone and Montoya-Weiss (1993) suggested that a well-formulated launch plan is essential for maximising acceptance of the new product in the target market. A launch plan involves identifying target markets, establishing marketing mix roles, forecasting financial outcomes and controlling the project. Launch planning begins with the development of marketing strategy and continues in parallel with product development and testing.

A launch plan includes decisions about launch strategy as well as tactics. Because products differ in the degree of newness, each new product launch plan must produce a different demand outcome. In a *pioneering launch strategy*, a launch plan needs to simulate adoption and lead to diffusion. A differentiated new product in an established market requires a plan to simulate trial purchase as a precursor to adoption, while products positioned as upgrades to existing product offerings achieve customer migration from the competition (Guiltinan, 1999).

Further reading
Calantone, R. G. and Montoya-Weiss, M. M. (1993) 'Product Launch and Follow On' in W.E. Soulder and J.D. Sherman (eds), *Managing New Technology Development*, New York, McGraw-Hill, pp. 217–48.

Guiltinan, Joseph P. (1999) 'Launch Strategy, Launch Tactics, and Demand Outcomes', *Journal of Product Innovation Management*, 16, 509–29.

Launch strategy

A launch strategy is the 'decisions and activities necessary to present a product to its market and begin to generate revenue from sales of the new product' (Hultink *et al.*, 1997). The launch strategy must be aligned with the strategy of the firm. The firm decides whether the technology or the markets drive **new product development (NPD)** projects, which NPD projects to pursue (that is, innovation, imitation or cost-reduction projects) and what the objectives are for the development project (Hultink *et al.*, 2000). Launch decisions can be classified as strategic if they are made early in the new product development project and are difficult or expensive to change later on. This includes decisions about the degree of innovation in a product and the scale of production entry. Strategic launch decisions determine what to launch, where to launch, when to launch and why to launch. Strategic launch variables are product strategy (product innovativeness, relative product newness and cycle time), market strategy (target markets, stages of product life cycle and target market growth), competitive position (number of competitors, product advantage) and firm strategy (innovation strategy, drivers of growth).

Launch strategy is complemented by launch tactics that generally involve marketing mix adjustments that govern the 'how' of the launch. Marketing mix (tactical) launch variables are product (breadth of product line), promotion (advertising, promotion), pricing and distribution (distribution intensity, sales forecast effort). The elements of the marketing mix are pricing, product and branding, advertising and promotion, and distribution. Marketing mix decisions are more easily modified late in the development cycle than are strategic launch decisions.

A new product may be highly innovative and superior to competitive products, but still fail due to a poor launch (Campbell, 1999). For example, product categories with network effects cause unique, competitive dynamics and change customer response patterns. Products with network effects call for product launch strategies that differ from what might normally be effective (Lee and O'Connor, 2003). In the *network effect*, the value of a product to its users increases as more people use it (Katz and Shapiro, 1985).

Lee and O'Connor (2003) found that some proven and tested normal launch tactics such as skimming pricing, niche targeting and secrecy prior to introduction may be harmful to the successful launch of networked products, while others such as the bundling strategy and pre-announcing strategy remain relevant.

See also: **launch decision, launch plan**

Further reading
Campbell, T. (1999) 'Back in Focus', *Sales and Marketing Management*, 151(2), 56–61.
Hultink, Erik Jan, Griffin, Abbie, Hart, Susan and Robben, Henry S. J. (1997) 'Industrial New Product Launch Strategies and Product Development Performance', *Journal of Product Innovation Management*, 14, 243–57.
Hultink, Erik Jan, Hart, Susan, Robben, Henry S. J. and Griffin, Henry S. J. (2000) 'Launch Decisions and New Product Success: An Empirical Comparison of Consumer and Industrial Products', *Journal of Product Innovation Management*, 17(1), 5–23.
Katz, M. and Shapiro, C. (1985) 'Network Externalities, Competition, and Compatibility', *American Economic Review*, 75(3), 424–40.
Lee, Yikuan and O'Connor, Gina Colarelli (2003) 'New Product Launch Strategy for Network Effects Products', *Journal of the Academy of Marketing Science*, 31(3), 241–55.

Lead users

Lead users are at the leading of edge of markets. They experience needs that will become prevalent in the marketplace in the future (Schreier and Prügl, 2008).

Eric von Hippel (1986) introduced the concept of lead users being the most suitable innovators, as they innovate to find solutions to unfilled needs and derive benefit from innovating. Commercial firms engage lead users to develop attractive **user innovations**. Lead users are a highly promising source of innovation for new product development tasks (Schreier and Prügl, 2008).

Firms collaborating with lead-user customers report an increased rate of new product success. For example, lead-user concepts developed at 3M showed an average of eight times the sales potential of traditionally developed concepts (Lilien *et al.*, 2002). Ordinary users do not have the technical knowledge of lead users to contribute ideas suitable for entry into the **new product development**

process (Magnusson, 2009). Instead, companies can use them to learn about user needs.

Further reading
Lilien, Gary L., Morrison, Pamela D., Searls, Kathleen, Sonnack, Mary and Von Hippel, Eric (2002) 'Performance Assessment of the Lead User Idea-Generation Process for New Product Development', *Management Science*, 48(8), 1042–59.
Magnusson, Peter R. (2009) 'Exploring the Contributions of Involving Ordinary Users in Ideation of Technology-Based Services', *Journal of Product Innovation Management*, 26(5), 578–93.
Schreier, Martin and Prügl, Reinhard (2008) 'Extending Lead-User Theory: Antecedents and Consequences of Consumers' Lead Userness', *Journal of Product Innovation Management*, 25(4), 331–46.
Von Hippel, E. (1986) 'Lead Users: A Source of Novel Product Concepts', *Management Science*, 32(7), 791–806.

Lean launch

The concept of lean involves continuing efforts to eliminate waste. A lean launch refers to a new product roll-out that involves a limited inventory commitment. It is based on the principle of **postponement**, supported by a response-based logistics system that is capable of rapidly responding to initial sales success (Bowsersox *et al.*, 1999). A lean launch enables companies to achieve profit from partial or niche successes.

Managing a lean launch involves planning lean inventory requirements to support successful product sales while reducing the quantity of failed product, achieving flexibility and responsiveness to demand through agile manufacturing and postponement, and use of information communications technologies to achieve successful replenishment.

Further reading
Bowersox, Donald J., Stank, Theodore P. and Daugherty, Patricia J. (1999) 'Lean Launch: Managing Product Introduction Risk through Response-Based Logistics', *Journal of Product Innovation Management*, 16, 557–68.

Lean product development

Lean product development attempts to increase productivity by lowering the cost structure of the development process. It does so by simplifying work units and processes, and developing the capability for rapid changeover from one process to another (Githens, 2005). Achieving lean product development sometimes involves defining and stabilising a robust but flexible, common technology platform that allows customers to select the features they want to add to that platform. Lean enterprises create a highly efficient system designed for accepting incremental product launches. Dell Computer and Toyota exemplify lean development.

Liker and Morgan (2006) outline four process principles of lean product development using the Toyota lean product development system as an example:

1. Establish the customer-defined value to differentiate value added from waste.

2. Thoroughly explore alternative solutions early in the development process with maximum design space.
3. Create a levelled product development process flow with a knowledge work job shop to level workload, meet customer demand rate and minimise queues. Synchronise work processes across functional departments and virtually eliminate rework.
4. Reduce variation, and create flexibility and predictable outcomes with rigorous standardisation.

Lean product development is often confused with **agile development**, as both attempt productive improvements.

Further reading
Githens, G. (2005) 'Agile and Lean Development: Old Wine in New Jugs?', *Visions*, 29(4), 7.
Liker, Jeffrey K. and Morgan, James M. (2006) 'The Toyota Way in Services: The Case of Lean Product Development', *Academy of Management Perspectives*, 20(2), 5–20.

Linear model of innovation

While Joseph Schumpeter is a seminal person in **technological innovation**, *Rupert MacLaurin* constructed one of the first taxonomies for measuring technological innovation (Godin, 2008). The linear model of innovation proposed by MacLaurin (1947) is one of the first theoretical frameworks developed for understanding science and technology and its relationship to the economy. MacLaurin (1947) suggests that technological innovation is a sequential process composed of four distinct stages: fundamental research, applied research, engineering development and production engineering. To him, science and technology can be broken down into five distinct stages: fundamental research, applied research, engineering development, production engineering and service engineering (Godin, 2008).

The MacLaurin model is now largely discredited, however, and has been superseded by systemic models of innovation.

Further reading
Godin, Benoît (2008) 'In the Shadow of Schumpeter: W. Rupert Maclaurin and the Study of Technological Innovation', *Minerva: A Review of Science, Learning & Policy*, 46(3), 343–60.
MacLaurin, W. Rupert (1947) 'Federal Support for Scientific Research', *Harvard Business Review*, Spring,
385–96.

Living labs

Living labs is a new concept for **research and development (R&D)** and **innovation** in support of the Lisbon strategy for jobs and growth in Europe. It brings together stakeholders in co-creative ways and stresses human-centric involvement versus a technology-centric environment in the development of new ICT-based services and products. The aim is to include a diverse mix of user representatives in all the development stages of products and prototypes, before the innovations are launched into the market.

In the living labs, users are involved throughout the **innovation process**. Involvement begins with people in the streets as well as users and user communities as contributors and co-creators of innovations. Users get the opportunity to gain an understanding of a new product or service in their everyday life because living lab activities occur throughout the day. This immersion allows potential users to influence the final design of products through close cooperation with the company.

The living lab network supports the 'innovation life cycle' for all stakeholders in the European innovation system: end-users, small and medium enterprises (SMEs), corporations, public sector and academia. The Portuguese government found living labs as a research methodology consistent with its policy measures to design and implement a method to foster an innovation-friendly market. Nokia also found the concept an excellent opportunity to speed up its innovation.

See also: **co-creation, user-centred design**

Further reading

Niitamo, V. P., Kulkki, S., Eriksson, M. and Hribernik, K. A. (2006) 'State-of-the-Art and Good Practice in the Field of Living Labs', *Proceedings of the 12th International Conference on Concurrent Enterprising: Innovative Products and Services through Collaborative Networks*, Milan, Italy, 349–57.

Schumacher, J. and Feurstein, K. (2007) 'Living Labs – The User as Co-creator', *ICE*. euniversityforum.org, http://www.openlivinglabs.eu/concept.html (accessed on December 18, 2009).

Localisation

Localisation involves tailoring a product or marketing strategy so that it meets the needs of customers in specific foreign countries. To better reach potential customers abroad, companies make changes to their products and services in view of differences in culture, language and legal regulations. Localisation is any endeavour that involves the management of multilingual activity (Fischer, 1998). In the software industry, localisation is a common practice: *software localisation* is the process of transforming programs (and their accompanying printed documentation) from one language version to another.

Further reading

Fischer, Christina (1998) 'Global Necessities Bring Localization To the Fore- – with Broader Scope', *Seybold Report on Publishing Systems*, 28(3), 23–33.

Lieu, Tina (1997) 'Software Localization: The Art of Turning Japanese', *Computing Japan*, 4(12), 23–7.

Low-end disruption

Disruption refers to the extent to which an emerging customer segment, not the mainstream customer segment, sees value in the innovation at the time of introduction, which over time disrupts the products that mainstream customers use (Abernathy and Clark, 1985; Adner, 2002; Govindarajan and Kopalle, 2004, 2006).

A **disruptive innovation** introduces a different set of features, performance and price attributes relative to the existing product. It is an unattractive offer for mainstream customers at the time of product introduction because of inferior performance on the attributes these customers value. However, the new product may perform better on an alternative dimension and create a new market (it may be easier to use or offered at lower cost, for example) (Schmidt and Druehl, 2008).

Low-end disruption occurs when a new entrant disrupts a market 'by initially targeting the least demanding customers in the market with an adequate and lower-priced product' (Christensen, 2003). Low-end disruptors nearly always win when the product performance is below the levels customers can use or absorb. Low-end disruption does not create entirely new markets, but can create new growth businesses resulting in new entrants challenging existing players (*ibid.*). A recent low-end disruption occurred with digital photography, which posed a major challenge to Kodak (Dewald and Bowen, 2010), the incumbent player in the chemistry-based photographic industry.

See also: **disruptive innovation, new market disruption**

Further reading
Abernathy, William J. and Clark, Kim B. (1985) 'Mapping the Winds of Creative Destruction', *Research Policy*, 14(1), 3–22.
Adner, Ron (2002) 'When Are Technologies Disruptive? A Demand-Based View of the Emergence of Competition', *Strategic Management Journal*, 23(8), 667–88.
Christensen, Clayton (2003) 'HBSP Audio Conference – Beyond the Innovator's Dilemma', http://hbswk.hbs.edu/archive/3709.html (accessed on December 3, 2010).
Dewald, Jim and Bowen, Frances (2010) 'Storm Clouds and Silver Linings: Responding to Disruptive Innovations Through Cognitive Resilience', *Entrepreneurship: Theory & Practice*, 34(1), 197–218.
Govindarajan, Vijay and Kopalle, Praveen (2004) 'Can Incumbents Introduce Radical and Disruptive Innovations?' Working Paper #04-001, Cambridge, MA, Marketing Science Institute.
Govindarajan, Vijay and Kopalle, Praveen K. (2006) 'The Usefulness of Measuring Disruptiveness of Innovations Ex Post in Making Ex Ante Predictions', *Journal of Product Innovation Management*, 23(1), 12–18.
Schmidt, G. M. and Druehl, C. T. (2008) 'When Is a Disruptive Innovation Disruptive?', *Journal of Product Innovation Management*, 25, 347–69

L

Management role to support NPD

New product innovation is a cross-functional effort. A cross-functional effort brings together various functional capabilities needed for executing new product design and development. Top management support is necessary to achieve the successful integration of members from different functions into a cohesive **new product development (NPD) team** (Hitt *et al.*, 1999).

The primary role of top management is to provide leadership. Leadership orchestrates shared values and **vision** and focuses organisational efforts towards clear goals. Management can communicate the vision throughout the organisation. It can focus the organisation's attention on achieving innovation goals and loosely couple functions involved in new product development and innovation through shared values. Subtle leadership is critical to the success of 'more innovative' projects. Brown and Eisenhardt (1995) argue that subtle control by senior managers leads to a common understanding of goals and a coherent product concept. A coherent product concept produces internal consistency in new product design. It also ensures that new product design is compatible by looking at it in the initial stages of the NPD process.

NPD activities in which managers can show support for new product innovation include establishing a **product strategy, portfolio management, new product development process** and initiation of NPD team with a **product innovation charter**.

Further reading
Brown, S. L. and Eisenhardt, K. M. (1995) 'Product Development: Past Research, Present Findings, and Future Directions', *Academy of Management Review*, 20, 343–78.
Hitt, Michael A., Nixon, Robert D., Hoskisson, Robert F. and Kochhar, Rahul (1999) 'Corporate Entrepreneurship and Cross-Functional Fertilization: Activation, Process and Disintegration of a New Product Design Team', *Entrepreneurship: Theory & Practice*, 23(3), 145–67.

Market definition

The definition of a 'market' in common trade is different from its regulatory definition. In common trade, a market 'is a large group of potential customers, with common needs or problems, who purchase a common class of products and/or services for similar use or application'. Less commonly understood by innovation professionals is the regulatory definition.

In US antitrust law, the focus is on 'monopoly power' in the market such as mergers and economic concentration. The regulatory definition of a market is 'the methods contained in the 1982 Horizontal Merger Guidelines released by the US

Justice Department and indices that roughly correspond to those identified in Brown Shoe [an early horizontal merger case]' (Pleatsikas and Teece, 2001). Some economists define a market by utilising interchangeability of use or cross-elasticity of demand for products and services, price sensitivity of sales, industry or public recognition of submarkets (such as high-technology manufacturing or environmental technologies). They also utilise a product's unique characteristics and uses, specialised vendors and production facilities, distinct customers and prices as additional factors to define markets (*ibid*.). However, a narrow focus on these market factors involves a weakness. It excludes consideration of substitute products and could result in the assertion of narrow technology markets and overestimation of market power. *Market power* is defined by the Guidelines as the 'ability to profitably maintain prices above competitive levels for a significant period of time' (Glick and Campbell, 2007).

Pleatsikas and Teece (2001) argue that traditional antitrust analysis is insufficient to define markets and evaluate market power in knowledge-intensive high-technology industries. In such markets, entry with duplicate or substitutable products is often difficult because of trade secrets or **intellectual property** protections (*ibid*.). It may even be competitively unwise to enter high-technology markets with duplicate products since product **differentiation** is a source of **competitive advantage**.

Unless a market is defined correctly, analysis of market concentration is unreliable in assessing market power (Hyman and Kovacic, 2004). The US Department of Justice makes a distinction between markets for technologies and markets for innovations (USDOJ, 1995). Markets for technologies refer to existing markets for licensed intellectual property. Markets for innovation are markets for future technologies (such as contract research, R&D joint ventures, technology alliances and technical services) (Rao, 2005).

Further reading
Glick, Mark A. and Campbell, Donald (2007) 'Market Definition and Concentration: One Size Does not Fit All', *Antitrust Bulletin*, 52(2), 229–37.
Hyman, David A. and Kovacic, William E. (2004) 'Monopoly, Monopsony, and Market Definition: An Antitrust Perspective on Market Concentration among Health Insurers', *Health Affairs*, 23(6), 25–8.
Kauper, Thomas E. (1983) 'The 1982 Horizontal Merger Guidelines: Of Collusion, Efficiency, and Failure', *California Law Review*, 71(2), 497.
Markovits, Richard S. (2002) 'On the Inevitable Arbitrariness of Market Definitions', *Antitrust Bulletin*, 47(4), 571–601.
Pleatsikas, Christopher and Teece, David (2001) 'The Analysis of Market Definition and Market Power in the Context of Rapid Innovation', *International Journal of Industrial Organization*, 19, 665–93.
Rao, P. M. (2005) 'Sustaining Competitive Advantage in a High-Technology Environment: A Strategic Marketing Perspective', *Advances in Competitiveness Research*, 13(1), 33–47.
U.S. Department of Justice (USDOJ) (1995). *Antitrust Guidelines for the Licensing of Intellectual Property*, Washington, DC, Government Printing Office.

M

Market orientation

Narver and Slater (1990) define market orientation as 'an organizational culture that most effectively and efficiently creates the necessary behavior for the creation of superior value for buyers and superior firm performance'. It was conceived as a customer-centric concept (Coley *et al.*, 2010). Sheth and Uslay (2007) claim that a customer-centric framework 'is limiting for the conceptualization of marketing' because the consumer is a separate market factor from the producer's customer. Some researchers believe that a strong market orientation is required for **radical innovations** (Day, 1994), while others believe it leads to **incremental innovations** (Lukas and Ferrell, 2000). Kohli and Jaworski (1990) view market orientation as a process of implementing the marketing concept.

Market orientation consists of three behavioural components: customer orientation, competitor orientation and interfunctional coordination (*ibid.*). Customer-orientation behaviours are developing sufficient knowledge of one's target buyers and then continuously creating greater value for them. In competitor-orientation behaviour, a seller understands competitors' strengths, weaknesses, opportunities and threats (**SWOT analysis**) and their long-term capabilities and strategies. Interfunctional orientation is an alignment of functional areas, incentives and the creation of interfunctional dependency. It is also being sensitive and responsive to the perceptions and needs of all other departments in the **business**.

A market orientation may not be the best orientation for all companies. Kahn (2001) cautions that implementing a market orientation may or may not improve product development performance in apparel and textile organisations. This may be due to the fact that the greatest challenge for creating a market-oriented firm is to determine exactly which behaviours will align with the market. The business performance of market-oriented firms can be improved by satisfactorily responding to questions such as: What to change? What to change to? How to cause a change? (Kaur and Gupta, 2010).

Further reading
Coley, Linda Silver, Mentzer, John T. and Cooper, Martha C. (2010) 'Is "Consumer Orientation" A Dimension of Market Orientation in Consumer Markets?', *Journal of Marketing Theory & Practice*, 18(2), 141–54.
Day, George S. (1994) 'The Capabilities of Market-Driven Organizations', *Journal of Marketing*, 58, 37–52.
Day, George S. (1999) *The Market Driven Organization: Understanding, Attracting, and Keeping Valuable Customers*, New York, Free Press.
Kahn, Kenneth B. (2001) 'Market Orientation, Interdepartmental Integration, and Product Development Performance', *Journal of Product Innovation Management*, 18(5), 314–23.
Kaur, Gurjeet and Gupta, Mahesh C. (2010) 'A Perusal of Extant Literature on Market Orientation – Concern for its Implementation', *Marketing Review*, 10(1), 87–105.
Kohli, A. K. and Jaworski, B. J. (1990) 'Market Orientation: The Construct, Research Propositions, and Managerial Implications', *Journal of Marketing*, 54(2), 1–18.
Lukas, Bryan A. and Ferrell, O. C. (2000) 'The Effect of Market Orientation on Product Innovation', *Journal of the Academy of Marketing Science*, 28(2), 239–48.
Narver, J. C. and Slater, S. F. (1990) 'The Effect of a Market Orientation on Business Philosophy', *Journal of Marketing*, 54, 20–35.
Sheth, Jagdish and Uslay, Can (2007) 'Implications of the Revised Definition of Marketing: From Exchange to Value Creation', *Journal of Public Policy & Marketing*, 26(2), 302–7.

M

Market research

The primary function of market research is to define the problem and identify a market opportunity to assist the company in making decisions. Market research includes research activities for sensing, learning from and understanding customers, competitors and macro-level market forces. Market research can be explorative or confirmatory (McQuarrie, 2006). *Explorative research* is discovery research. It is qualitative (**focus groups**, customer visits). *Confirmatory research* is quantitative (**conjoint analysis**, surveys). **Best-practice** companies involve customers throughout the development process and measure customer response throughout the **product development process**.

There are four levels of sophistication for market research (Kahn and Barczak, 2006). According to Kahn and Barczak (2006), level-one companies do not conduct market research; they rely on anecdotal evidence. Level-two companies use market research reactively. They rely heavily on pilot testing or product testing to obtain customer feedback. Level-three companies use **concept tests**, product tests (**beta testing**) and market tests (**gamma testing**) across projects. **Incremental innovation** projects receive less testing than 'more innovative' projects. Level-four companies build market research into their **new product development (NPD) process**.

Best-practice companies use both qualitative and quantitative research in their NPD process. Non-best-practice firms overvalue quantitative research and under-value qualitative research (McQuarrie, 2006). Reliance on quantitative research to fund **radical innovation projects** leads to failure.

See also: **ethnography research, market testing, strategy at business unit level**

Further reading
Cooper, Robert G., Edgett, Scott J. and Kleinschmidt, Elko J. (2002) *Improving New Product Development Performance and Practices, Benchmarking Study*, Houston, TX, American Productivity & Quality Center.
Kahn, Kenneth B. and Barczak, Gloria (2006) 'Perspective: Establishing an NPD Best Practices Framework', *Journal of Product Innovation Management*, 23(2), 106–16.
McQuarrie, Edward F. (2006) *The Market Research Toolbox*, 2nd edn, Thousand Oaks, CA, Sage.

Market testing

Market testing of new products involves the trial reproduction on a small scale of the planned full-scale marketing programme for a new product (Enright, 1958). Its purposes are to determine the acceptability of a new product, the effectiveness of marketing plan, probable marketing success and to detect any significant problems overlooked in planning the marketing programme (*ibid.*). However, the majority of new consumer products fail in the test market because they do not contain a demonstrable consumer advantage over competitive brands (Cadbury, 1975).

Both consumer and manufacturing companies use market testing to evaluate the new product together with its marketing plan. Manufacturing companies market test goods and services. Consumer companies market test their packaged goods, durable goods and services. Three categories of market tests used are the pseudo sale method, control sale method and full sale method (Crawford and Di Benedetto, 2008).

The pseudo sale method includes speculative sale and simulated test market (STM). The control sale method includes informal selling, direct marketing and mini-marketing. The full sale method includes test marketing, roll-out by application, roll-out by influence, roll-out by geography and roll-out by trade channel (*ibid.*).

Pseudo sale is cheap and fast. Many manufacturing companies use 'pseudo sale' to obtain quick market knowledge and follow it up with applications roll-out. A consumer goods company may use an STM to get a quick read of the market followed by a geographical roll-out, or an STM followed by a mini-market and then full launch (*ibid.*). With advances in information and communications technology (ICT), companies can gain deep insights at the level of the household and the level of the firm.

Further reading
Cadbury, N. D. (1975) 'When, Where, and How to Test Market', *Harvard Business Review*, 53(3), 96–105.
Crawford, Merle and Di Benedetto, Anthony (2008) *New Products Management,* 9th edn, New York, McGraw-Hill.
Enright, Ernest J. (1958) 'Market Testing', *Harvard Business Review*, 36(5), 72–80.
Klompmaker, Jay E., Hughes, G. David and Haley, Russel I. (1976) 'Test Marketing in New Product Development', *Harvard Business Review*, 54(3), 128–38.
Silk, Alvin J. and Urban, Glen L. (1978) 'Pre-Test-Market Evaluation of New Packaged Goods: A Model and Measurement Methodology', *Journal of Marketing Research*, 15(2), 171–91.

Marketing innovation

The American Marketing Association (AMA) defines marketing as 'an organizational function and a set of processes for creating, communicating and delivering value to customers and for managing customer relationships in ways that benefit the organization and its stakeholders' (Keefe, 2004). The *Oslo Manual* (2005) defines marketing innovation as 'the implementation of a new marketing method involving significant changes in product design or packaging, product placement, product promotion or pricing'. This is the *4Ps model for marketing strategies*: product, price, placement and promotion.

Product design changes are changes in form and appearance of the product. They do not alter the functionality of a product or its user characteristics. They also include changes in packaging. This is significant, because packaging determines the appearance of products in industries such as foods, beverages and detergents. Product placement methods are methods used to sell goods and services. Promotion methods include new marketing campaigns, advertising in traditional and new media or branding efforts to promote a firm's products and services. Pricing includes pricing strategies to promote a firm's products or services.

Further reading
Bloch, Carter (2007) 'Assessing Recent Developments in Innovation Measurement: The Third Edition of the Oslo Manual', *Science & Public Policy*, 34(1), 23–34.

Keefe, L. (2004) 'What Is the Meaning of "Marketing"?', *Marketing News*, September 15, 17–18.

OECD/Eurostat (2005) *Guidelines for Collecting and Interpreting Innovation Data – The Oslo Manual*, 3rd edn, Paris, OECD.

Marketing launch

See **launch**

Mass customisation

Mass customisation is 'the technologies, and systems used to deliver goods and services that meet individual customers' need at mass production efficiency level' (Tseng and Jiao, 2001). According to Frank Piller, it is the practice of creating products and services geared specifically for individual customers (Tseng and Piller, 2003). Strategic and tactical mechanisms for mass customisation include modular product design, use of a finite set of production processes to provide a large number of product configurations, delayed differentiation and customer **co-design** (Kumar and Williams, 2006).

There are four types of mass-customisation companies (Gilmore and Pine, 1997): collaborative customisers work with customers to choose the optimal product; adaptive customisers let customers do the customising themselves; cosmetic customisers sell the same basic product to different segments; transparent customisers do not tell their customers that they are customising the product for them.

Mass-customised products do not carry the price premiums connected with craft customisation (when craftspeople customise the product for different customers). Customers pay a premium price for the added satisfaction they derive from individualisation. The added value of an individualised solution for an individual customer is the defining characteristic of mass customisation.

Adidas Salomon AG is a leader in mass customisation. It has integrated customers into its requirements-elicitation process and customers co-design their own products (Berger and Piller, 2003).

Further reading
Berger, Christoph and Piller, Frank T. (2003) 'Customers as Co-Designers', *Manufacturing Engineer*, 82(4), 42–5.
Gilmore, James H. and Pine, B. Joseph, II (1997) 'The Four Faces of Mass Customization', *Harvard Business Review*, January–February, 91–101.
Kumar, Asbok and Williams, H. James (2006) 'Mass Customization Is Strategic Imperative', *Grand Rapids Business Journal*, 24(36), 36.
Tseng, M. M. and Jiao, J. (2001) 'Mass Customization', in G. Salvendy (ed.), *Handbook of Industrial Engineering*, 3rd edn, New York, John Wiley & Sons, pp. 684–709.
Tseng, M. M. and Piller, F. T. (eds) (2003) *The Customer Centric Enterprise: Advances in Mass Customization*, Berlin, Springer.

Mergers and acquisitions (M&A)

Mergers and acquisitions (M&A) are traditionally defined as the purchase of entire companies or specific assets by another company. Ahern and Weston (2007) added joint ventures, alliances and divestitures to the M&A definition. Mergers enable a firm to cope with change more rapidly than sole dependence on internal organic growth. Major shocks to operating environments create a greater potential role for M&As. For example, price instability in the oil industry (Weston *et al.*, 1999), deregulation in the banking industry (Becher, 2000) and the increased cost of discovery of molecules in the pharmaceutical industry (Weston, 2001) are associated with increased M&A activity. Other drivers of M&A include technology change, globalisation, commoditisation, low growth, chronic excess capacity (consolidation), fragmentation and augmented capabilities (Ahern and Weston, 2007).

In successful M&A, a combination of assets has greater value (that is, synergy) than the sum of its individual parts. Successful acquirers emphasise cultural due diligence, operational due diligence and financial due diligence. In the high-tech industry, Cisco became a role model for other high-tech companies by succeeding repeatedly in using acquisitions to reshape itself and fill gaps in product lines (Miller, 2000).

Many companies fail to create shareholder value after M&A. In 1999, KPMG reported that more than 8 in 10 companies failed to create shareholder value after M&A. One failure factor may be due to the acquiring company not selecting the right brand strategy for the combined brand portfolio (Yohn, 2010).

See also: **due diligence for technology deals**

Further reading
Ahern, Kenneth R. and Weston, J. Fred. (2007) 'M&As: The Good, the Bad, and the Ugly', *Journal of Applied Finance*, 17(1), 5–20.
Becher, D. A. (2000) 'The Valuation Effects of Bank Mergers', *Journal of Corporate Finance*, 6(2), 189–214.
Miller, Roger (2000) 'How Culture Affects Mergers and Acquisitions', *Industrial Management*, 42(5), 22–6.
Weston, J. F. (2001) 'Mergers and Acquisitions as Adjustment Processes', *Journal of Industry, Competition, and Trade*, 1(I), 395–410.
Weston, J. F., Johnson, B. and Siu, J. A. (1999) 'Mergers and Restructuring in the World Oil Industry', *Journal of Energy Finance and Development*, 4(2), 149–83.
Yohn, Denise Lee (2010) 'M&A Brand Strategies', *Mergers & Acquisitions: The Dealermaker's Journal*, 45(8), 46–7.

Middle manager as innovator

Rosabeth Moss Kanter (2004) studied the effectiveness of middle managers working in large corporations. Her studies revealed that middle managers who foster **innovation** share personal traits such as 'thoroughness, persistence, discretion, persuasiveness, and comfort with change'. Effective middle managers, just below officer level, design systems, carry out plans and redirect their staff's activities to further business goals (*ibid.*).

According to Kanter (2004), innovative middle managers know that uncertainties are reduced as the innovation project journeys through various stages in the **innovation process** and they choose projects carefully with a long-term view. Innovative middle managers are also detailed, practise participative management and are persuasive (*ibid.*). They prepare carefully for meetings and present information professionally. Their insight into organisational politics allows them a sense of the **champions** needed at different stages in the innovation process (*ibid.*).

Innovative middle managers flourish under senior managers who consciously encourage innovation and achievement.

Further reading
Kanter, Rosabeth Moss (2004) 'The Middle Manager as Innovator', *Harvard Business Review*, 82(7/8), 150–61.

Models of innovation

See **corporate model of innovation, entrepreneurial model of innovation**

Modular product design

Modularity (or modularisation) is an 'approach for organizing complex products and processes efficiently, by decomposing complex tasks into simpler portions' (Baldwin and Clark, 1997). Simpler portions can be managed independently and yet they operate together as a whole. The goal of modular design is to reduce cost by grouping components that have a high dependence on one another. Modular design, early in the design process, enables efficient **design for assembly (DFA)** later in the process.

Product modularity allows a wide variety of products to be constructed from a smaller set of different components (Ulrich and Tung, 1991). This approach promotes standardisation and reuse of existing modules for developing new products (Pahl and Beitz, 1996). Quality can be built into modular product during the conceptual design phase by utilising quality principles such as **axiomatic design** and **robust design** (Nepal *et al.*, 2006).

Modularity is the foundation of modular product design. Modular product design methods allow designers to achieve the optimal module composition with maximum modularity. The benefits of modularity range from development to production (Ulrich and Tung, 1991).

Baldwin and Clark (1997) attribute the success of the computer industry to modularity. It allows flexibility (that is, reconfiguration and customisation options), which enables customers to customise options to build a final product that suits their needs (*ibid.*). Promoting modularity can be both beneficial and hazardous to capturing value from **innovation** (Pisano and Teece, 2007). It is beneficial if the firm retains competence in (and control over) the systems integration function (*ibid.*).

M

See also: **profiting from innovation**

Further reading
Baldwin, C. Y. and Clark, K. B. (1997) 'Managing in an Age of Modularity', *Harvard Business Review*, 75(5), 84–93.
Gershenson, J. K., Prasad, G. J. and Zhang, Y. (2003) 'Product Modularity: Definitions and Benefits', *Journal of Engineering Design*, 14(3), 295–313.
Lai, X. and Gershenson, J. K. (2008) 'Representation of Similarity and Dependency for Assembly Modularity', *International Journal of Advanced Manufacturing Technology*, 37(7/8), 803–27.

Nepal, Bimal, Monplaisir, Leslie and Singh, Nanua (2006) 'A Methodology for Integrating Design for Quality in Modular Product Design', *Journal of Engineering Design*, 17(5), 387–409.

Pahl, G. and Beitz, W. (1996) *Engineering Design: A Systematic Approach*, 2nd edn, London, Springer.

Pisano, Gary P. and Teece, David J. (2007) 'How to Capture Value from Innovation: Shaping Intellectual Property and Industrial Architecture', *California Management Review*, 50(1), 278–96.

Ulrich, K. and Tung, K. (1991) 'Fundamentals of Product Modularity', *Issues in Design/Manufacture Integration*, 39, 73–7.

M

National innovation systems (NIS)

Richard Nelson, Columbia University, and Nathan Rosenberg, Stanford University, define a national innovation system (NIS) as 'the set of institutions whose interactions determine the innovative performance of national firms' (Hill, 2007). A country's national innovation system comprises a set of organisations and institutions and linkages between them, which support the generation, diffusion and application of scientific and technological knowledge and innovation. Public and private universities, large industrial research laboratories and federal R&D agencies sustain the US NIS.

The blueprint for building the NIS was *The Vannevar Bush Report*, published in 1945. The central point of this report, 'Science – The Endless Frontier', was that 'generous public support of fundamental research in the sciences yields enormous benefits to the nation' and 'scientific progress insures health, prosperity, and security as a nation in the modern world'. The Bush Report became the post-Second World War blueprint for the US Science and Technology Policy (STP). Before the Bush Report, this was *ad hoc*.

The USA built key elements of its NIS based on the premise that 'deep understanding of scientific principles is the basis for technological progress'. In the public sector, the USA established the *National Science Foundation* in 1950. This embraced the role of US universities as teachers and investigators of basic research and reformed the country's engineering education system. In the private sector, corporations invested in scientific research and development (R&D) for commercial purposes (that is, applied research). Private corporations designed their R&D labs like university campuses, built them away from corporate headquarters and staffed them with eminent scientists from the natural sciences and engineering. The educational system and research in physical, mathematical, engineering and biological sciences became the backbone of US innovation capability and the country led in innovation across multiple science, technology and engineering (STE) fields such as aerospace, electronics, pharma and nanotechnology.

In the early twenty-first century, the USA recognised the need to train well-educated scientists in order to retain its innovation leadership. Well educated means scientists with creativity, autonomy, problem-solving skills, industriousness and a passion for knowledge. During 2005–07, the USA reshaped its *innovation policy*. Two reports, the National Academies' *Rising Above the Gathering Storm* and the Council on Competitiveness's *Innovate America,* were published in 2005. They played a crucial role in developing President Bush's American Competitiveness Initiative and the America COMPETES Act in 2007.

The Innovate America Report (2005) represents a paradigm shift in US innovation policy. It redefined America's tasks in the twenty-first century as distinctly different from those in the second half of the twentieth century. America's new task 'is to transition from optimizing organizations for efficiency and quality to optimize its entire society for innovation'. In addition, the National Innovation Initiative (NII) of the Council on Competitiveness defined innovation 'as the intersection of invention and insight, leading to the creation of social and economic value'. While the Bush Report (1945) focused America's efforts on economic gain as an innovation output, the NII refocuses innovation efforts on the creation of both economic and social values.

See also: **national system of innovation**

Further reading
Augustine, Norman R. (2005) 'Rising above the Gathering Storm: Energizing and Employing America for a Brighter Economic Future'. http://www7.nationalacademies. org/ocga/testimony/gathering_storm_energizing_and_employing_america2.asp (accessed on December 5, 2010).
Bush, Vannevar (1945) *Science – The Endless Frontier*, United States Government Printing Office, Washington, DC, http://www.nsf.gov/od/lpa/nsf50/vbush1945.htm (accessed on December 5, 2010).
Galli, R and Teubal, M. (1997) 'Paradigmatic Shifts in National Innovation Systems' in C. Edquist (ed.), *Systems of Innovation. Technologies, Institutions and Organizations*, London, Pinter, pp. 342–70.
Hill, Christopher T. (2007) 'The Post-Scientific Society', *Issues in Science & Technology*, 24(1), 78–84.
'Innovate America', National Innovation Initiative Summit and Report, May 2005, Council on Competitiveness, http://www.compete.org/publications/detail/202/innovate-america/ (accessed on December 5, 2010).
Schibany, Andreas and Streicher, Gerhard (2008) 'The European Innovation Scoreboard: Drowning by Numbers?', *Science and Public Policy*, 35(10), 717–32.
The America Creating Opportunities to Meaningfully Promote Excellence in Technology, Education, and Science Act (COMPETES). H.R. 2272 (July 31, 2007), http://science. house.gov/legislation/leg_highlights_detail.aspx?NewsID=1938 (accessed on December 5, 2010).

National innovative capacity

Furman *et al.* (2000) define national innovative capacity as 'an economy's potential, at a given point in time, for producing a stream of commercially relevant innovations'. National innovative capacity depends on the common innovation infrastructure, cluster-specific environment for innovation and quality of linkages (*ibid.*):

- The common innovation infrastructure includes human and financial resources devoted to scientific and technological advances, policies to foster innovative activity and technological sophistication.
- Clusters offer important externalities for innovation. For example, interactions between firms located within a cluster allow firms to perceive both the need and the opportunity for innovation and the ability to act rapidly to turn new opportunities into reality.
- Strong clusters reinforce the common infrastructure and also benefit from it. Without strong linkages, a nation's scientific and technical advances can diffuse to other countries before exploitation at home.

The National Innovation Capacity Index measures and reports national innovation capacities annually as part of the Global Competitiveness Report (Porter and Stern, 2002). Georgia Tech's 2007 national 'High Tech Indicators' (HTI), based on export of high-tech products, ranked China as the number one country in technological standing. However, the 2006–07 Global Competitiveness Index (GCI/World Economic Forum, 2006) ranked China 54th among the 125 countries it measured. The differences in these indicators measuring the innovative capacity of nations provide multiple perspectives that complement each other (Porter *et al.*, 2009).

Further reading
Furman, Jeffrey L.. Porter, Michael E. and Stern, Scott (2000) 'Understanding the Drivers of National Innovative Capacity', *Academy of Management Proceedings & Membership Directory*, A1–A6.
Porter, Alan L., Newman, Nils C., Roessner, J. David, Johnson, David M. and Jin, Xiao-Yin (2009) 'International High Tech Competitiveness: Does China Rank Number 1?', *Technology Analysis & Strategic Management*, 21 2), 173–93.
Porter, M. E. and Stern, S. (2002) 'National Innovative Capacity' in *The Global Competitiveness Report* 2001–2002, New York, Oxford University Press, pp. 102–18.

National system of innovation (NSI)

The concept of a national system of innovation (NSI) first appeared in the literature by Freeman (1987) in a study, at national level, about the performance of the Japanese economy. Lundvall (1992) employed the concept initially to describe the interaction and importance of learning processes in a knowledge-based economy.

Innovation is an 'interactive process' in which agents and organisations communicate, cooperate and establish long-term relationships. Freeman (1995) defined a national system of innovation as 'a number of distinct technology-based systems, each of which is geographically and institutionally localized within the nation but with links into the supporting national and the international system'. The NSI concept is historical in that both the character of the **innovation process** and the role of national borders change over time (Freeman, 1995).

See also: **national innovation systems**

Further reading
Freeman, C. (1995) 'The National System of Innovation in Historical Perspective', *Cambridge Journal of Economics*, 19, 5–24.
Freeman, C. (ed.) (1987) *Technology Policy and Economic Performance: Lessons from Japan*, London, Pinter Publishers.
Lundvall, B.-A. (ed.) (1992) *National Systems of Innovation: Towards a Theory of Innovation and Interactive Learning*, London, Pinter.
Lundvall, Bengt-Ake (1998) 'Why Study National Systems and National Styles of Innovation?', *Technology Analysis & Strategic Management*, 10(4), 407–21.

Network externalities

Jeff Rohlfs (1974) at Bell Labs defined network externalities as 'effects on a user of a product/service of others using the same or compatible products/services'. *Network effect* is another name for network externalities. These occur when the perceived value of an innovation depends on the number of consumers who have adopted that innovation. In the video game markets, developers develop new applications for the popular Xbox or PlayStation and more independent developers develop more applications for the popular BlackBerry and iPhone, making them even more popular (Wattal *et al.*, 2010).

There two types of network externalities: direct and indirect. Direct network externalities affect the value for the user of a product based on the number of users of the same product. Users experience indirect network externalities based on the availability of related or compatible products. For example, social networking websites such as Facebook and LinkedIn exhibit direct network externalities because the value for each user increases as more users join the network. Video gaming consoles, such as the Xbox and PlayStation, exhibit indirect network externalities because the value of the console increases as more game titles become available for it (*ibid.*).

Network externalities can be positive or negative. Positive externalities exist if more users make a product more valuable, as with the telephone. Negative externalities exist if the value of the product or services diminishes as more people start using it, such as is the case with oversubscribed wireless networks becoming congested with more subscribers and overloading the bandwidth. Wattal *et al.* (2010) found stronger network effects for younger generations and for women over men. Song *et al.* (2009) recommend that managers of innovations whose market potential is affected by network externalities must manage consumer awareness and understanding of the relationship between network externality variables and innovation potential to achieve positive results.

Further reading
Rohlfs, J. (1974) 'A Theory of Interdependent Demand for a Communication Service', *Bell Journal of Economics*, 5, 16–37.
Song, Michael, Parry, Mark E. and Kawakami, Tomoko (2009) 'Incorporating Network Externalities into the Technology Acceptance Model', *Journal of Product Innovation Management*, 26, 291–307.
Wattal, Sunil, Racherla, Pradeep and Mandviwalla, Munir (2010) 'Network Externalities and Technology Use: A Quantitative Analysis of Intraorganizational Blogs', *Journal of Management Information Systems*, 27(1), 145–73.

N

New concept development model

See **front end of innovation**

New market disruption

See **disruptive innovation**

New product development

The meaning of the term 'new product' differs if it is sought from the perspective of the firm or the marketplace. New products can be categorised in terms of their newness to the market and to the firm. There are six categories of new products (Crawford and Di Benedetto, 2008):

1. *New-to-the world products or really-new products*. The Polaroid camera, Sony Walkman and Palm Pilot are examples of the new-to-the-world products. This category accounts for about 10 per cent of new products.
2. *New-to-the-firm products*. These are not new to the world but only new to the firm. Procter & Gamble's first shampoo or coffee, Hallmark gift items and Canon's laser printer are examples. This category accounts for about 20 per cent of new products.
3. *Additions to existing products*. P&G's Tide liquid detergent, Bud Light beer and Apple's iMac are examples. This category accounts for about 20 per cent of new products.
4. *Improvements and revisions to existing products*. P&G's Ivory soap and Tide powder detergent are examples. This category accounts for about 26 per cent of new products.
5. *Repositionings*. Arm & Hammer baking soda repositioning as a drain deodorant, aspirin repositioning as beneficial for the heart and Marlboro cigarettes repositioning for women are examples. This category accounts for about 7 per cent of new products.
6. *Cost reductions*. These are new products that offer similar performance but at a lower cost through a new design or a new manufacturing process. This category accounts for about 11 per cent of new products.

New product development (NPD) consists of a basic **new product development process** plus **the front end of product development**, also known as the **fuzzy front end of product development**. The NPD process is executed by cross-functional new product development teams. The Product Development and Management Association (PDMA) has conducted benchmarking studies in NPD, the **comparative performance assessment studies (CPAS).** Using the CPAS results, PDMA has developed a **new product development best practices framework**. According to PDMA, best-in-class NPD organisations work half as hard as average-performing companies while achieving twice the success rate for new products.

Further reading
Crawford, Merle and Di Benedetto, Anthony (2008) *New Products Management,* 9th edn, New York, McGraw-Hill.

N

New product development process

New product development (NPD) processes 'involve a series of stages aimed at delivering a functional, commercial benefit to customers' (Calantone *et al.*, 1995). Crawford and Di Benedetto (2008) outline a basic NPD process consisting 'of opportunity identification and selection, concept generation, concept/project evaluation, development, and testing and launch phases'. The first three phases, opportunity identification and selection, concept generation and concept/project evaluation, are commonly referred the **fuzzy front end (FFE).**

- *Opportunity identification and selection.* This is the process of creatively identifying opportunities. It can be based on a company skill or resource, or a customer problem.
- *Concept generation.* A concept is an idea that has not been not fulfilled. Ideation is a creative process that requires three inputs: the form, the technology and the need/benefit. Any of the three inputs can start the creative process and the generation of ideas or concepts. IDEO's Deep Dive process is an example of best-in-class concept generation.
- *Concept/project evaluation.* Concept/project evaluation activity is sometimes called screening or pre-technical evaluation. It varies widely depending on the firm and the concept/project and moves from quick looks to complete discounted cash flow and net present value analysis. If a concept is a 'go', the decision turns into project evaluation. This involves preparing a statement of what benefits are needed from the new product, the first list of customer needs. A more generic term is the product description, product definition or product protocol.
- *Development.* Development is the main body of effort. It consists of development of the product or service, the marketing plan and a **business plan**.
- *Testing and launch (commercialisation).* The term 'launch' or '**commercialisation**' is used to describe that time or that decision where the firm chooses to market a product (the 'go or no/go' decision).

The **Stage-Gate™ process** is a step-wise new product development process (Cooper and Kleinschmidt, 1991). Cooper and Edgett (2005) note that this process is a 'detailed and operational playbook' containing a list of activities, best practices and the required deliverable at each stage. Adherence to a structured NPD process allows managers to manage risk and increase the efficiency of NPD (Calantone and Di Benedetto, 1988). However, the process often results in lower-risk, immediate-reward and incremental projects (McDermott and O'Connor, 2002), inhibiting companies from achieving **radical innovation** (Pétrie, 2008).

Cooper (2008) debunked the myths and misconceptions about his Stage-Gate™ process. He distinguished between the original, simple process and the more advanced second-generation process. According to him, the basic Stage-Gate™ process consists of 'a series of stages, where the project team undertakes the work, obtains the needed information, and does the subsequent data integration and analysis, followed by gates, where go/kill decisions are made to continue to invest in the project' (*ibid.*). For Cooper, the basic process is a set of information-gathering stages followed by go/kill decision gates. In contrast, the second-generation process consists of a five-stage, five-gate system along with discovery and post-launch review. Scoping, building the **business case development**, testing and validation, and **launch** are the five stages. He further argued that three of the stages (or half the second-generation model) happen before development begins. He reiterated that the **fuzzy front end** activities (that is, ideation, scoping the project, defining the product and building the business case) are the most critical part of his second-generation Stage-Gate™ process (*ibid.*).

The PRTM phase-gate new product development (NPD) process, also known as the **Product and Cycle Time Excellence (PACE®) process**, is widely used in the high-technology industry. It proceeds in a sequence of six phases, each of which includes steps and milestones (Kumar and Krob, 2007). The requirements for each phase are documented in checklists and reviewed during phase reviews held at the completion of each phase (*ibid.*). The review is held with the appropriate **Product Approval Committee (PAC)** to approve advancement of the product and to the next phase (*ibid.*). PACE® facilitates the development of products and manufacturing processes in a more defined and organised manner and also reduces confusion in the early stages of the traditional NPD process (*ibid.*). However, Harmancioglu *et al.* (2007) argue that in high-tech industries there is unpredictability of customer requirements and competitor strategies, and so high-tech companies require the use of step-wise NPD processes, establishment of a stable business case, full integration of customers and higher specialisation (as opposed to cross-functional collaboration).

In many organisations, NPD managers fail to utilise the full potential of **design** in the NPD process. *Industrial design* is essential for the creation of products that satisfy user needs and aspirations, and NPD managers can leverage industrial design to differentiate themselves from the competition (Goffin and Micheli, 2010).

Further reading

Calantone, R. J. and Di Benedetto, C. A. (1988) 'An Integrative Model of the New Product Development Process: An Empirical Validation', *Journal of Product Innovation Management*, 5(3), 201–15.

Calantone, R. J., Vickery, S. K. and Droge, C. (1995) 'Business Performance and Strategic New Product Development Activities: An Empirical Investigation', *Journal of Product Innovation Management*, 12(3), 214–23.

Cooper, Robert G. (2008) 'Perspective: The Stage-Gate Idea-to-Launch Process-Update, What's New and NextGen Systems', *Journal of Product Innovation Management*, 25(3), 213–32.

Cooper, R. G. and Edgett, S. J. (2005) *Lean, Rapid and Profitable New Product Development*, Ancaster, ON, Product Development Institute.

Cooper, R. G. and Kleinschmidt, E. J. (1991) 'New Product Processes at Leading Industrial Firms', *Industrial Marketing Management*, 20(2), 137–47.

Crawford, Merle and Di Benedetto, Anthony (2008) *New Products Management*, 9th edn, New York, McGraw-Hill.

Goffin, Keith and Micheli, Pietro (2010) 'Maximizing the Value of Industrial Design in New Product Development', *Research Technology Management*, 53(5), 29–37.

Harmancioglu, Nukhet, McNally, Regina C., Calantone, Roger J. and Durmusoglu, Serdar S. (2007) 'Your New Product Development (NPD) Is Only as Good as Your Process: An Exploratory Analysis of New NPD Process Design and Implementation', *R&D Management*, 37(5), 399–424.

Kumar, Sameer and Krob, William (2007) 'Phase Reviews versus Fast Product Development: A Business Case', *Journal of Engineering Design*, 18(3), 279–91.

McDermott, C. M. and O'Connor, G. C. (2002) 'Managing Radical Innovation: An Overview of Emergent Strategy Issues', *Journal of Product Innovation Management*, 19(6), 424–38.

Pétrie, A. (2008) 'Developing Products with a Holistic Process', *Design Management Review*, 19(3), 68–73.

N

New product development (NPD) team

A team refers to a group of people brought together to perform a specific task. A team has a specific purpose that the team itself delivers. Teams have norms, informal standards that inform members how to behave and what to expect.

Cross-functional core teams are the most successful team types in new product development (NPD). A cross-functional team consists of individuals with varying backgrounds, points of view and skills, and produces better results to those an individual could provide alone. Key responsibilities of a cross-functional NPD core team member include ensuring that functional knowledge is represented on the project and sharing responsibility for team results. Core NPD team members should be chosen based on the skills and information needed and willingness to commit to the team, and the ideal size is 6–10 members.

According to Wheelwright and Clark (1995) there are four types of teams: functional, lightweight, heavyweight and autonomous (tiger) teams. A heavyweight team is appropriate for a new category entry project and autonomous teams when the problem is to break from the past, enter a new market or do radical technology or commercial projects (new-to-the-world projects).

The NPD task is often conceptualised as the search for a **design** with appropriate functionality (Leenders *et al.*, 2007). The task may be the creation of new solutions or variation from or modification to existing solutions. Reordering and recombination of existing elements, developing, refining, testing and implementing new solutions are some of the key responsibilities of NPD teams. Effective interaction and knowledge sharing among NPD team members are essential to the success of their cross-functional effort. Team trust is the most significant factor for effective team interaction and performance. Successful teams translate their common goal into specific, measurable and realistic performance goals.

Traditional NPD teams are co-located while *virtual NPD teams* are dispersed and use a variety of information and communication technology (ICT) tools to work across space, time and organisational boundaries (Montoya *et al.*, 2009). A virtual NPD team 'is a group of people who interact through interdependent tasks guided by common purpose' (Powell *et al.*, 2004). It is necessary to understand the communication and task processes in virtual NPD teams and how they cope with the opportunities and challenges of cross-boundary work (Montoya *et al.*, 2009).

Cross-functional teams are necessary when a major product change is desired or a technology breakthrough is needed (new-to-the-world projects). A recent phenomenon in new product development is not to create value for customers, but seek to co-create value with them.

See also: **co-creation, types of development teams, structures and roles**

Further reading
Leenders, Roger Th. A. J., van Engelen, Jo M. L. and Kratzer, Jan (2007) 'Systematic Design Methods and the Creative Performance of New Product Teams: Do They Contradict or Complement Each Other?', *Journal of Product Innovation Management*, 24(2), 166–79.
Montoya, Mitzi M., Massey, Anne P., Hung, Yu-Ting Caisy and Crisp, C. Brad. (2009) 'Can You Hear Me Now? Communication in Virtual Product Development Teams', *Journal of Product Innovation Management*, 26(2), 139–55.

Powell, A., Piccoli, G. and Ives, B. (2004) 'Virtual Teams: A Review of Current Literature and Directions for Future Research', *Database for Advances in Information Systems*, 35(1), 6–36.

Wheelwright, S. C. and Clark, K. B. (1995) *Leading Product Development*, New York, Free Press.

New product strategy

New products are essential for a firm's continued success. A firm's **strategy** describes how it intends to create value for its shareholders, customers and citizens (Kaplan and Norton, 2004). A new product strategy must exist and operate within the boundaries of the firm's strategy to avoid wasted time, effort and resources as well as employee confusion and disagreement (Crawford, 1972). It should identify specific customer segments that are targeted for growth and profitability. For example, Southwest Airlines offers low prices to satisfy and retain price-sensitive customers, while Neiman Marcus targets customers with high disposable incomes (who are willing to pay more for high-end merchandise) (Kaplan and Norton, 2004). Strategy is the most important factor contributing to the success of new product development. It is senior management's role to define strategy.

Henry Mintzberg (1987) used the metaphor of a potter to describe managers as craftsmen and strategy as their clay to describe the process of strategy making. **New product strategy** can emerge in response to an evolving situation, or it can be orchestrated carefully, through a process of formulation followed by implementation. A number of paradigms exist for guiding new product strategy:

- Ansoff and Stewart (1967) developed a four-point scale of strategy orientation of first-to-market, follow-the-leader, applications-engineering and me-too products based on the timing of entry of a technological firm into an emerging industry.
- Miles and Snow (1978) created four strategy types of defenders, prospectors, analysers and reactors based on rate at which a firm changes its products or markets in response to its environment.
- Cooper (1985) introduced five strategic arenas of the markets, industry sectors, applications, product types and technologies on which firms focus their new product efforts.
- Porter (1985) suggested three generic strategies of cost leadership, **differentiation** and market focus based on his five forces framework for achieving profitability and sustaining a **competitive advantage**.

Zhang *et al.* (2009) operationalised a subsidiary's product development strategy as three strategic focuses (breakthrough, platform and incremental focuses). Each focus dictates a different allocation of resources across NPD projects with a high, moderate or low level of innovativeness (Cooper and Kleinschmidt, 1987). Many firms use the product-market matrix to identify strategic arenas on which to focus new product development. Each cell in the matrix points to new product opportunities (Cooper and Edgett, 2010).

N

Further reading
Ansoff, H. I. and Stewart, J. M. (1967) 'Strategies for a Technology-Based Business', *Harvard Business Review*, 45(6), 71–83.

Cooper, R. G. (1985) 'Overall Corporate Strategies for New Product Programs', *Industrial Marketing Management*, 14, 179–93.

Cooper, Robert G. and Edgett, Scott J. (2010) 'Developing a Product Innovation and Technology Strategy for Your Business', *Research Technology Management*, 53(3), 33–40.

Cooper, Robert G. and Kleinschmidt, Elko J. (1987) 'New Products: What Separates Winners from Losers?', *Journal of Product Innovation Management*, 4(September), 169–84.

Crawford, C. Merle (1972) 'Strategies for New Product Development: Guidelines for a Critical Company Problem', *Business Horizons*, 15(6), 49–58.

Kaplan, Robert S. and Norton, David P. (2004) 'Strategy Maps', *Strategic Finance*, 85(9), 27–35.

Miles, R. and Snow, C. (1978) *Organizational Strategy, Structure, and Process*, New York, McGraw-Hill.

Mintzberg, Henry (1987) 'Crafting Strategy', *Harvard Business Review*, 66(4), 66–75.

Porter, Michael E. (1985) *Competitive Advantage*, New York: Free Press.

Zhang, Junfeng, Di Benedetto, C. Anthony and Hoenig, Scott (2009) 'Product Development Strategy, Product Innovation Performance, and the Mediating Role of Knowledge Utilization: Evidence from Subsidiaries in China', *Journal of International Marketing*, 17(2), 42–58.

New technology ventures

See **new ventures**

New ventures

New ventures or **new technology ventures** (**NTV**s) are new entities funded and formed with three elements (product markets, capabilities, future vision) (Ambos and Birkinshaw, 2007). The creation of new ventures is a process by which 'entrepreneurs come to imagine the opportunity for novel ventures, refine their ideas, and, after an initial investment, justify their ventures to relevant others to gain much-needed support and legitimacy' (Alvarez & Barney, 2007). Cornelissen and Clarke (2010) argue that inductive analogical or metaphorical reasoning generates a platform for the creation and commercialisation of novel ventures.

An NTV can be launched as an independent entity, a new profit centre within a company or as a joint venture.

Song *et al.* (2008) claim that NTV performance depends on three factors:

1. Market and opportunity (that is, market scope).
2. Entrepreneurial team (with industry and marketing experience).
3. Resources (including financial resources, firm age, patent protections, size of the founding team and supply chain integration).

New ventures often lack adequate financial and managerial resources and effective, strategic decision making (Atuahene-Gima and Li, 2004). They face the 'liability of newness' with a high propensity to fail (Stinchcombe, 1965). Song *et al.* (2008) report that NTVs have a limited survival rate, with 21.9 per cent survival after five years.

Further reading
Alvarez, S. A. and Barney, J. B. (2007) 'The Entrepreneurial Theory of the Firm', *Journal of Management Studies*, 44, 1057–63
Ambos, Tina C. and Birkinshaw, Julian (2007) 'How Do New Ventures Evolve? The Process of Charter Change in Technology Ventures', *Academy of Management Annual Meeting Proceedings*, 1–6.
Atuahene-Gima, Kwaku and Li, Haiyang (2004) 'Strategic Decision Comprehensiveness and new Product Development Outcomes in New Technology Ventures', *Academy of Management Journal*, 47(4), 583–97.
Cornelissen, J. P. and Clarke, J. S. (2010) 'Imagining and Rationalizing Opportunities: Inductive Reasoning, and the Creation and Justification of New Ventures', *Academy of Management Review*, 35(4), 539–-57.
Song, Michael, Podoynitsyna, Ksenia, Vander Bij, Hans and Halman, Johannes I. M. (2008) 'Success Factors in New Ventures: A Meta-analysis', *Journal of Product Innovation Management*, 25(1), 7–27.
Stinchcombe, A. L. (1965) 'Organizations and Social Structure', in J. G. March (ed.), *Handbook of Organizations*, Chicago, IL, Rand McNally.

Next practice

The term 'next practice' is all about **innovation** (Prahalad, 2010). C. K. Prahalad coined it to suggest that companies aiming at **competitive advantage** through innovation need to be focused on 'next practices' instead of 'best practices'. **Best practice** is about what worked in the past. Next practice improves on prior experience. Best practice asks what worked, whereas next practice 'imagines what the future will look like; identifying the mega-opportunities that will arise; and building capabilities to capitalize on them' (Prahalad, 2010). Steve Jobs of Apple and Ratan Tata of Tata Motors are two well-known practitioners of next practice philosophy.

Nidumolu *et al.* (2009) encourage firms to develop next practice platforms with a five-stage sustainable innovation process: viewing compliance as opportunity, making value chains sustainable, designing sustainable products and services, developing new business models, and creating next practice platforms (Nidumolu *et al.*, 2009).

Further reading
Nidumolu, Ram, Prahalad, C. K. and Rangaswami, M. R. (2009) 'Why Sustainability Is Now the Key Driver of Innovation', *Harvard Business Review*, 87(9), 56–64.
Prahalad, C. K. (2010) 'Best Practices Get You Only So Far', *Harvard Business Review*, 88(4), 32.

N

Open innovation

Open innovation is a concept developed by Henry Chesbrough (2003). In open innovation, firms seek out research and development opportunities from outside. This contrasts with closed innovation, in which new product development and new product marketing activities occur within the firm.

Closed innovation firms rely solely on their own resources for innovation. They believe that they must discover, develop and commercialise innovation in order to benefit from it. Closed innovation firms also believe in controlling intellectual property (IP), using it to disadvantage competitors (preventing competitors from benefiting from IP).

In contrast, open innovation firms combine internal and external ideas to develop new technologies and internal and external paths to market. They leverage both internal and external expertise and R&D capability to create significant value. Their focus is to benefit from innovation, wherever it may occur. They want to be the first to market with a better **business model** rather than being restricted to internal innovation for market leadership.

Open innovation stresses a paradigm shift from closed innovation to the effect that not all good ideas are developed within the company, and not all ideas should automatically be developed within the firm's own boundaries. For instance, Deutsche Telekom enhanced its innovation capacity by opening up its traditional development process and embracing external creativity and knowledge resources (Rohrbeck *et al.*, 2009).

Ilcisin and Starkloff (2010) recommend the following practices for implementing open innovation in technology companies:

1. Identify customer needs. Disruptive or game-changing opportunities would be hard to implement without inflows and outflows of knowledge.
2. Research and document the opportunity and customer need.
3. Determine the level of partnership. Is it a licensing agreement or a joint development opportunity?
4. Assign core teams from the companies involved and allow the teams to get acquainted with each other in meaningful ways.
5. Develop a shared **vision** and use it to resolve questions around development expertise, manufacturing, marketing, sales and support.

Herstad *et al.* (2010) favour reframing of national innovation policies in global open innovation. Global open innovation includes distributed knowledge networks, technology flows across industrial sectors and open innovation processes. These authors argue for a 'loosening up' of policy emphasis on 'containing interaction within national boundaries', as it is a weakness in globalisation. For global competition, the challenge for national innovation policy is to support the domestic embedding of internationally linked industries.

See also: **co-creation**

Further reading
Chesbrough, H. (2003) *Open Innovation: The New Imperative for Creating and Profiting from Technology,* Boston, MA, Harvard Business School Press.
Herstad, Sverre J., Bloch, Carter, Ebersberger, Bernd and Van De Velde, Els (2010) 'National Innovation Policy and Global Open Innovation: Exploring Balances, Tradeoffs and Complementarities', *Science & Public Policy,* 37(2), 113–24.
Ilcisin, Kevin J. and Starkloff, Eric (2010) 'Using Open Innovation to Extend R&D Investments' *EE: Evaluation Engineering,* 49(8), 18–21.
Rohrbeck, Ren, Hölzle, Katharina and Gemünden, Hans Georg (2009) 'Opening Up for Competitive Advantage – How Deutsche Telekom Creates an Open Innovation Ecosystem', *R&D Management,* 39(4), 420–30.

Open-source innovation model

In the open-source innovation model, innovation thrives in a market without traditional intellectual property (IP). The intellectual property regime prevents competitors from selling the same products because of exclusive rights granted to innovators under intellectual property law. In open-source development, users develop software programs to overcome personal and shared technical problems. They freely reveal their innovations, forfeiting private returns from selling the software.

Raasch *et al.* (2009) define open-source innovation (OSI) as 'characterized by free revealing of information on a new design with the intention of collaborative development of a single design or a limited number of related designs for market or nonmarket exploitation'. In OSI software development occurs in a private–collective model of innovation that offers society the best of private investment and collective action (von Hippel and von Krogh, 2003, 2006). The 'collective' model assumes that innovators collaborate in order to provide public benefit when there is a market failure. The 'private investment' model assumes that innovators receive returns on innovation by selling private goods.

Red Hat is a private company founded on a collective model of innovation. In this model, knowledge is first passed from the developers to Red Hat Certified Software Engineers. Red Hat customers pay fees to access knowledge (service and support) from Red Hat engineers or a self-service database of technical information and updates. In the process, knowledge spreads and others gain the expertise. The base of knowledge expands and the selling price of the expertise drops. Innovations are created; new knowledge is generated; and the cycle repeats.

A note of caution is in order for for-profit companies using open-source software in their products. They need to pay particular attention to infringement risks, as evidenced by Red Hat *v.* Firestar, a patent dispute between Red Hat and Firestar Software (Walsh and Tibbetts, 2010).

Further reading
Boldrin, Michele and Levine, David K. (2007) 'Open-Source Software: Who Needs Intellectual Property?', *The Freeman*, 57(1), 26.
Raasch, Christina, Herstatt, Cornelius and Balka, Kerstin (2009) 'On the Open Design of Tangible Goods', *R&D Management*, 39(4), 382–93.
von Hippel, Eric and von Krogh, George (2003) 'Open Source Software and the Private-Collective Innovation Model: Issues for Organization Science', *Organization Science*, 14(2), 209–36.
von Hippel, E. and von Krogh, G. (2006) 'Free Revealing and the Private-Collective Model for Innovation Incentives', *Research and Development Management*, 36(3), 295–306.
Walsh, Edmund J. and Tibbetts, Andrew J. (2010) 'Reassessing the Benefits and Risks of Open Source Software', *Intellectual Property & Technology Law Journal*, 22(1), 9–13.

Opportunity analysis

Opportunity analysis is conducted at the **front end of innovation**. It consists of assessing an opportunity to verify that it is worth pursuing. Opportunity analysis recognises that technological and market uncertainty exists at the **front end of innovation**.

Companies use *market opportunity analysis (MOA)* to determine the feasibility of entering or expanding operations in existing markets. The MOA involves assessing the internal capabilities and market potential for proposed goods and services, competitive players and unmet **customer needs**. For example, small-city airports can use MOA to evaluate their capabilities to attract and serve customers and outperform competitors (Golicic *et al.*, 2003).

Further reading
Golicic, Susan L., McCarthy, Teresa M. and Mentzer, John T. (2003) 'Conducting a Market Opportunity Analysis for Air Cargo Operations', *Transportation Journal*, 42(4), 5–15.
Koen, Peter A. (2003) 'Front End of Innovation: Effective Methods, Tools and Techniques in Proceedings of Managing the Front End of Innovation', PDMA and IR Conference, May, Boston, MA.

Opportunity identification

An opportunity may be a specific problem or an unmet **customer need**. A firm may design a solution to realise the opportunity, respond to a threat or develop a technological solution to a specific problem. An opportunity can be incremental or breakthrough. Opportunity identification at the **front end of innovation** consists of activities and methods that a firm uses to identify opportunities (Koen, 2003).

For instance, a food manufacturing company identified an opportunity to develop low-fat products. It invested in low-fat technology development by recognising the problem of obesity as an opportunity.

Entrepreneurs identify opportunities to replicate a product, innovate a product, acquire an income stream or educate consumers (Chandler *et al.*, 2005). In small companies, entrepreneurial experience affects the ability of entrepreneurs to identify opportunities. Experience within an industry leads to **incremental innovations**; radical changes come from outside the industry.

Further reading
Chandler, Gaylen N., Lyon, Douglas W. and Detienne, Dawn R. (2005) 'Antecedents and Exploitation of Outcomes of Opportunity Identification Processes', *Academy of Management Proceedings*, 1–6.
Koen, Peter A. (2003) 'Front End of Innovation: Effective Methods, Tools and Techniques in Proceedings of Managing the Front End of Innovation', PDMA and IR Conference, May, Boston, MA.

Organisational innovation

Amabile *et al.* (1996) define innovation as the successful implementation of creative ideas within an organisation. Organisational innovation is the development of new or improved products or services and their successful launch to the market (Gumusluo lu and Ilsev, 2009). Another definition of organisational innovation is that it involves adoption of an idea, material artifact or behaviour that is new to the organisation adopting it (Daft, 1978).

Transformational leaders transform followers' personal values and self-concepts and move them to higher levels of needs and aspirations (Jung, 2001). They also raise the performance expectations of their followers (Bass, 1995).

Gumusluo lu and Ilsev (2009) encourage managers to become transformational leaders in order to boost organisational innovation. Their research highlights the importance of external support in the organisational innovation process. They found that externally received technical and financial support played a more prominent contextual influence in accelerating innovation in micro- and small-sized companies than an innovation-supporting internal climate. They encourage managers of micro- and small-sized companies to build relationships with external institutions that provide technical and financial support to achieve the high innovation potential of their firms. Relationships can be both a formal and an informal exchange of knowledge. External relations have also been shown to have a positive influence on innovation in small, knowledge-intensive firms in Norway. Jenssen and Nybakk (2009) found that market innovation is positively affected by senior management interaction with other firms, and product innovation by interaction with external research and development (R&D).

See also: **strategic management**

Further reading
Amabile, T. M., Conti, R., Coon, H., Lazenby, J. and Herron, M. (1996) 'Assessing the Work Environment for Creativity', *Academy of Management Journal*, 39(5), 1154–84.
Bass, B. M. (1995) 'Transformational Leadership', *Journal of Management Inquiry*, 4(3), 293–8.
Daft, R. (1978) 'Dual Core Model of Organizational Innovation', *Academy of Management Journal*, 21(2), 193–210.
Gumusluo lu, Lale and Ilsev, Arzu (2009) 'Transformational Leadership and Organizational Innovation: The Roles of Internal and External Support for Innovation', *Journal of Product Innovation Management*, 26(3), 264–77.
Jenssen, Jan Inge and Nybakk, Erlend (2009) 'Inter-Organizational Innovation Promoters in Small, Knowledge Intensive Firms', *International Journal of Innovation Management*, 13(3), 441–66.
Jung, D. I. (2001) 'Transformational and Transactional Leadership and Their Effects on Creativity in Groups', *Creativity Research Journal*, 13(2), 185–95.

O

Organisational knowledge-creation theory

Organisational knowledge is an organisation's internal representation of the world (Daft and Weick, 1984). It comprises codified and non-codified organisational information and individual cognitions and memories (Grant, 1996). The knowledge of an organisation determines what actions its members are capable of taking and how they coordinate and integrate their efforts (Madsen and Desai, 2010).

Nonaka (1994) proposes a theory of organisational knowledge creation, postulating that 'a continuous dialogue between tacit and explicit knowledge' creates new organisational knowledge. Five patterns of dialogue are interactions, socialisation, combination, internalisation and externalization (*ibid.*). Explicit knowledge is codified knowledge, which can be transmitted in formal, systematic language, while tacit knowledge is personalised knowledge. Tacit knowledge is difficult to transmit and is deeply rooted in action, commitment and involvement, within a given organisational context.

Socialisation represents interaction between individuals through observation, imitation or apprenticeships. Explicit knowledge can be combined through meetings, via personal conversations or using information systems to communicate. Internalisation refers to the conversion of explicit knowledge into tacit knowledge, whereas externalisation refers to the conversion of tacit knowledge into explicit knowledge.

Further reading
Daft, R. L. and Weick, K. E. (1984) ,Toward a Model of Organizations as Interpretation Systems', *Academy of Management Review*, 9, 284–95.
Grant, R. M. (1996) 'Toward a Knowledge-based Theory of the Firm', *Strategic Management Journal*, 17, 109–22.
Madsen, Peter M. and Desai, Vinit (2010) 'Failing to Learn? The Effect of Failure and Success on Organizational Learning in the Global Orbital Launch Vehicle Industry', *Academy of Management Journal*, 53(3), 451–76.
Nonaka, Ikujiro (1994) 'A Dynamic Theory of Organizational Knowledge Creation', *Organization Science*, 5(1), 14–37.

Organisational learning

Organisational learning is an *adaptive process*. Organisations develop and use knowledge in innovation. Knowledge development includes the *exploration* of new possibilities and the *exploitation* of old certainties. Organisations favour exploitation because, through rapid exploitation, they produce results in the short run. However, short-term exploitation becomes self-destructive in the long run (March, 1991).

An organisation focused on the short term may be practising *survival learning* or *adaptive learning* (Senge, 2006). An adaptive organisation socialises new employees only to the languages, beliefs and practices that constitute adaptive learning; it is not responsive to new ideas or change. So the beliefs reflected in the adaptive learning organisation and those held by its people remain identical and fixed, regardless of changes in a dynamic market or changing **customer needs** and preferences. Thus a difference begins to grow between the capabilities of the firm and the capabilities it requires to meet emerging customer needs. As a result, the value-adding capability of the adaptive learning organisation and individuals within it is systematically degraded over time in a mutual learning situation (March, 1991).

To enhance innovation capacity, an adaptive organisation must become a *learning organisation*. This requires a combination of *adaptive learning* with '*generative learning*' or '*exploratory learning*' (Senge, 2006). IBM, Hewlett-Packard and Apple implemented a mix of adaptive and exploratory learning processes in their organisations to enhance their creative and innovative abilities (*ibid.*).

While organisations can learn from both successes and failures, Madsen and Desai (2010) found that they learn more effectively from failures than successes. They also found that knowledge derived from failures depreciates more slowly, and that the magnitude of failure influences how effectively an organisation learns from various forms of experience. In the orbital vehicle launch industry, for example, the primary driver of organisational learning and improvement was large failures, rather than successes or small failures (*ibid.*).

Further reading
Madsen, Peter M. and Desai, Vinit (2010) 'Failing to Learn? The Effect of Failure and Success on Organizational Learning in the Global Orbital Launch Vehicle Industry', *Academy of Management Journal*, 53(3), 451–76.
March, James G. (1991) 'Exploration and Exploitation in Organizational Learning', *Organizational Science*, 2(1), 71–87.
Senge, Peter M. (2006) *The Fifth Discipline: The Art and Practice of the Learning Organization*, New York, Doubleday.

Osborn–Parnes creative problem-solving (CPS) model

Creative problem solving (CPS) is a widely accepted and well-researched method for developing creative thinking skills (Torrance and Presbury, 1984). It is any activity during which an individual, team or organisation attempts to create novel solutions to ill-defined problems (Puccio, 1999). Osborn (1963) introduced the CPS model in his book *Applied Imagination*, providing a structured framework for creative thinking principles, tools and stages. The Osborn–Parnes creative problem solving model consists of six steps: identifying the goal, wish, or challenge; gathering data; clarifying the problem; generating ideas; selecting and strengthening solutions; and developing a plan for action (Hughes, 2003).

In its iterated form, the model includes three components (understanding the problem; generating ideas; and planning for action) broken down into six stages (Isaksen *et al.*, 1994). Understanding the problem includes 'Mess-Finding, Data Finding and Problem Finding. Generating Ideas is a single step; it is called Idea-Finding, while Planning for Action includes Solution-Finding and Acceptance Finding' (*ibid.*).

Frito-Lay reported millions of dollars in manufacturing cost savings during the first six years of applying the CPS model to its operations (personal communication).

Further reading
Hughes, G. David (2003) 'Add Creativity to Your Decision Process', *Journal for Quality & Participation*, 26(2), 4–13.

Isaksen, S. G., Dorval, K. B. and Treffinger, D. J. (1994) 'Creative Approaches to Problem Solving', Dubuque, IA, Kendall/Hunt.

Osborn, A. F. (1963) *Applied Imagination: Principles and Procedures of Creative Problem-Solving*, 3rd edn, New York, Charles Scribner's Sons.

Puccio, Gerard (1999) 'Creative Problem Solving Preferences: Their Identification and Implications', *Creativity & Innovation Management*, 8(3), 171–8.

Solomon, C. (1990) 'What an Idea: Creativity Raining' in S. Parnes (ed.), *Source Book for Creative Problem Solving*, Buffalo, NY, Creative Education Foundation Press pp. 473–85.

Tassoul, Marc and Buijs, Jan (2007) 'Clustering: An Essential Step from Diverging to Converging', *Creativity & Innovation Management*, 16(1), 16–26.

Torrance, E. P. and Presbury, J. (1984) 'The Criteria of Success Used in 242 Experimental Studies of Creativity', *The CreativeChild and Adult Quarterly*, 9, 238–43.

Outcome-driven innovation

Introduced by Anthony Ulwick (2005), outcome-driven innovation shifts focus from the customer to the outcome. Ulwick recommends that companies become market driven rather than customer led (Pinegar, 2006). He shifts attention from people themselves to the things they try to achieve. For example, a CD helps customers to store music. Focusing 'on the job of storing music supports the discovery and creation of new ways to help customers' to do their jobs better (Ulwick and Bettencourt, 2008). According to Ulwick (2005), this is the essence of innovation. In outcome-driven innovation, companies capture jobs (which activities are to be carried out), constraints (what will prevent customers from adopting a product/service) and outcomes (which metrics are used to evaluate the performance of jobs).

The outcome-driven innovation process consists of eight steps: formulating innovation strategy; capturing customer inputs; identifying opportunities; segmenting the market; defining targeting strategy; positioning current offerings; prioritising the development pipeline; and defining breakthrough concepts.

Ulwick's approach provides a credible alternative for companies that struggle to benefit from **voice of the customer** input into their innovation process.

Further reading
Pinegar, Jeffrey S. (2006) '*What Customers Want: Using Outcome-Driven Innovation to Create Breakthrough Products and Services* by Anthony W. Ulwick', *Journal of Product Innovation Management*, 23(5), 464–6.

Ulwick, A. (2005) *What Customers Want: Using Outcome-Driven Innovation to Create Breakthrough Products and Services*, New York, McGraw-Hill.

Ullwick, Anthony W. and Bettencourt, Lance A. (2008) 'Giving Customers a Fair Hearing', *MIT Sloan Management Review*, 49(3), http://www.strategyn.com/resources/journal-articles/free-download-giving-customers-fair-hearing/ (accessed on December 9, 2010).

PACE (product and cycle time excellence)

See **new product development process**

Paradigm

A paradigm is a philosophical or theoretical framework that is based on a set of assumptions. It is a set of relationships like a model, but is more abstract and less quantitatively defined than a model. In his book *The Structure of Scientific Revolutions*, Thomas Kuhn (1996) defines paradigms as 'universally recognized scientific achievements that for a time provide model problems and solutions to a community of practitioners'. In a broad sense, paradigm means 'the entire constellation of beliefs, values, techniques, and so on shared by the members of a given community' and in a narrow sense it means 'one sort of element in that constellation, the concrete puzzle-solution which, employed as models or examples, can replace explicit rules as a basis for the solution of the remaining puzzles of normal science' (*ibid.*). Newtonian physics and Marxist economics are two examples of paradigms.

For Kuhn, paradigms gain their status because they are more successful than their competitors in solving questions that the scientific community considers acute (*ibid.*). By his account, acceptance or rejection of the paradigm determines how scientists see the world (Wendel, 2008). To question assumptions is a first step towards innovation and replacement of the current paradigm with a new one.

Further reading
Kuhn, Thomas S. (1996) *The Structure of Scientific Revolutions,* 3rd edn, Chicago, University of Chicago Press.
Wendel, Paul (2008) 'Models and Paradigms in Kuhn and Halloun', *Science & Education,* 17(1), 131–41.

Parallel thinking

Parallel thinking techniques allow each thinker in a group to put forward his or her thoughts in parallel with the thoughts of others. Parallel thinking is suitable for innovative, constructive and creative work. It unbundles thinking and separates ego from performance. Parallel thinking focuses on cooperative exploration, instead of adversarial confrontation (De Bono, 1995).

For De Bono, thinking is a skill. The main difficulty of thinking is confusion. The *Six Thinking Hats* is a method that allows a thinker to focus his or her thinking more clearly (De Bono, 1985). It is a practical way to apply parallel thinking. In this method, six metaphorical hats are used to represent six different types of thinking – the white hat looks at the facts, the black hat makes judgements, the red hat considers feelings, the yellow hat looks at benefits, the green hat is creative and the blue hat assesses the thinking process (*ibid.*). The thinker puts on or takes off one of these hats to indicate the type of thinking being used. When used alone by an individual, the method helps in carrying out a well-rounded analysis of a problem. When done in a group, everybody wears the same hat, at the same time, to synchronise their thinking as they explore all sides of an issue. It leads to more creative thinking and improves communication and decision making (*ibid.*).

Further reading
De Bono, E. (1985) *Six Thinking Hats*, Boston, Little, Brown.
De Bono, E. (1995) *Parallel Thinking: From Socratic Thinking to de Bono Thinking*, Harmondsworth, Penguin.

Participatory design

The concept of participatory design originated from Kristen Nygaard's pioneering work in the 1970s. It refers to the design and development processes in which end-users are invited to participate and contribute like co-designers (Burr and Mathews, 2008). **Co-design** is prototypical practice explored through an open-ended exploration of possibilities (Björgvinsson, 2008).

Participatory design is also a means of ensuring that technologies support and enhance users' knowledge and skills, rather than redefining or eliminating people's jobs through the introduction of new technologies into workplaces (Burr and Mathews, 2008). Early proponents of participatory design developed methods for exploring design alternatives that were meaningful and engaging for all stakeholders (Björgvinsson, 2008). They also ensured that methods allowed for a shared understanding of current and future ways of organising the work. **User-centred design** developed from participatory design (Schuler and Namioka 1993).

Baek and Lee (2008) utilised two toolkits, Info Block and Info Tree, to evaluate the usability of the Yahoo! Kids (Korea) directory. They found that the thinking of children is determined more by knowledge association than by conventional logic. Children navigate websites differently to adults and the participation of children in the design process for children's websites is essential (Baek and Lee, 2008).

See also: **user-driven innovation**

Further reading
Baek, Joon-Song and Lee, Kun-Pyo (2008) 'A Participatory Design Approach to Information Architecture Design for Children', *CoDesign*, 4(3), 173–91.
Björgvinsson, Erling Bjarki (2008) 'Open-Ended Participatory Design as Prototypical Practice', *CoDesign*, 4(2), 85–99.
Burr, Jacob and Matthews, Ben (2008) 'Participatory Innovation', *International Journal of Innovation Management*, 12(3), 255–73.
Buur, Jacob and Bagger, K. (1999) 'Replacing Usability Testing with User Dialogue', *Communications of the ACM*, 42(5), 63–6.
Holmquist, L. E. (2004) 'User-Driven Innovation in the Future Applications Lab', *Proceedings of the Conference on Human Factors in Computing Systems*, Vienna, ACM Press.

P

Schuler, D. and Namioka, A. (eds) (1993) *Participatory Design: Principles and Practice*, Hillsdale, NJ, Lawrence Erlbaum Associates.

People and teams

Formal processes for **new product development (NPD)** are now the norm (Barczak and Wilemon, 2003) and people and teams are at the heart of NPD processes. Use of cross-functional teams in NPD suggests the evolution of firms from functional silos to cross-functional management. The fundamental problem of cross-functional NPD management is to make a diverse group of individuals from various functional areas work together as a team for a specific time to accomplish specific project objectives (Barczak and McDonough, 2003). Diverse teams bring more **creativity** to problem solving and product development. However, diversity also impedes the team's implementation capability as compared to homogeneous teams (Ancona and Caldwell, 1992). Cross-functional teams depend on the collaborative contribution of each team member to increase innovation and speed to market (Lovelace *et al.*, 2001). Greater negotiation and conflict-resolution skills within the team can diminish potential issues in cross-functional teams. NPD best-practice firms better support their people through cross-functional team communications (Barczak *et al.*, 2009). However, many firms struggle to implement cross-functional teams successfully (Barczak and Wilemon, 2003).

There are two types of teams in NPD: dedicated and parallel. Dedicated NPD team members devote 100 per cent of their time to the NPD project. Parallel team members report into their functional managers, performing work assigned to them by their functional departments while also performing team work (Gross, 1995). More than 80 per cent of Fortune 500 companies use parallel teams and 90 per cent of these fail (Wang and He, 2008).

NPD may reside in a strategic business unit (SBU) under the direction of the SBU manager and under the control of a function in the organisation, or in a standalone NPD department or venture group (Barczak *et al.*, 2009). NPD governance structures and organisational goals exert a significant influence on the innovation performance of NPD people and teams.

Further reading

Ancona, Deborah Gladstein and Caldwell, David F. (1992) 'Demography and Design: Predictors of New Product Team Performance', *Organization Science*, 3(3), 321–41.

Barczak, G. and McDonough, E. F. (2003) 'Leading Global New Product Development Teams'. *Research*Technology Management*, 46(6), 14–18.

Barczak, Gloria and Wilemon, David (2003) 'Team Member Experiences in New Product Development: Views from the Trenches', *R&D Management*, 33(5), 463–79.

Barczak, Gloria. Griffin, Abbie and Kahn, Kenneth B. (2009) 'Trends and Drivers of Success in NPD Practices: Results of the 2003 PDMA Best Practices Study', *Journal of Product Innovation Management*, 26(1), 3–23.

Blindenbach-Drissen, Floortje (2009) 'The Effectiveness of Cross-Functional Innovation Teams', *Academy of Management Annual Meeting Proceedings*, 1–6.

Gross, S. E. (1995) *Compensation for Teams: How to Design and Implement Team-Based Reward Systems*, New York, American Management Association.

Kahn, Kenneth B., and Barczak, Gloria (2006) 'Establishing an NPD Best Practices Framework', *Journal of Product Innovation Management*, 23(2), 106–16.

P

Lovelace, K., Shapiro, D. L. & Weingart, L. R. (2001) 'Maximizing Cross-Functional New Product Teams' Innovativeness and Constraint Adherence: A Conflict Communications Perspective', *Academy of Management Journal*, 44, 779–93.
Wang, Sijun and He, Yuanjie (2008) 'Compensating Nondedicated Cross-Functional Teams', *Organization Science*, 19(5), 753–65.

Personas

Personas originally came from the software industry and were popularised by Alan Cooper's 1999 book *The Inmates Are Running the Asylum: Why High Tech Products Drive Us Crazy and How to Restore Sanity* (Wasserman, 2006). Personas are archetypal users, or customers, that represent real user or customer types (Guenther, 2006). They allow companies to incorporate an in-depth understanding of what consumers want in an interaction with a company, product or website into the marketing development process (Honigman, 2008). They are a tool that makes users and customers visible and tangible participants in the **new product development (NPD) process.**

Where the user is typically described with dry statistics and long descriptions, personas have a name, face and story that bring to life their social and cultural context. The process of creating personas generates deeper insights into user, or customer, personalities through qualitative and visual descriptions.

Personas are ideally developed from the organisation's marketing research data and they represent typical customers within targeted segments (*data-driven personas*). They can also be developed from knowledge and assumptions held by stakeholders (*assumption personas*) (Beale and Sutton, 2008).

While personas can be used informally on an *ad hoc* project basis, when supported by senior executives they can play a strategic role in building a customer and user-centred organisation. They can be used as a cross-functional tool for **customer-driven innovation**, for communicating market research insights, brand values or product goals to internal and external stakeholders, and to help with key decisions throughout the project life cycle (Pruitt and Adlin, 2006).

Whereas personas are often used by design or creative teams in conjunction with **scenarios**, they can be used at any stage of the NPD life cycle. Unilever, Whirlpool, Best Buy, Staples, Discover, Ford, Chrysler and Microsoft are examples of companies using personas as a marketing tool (Wasserman, 2006).

See also: **customer targeting**

Further reading
Beale, Claire-Juliette and Sutton, Tricia (2008) 'Personas and Scenarios Workshop Launches PDMA Chapter in South Carolina', *PDMA Visions Magazine*, 32(3), 26–7.
Guenther, Kim (2006) 'Developing Personas to Understand User Needs', *Online*, 30(5), 49–51.
Honigman, Daniel B. (2008) 'Persona-Fication', *Marketing News*, 42(6), 8.
Pruitt, John and Adlin, Tamara (2006) *Persona Lifecycle: Keeping People in Mind Throughout Product Design*, San Francisco, CA, Morgan Kaufmann.
Wasserman, Todd (2006) 'Unilever, Whirlpool Get Personal with Personas', *Brandweek*, 47(34), 13.

Platform   **203**

Platform

Various researchers refer to platform in different contexts. It is a concept that is difficult to understand (Cusumano, 2010). Cusumano differentiates between two platforms in information technology (IT) businesses: product platform and industry platform. *Product platform* refers 'to a foundation or base of common components around which a company builds a series of related products' (*ibid.*). In his studies of IT businesses, he found that companies such as Microsoft, Apple and Google became successful when their products became industry-wide platforms (*ibid.*). *Industry-wide platforms* include Microsoft Windows, the personal computer, the browser and the internet (*ibid.*). An industry platform is part of a technology 'system' whose components come from different companies (or different departments of the same company), which Cusumano called *complementors*. Another critical distinction of an industry platform and ecosystem from the product platform is the creation of *network effects*. In network effects, the more external adopters in the ecosystem, the more valuable the platform and the *complementors* become (*ibid.*). Direct network effects occurs when there is technical compatibility or interface standard (e.g. Windows–Intel PC and Windows-based applications) or indirect effects when complementors adopt a specific set of technical standards to use or connect to the platform (*ibid.*). Intel counts as complementors every company that has adopted the Windows–Intel platform for the PC and markets or sells PC hardware and peripherals (Dell and Lexmark), plus five million software developers who write applications and networking or systems software (Cusumano and Gawer, 2002).

For Robertson and Ulrich (1998), the *product platform* is a collection of common assets shared by a set of products called the product family (*ibid.*). The platform consists of versatile components that are loosely coupled with specifications whose variety is preferred by the market (Kang and Hong, 2009). A well-designed product platform achieves both commonality (that is, components that do not contribute to distinctiveness are made commonly across the products to save cost) and distinctiveness (that is, there are versatile components that support market-preferred differentiation without changing themselves) in the design of product platforms (Robertson and Ulrich, 1998).

The use of *platform architecture* in product development and manufacturing is one way to implement **mass customisation**. *Process platforms* deliver product variety while achieving a near mass-production efficiency (Zhang, 2009). Process platforms plan and utilise similar, yet optimal, production processes to those used in existing manufacturing processes to fulfil diverse, customised products (*ibid.*).

Consoli and Patrucco (2008) propose *innovation platforms* as a specific case of technology infrastructure. Technology infrastructures are strategic activities across specialised actors within either the public or private sectors, or at their interface (Smith, 1997). Consoli and Patrucco (2008) found that innovation platforms enabled capacity and capability building for individuals, teams and organisations in the health-care and automotive sectors in the UK and Italy (*ibid.*).

Product platform management is the process of managing major product platforms throughout their life cycles. McGrath (2000) outlines key activities in product platform management as:

P

- Identify where a platform is in its life cycle.
- Synchronise the replacement of a major platform with a next-generation platform.
- Extend the life of a major platform.
- Understand what causes a platform's life cycle to decline (e.g. changes in the desirability of a technology).
- Regularly replace platforms with short life cycles.

Product platform management enables synchronisation of the introduction of a next-generation platform with the decline of the platform it replaces.

See also: **e-innovation**

Further reading

Consoli, Davide and Patrucco, Pier Paolo (2008) 'Innovation Platforms and the Governance of Knowledge: Evidence from Italy and the UK', *Economics of Innovation & New Technology*, 17(7/8), 701–18.

Cusumano, Michael (2010) 'Technology Strategy and Management: The Evolution of Platform Thinking', *Communications of the ACM*, 53(1), 32–4.

Cusumano, Michael A. and Gawer, Annabelle (2002) 'The Elements of Platform Leadership', *MIT Sloan Management Review*, 43(3), 51–8.

Kang, Chang Muk and Hong, Yoo Suk (2009) 'A Framework for Designing Balanced Product Platforms by Estimating the Versatility of Components', *International Journal of Production Research*, 47(19), 5271–95.

McGrath, Michael E. (2000) *Product Strategy for Technology Companies*, 2nd edn, New York, McGraw-Hill.

Robertson, D. and Ulrich, K. (1998) 'Planning for Product Platforms', *MIT Sloan Management Review*, 39(4), 19–31.

Smith, K. (1997) 'Economic Infrastructures and Innovation Systems', in C. Edquist (ed.), *Innovation Systems: Institutions, Organizations and Dynamics*, London, Pinter.

Zhang, Lianfeng (2009) 'Modeling Process Platforms Based on an Object-Oriented Visual Diagrammatic Modeling Language', *International Journal of Production Research*, 47(16), 4413–35.

Platform leadership

Platform leadership refers to a company's ability 'to drive innovation around a particular platform technology at the broad industry level' (Gawer and Cusumano, 2002). Platform leaders are companies driving industry-wide innovation for an evolving platform (comprising a system of separately developed pieces of technology) (*ibid.*). Platform leaders create an industry ecosystem whose value is greater than the sum of its parts (Cusumano and Gawer, 2002). A key challenge for platform leaders is to balance cooperation and coercion. They face challenges from within the industry ecosystem, from companies that want to be platform leaders, as well as from *complementors.* Intel and Microsoft are simultaneously complementors and platform leaders for the personal computer (*ibid.*).

Google is a successful platform builder and leader. It solved the online search problem with its proprietary search engine and encouraged website developers and users to develop new applications to complement its search engine. It built an innovative **business model**, redesigning the relationship between advertisers and internet users with 'clicks for pay'. Microsoft and Adobe also use innovative business models. Microsoft gives away its server for free but charges for Windows software. Adobe gives away free Acrobat Reader while charging for its servers and editing tools (Cusumano, 2010a).

Apple gained its leadership with an 'open-but not-open' platform strategy (Cusumano, 2010b). It uses proprietary technology in its platform, and controls the user experience and the type of applications, content or service contracts that can operate on its devices (*ibid.*). At the same time, it also opened up access to outside application developers and content providers; there were 225,000 applications for the iPhone by mid-2010 (*ibid.*). According to Cusumano (2010b), Apple's success can be attributed to the creation of synergies and network effects across its products and complementary services. Apple devices (iPod, iPhone, iPad) and the iTunes service have interoperability with Windows. Apple also provides key complementary platforms such as the iTunes Store, the Apple App Store and the iBooks store (*ibid.*).

Apple, Google and Facebook are examples of platform leaders that have successfully married new consumer devices and internet platforms with a variety of online services and content (*ibid.*).

See also: **platform**

Further reading
Cusumano, Michael (2010a) 'Technology Strategy and Management: The Evolution of Platform Thinking', *Communications of the ACM*, 53(1), 32–4.
Cusumano, Michael A. (2010b) 'Technology Strategy and Management Platforms and Services: Understanding the Resurgence of Apple', *Communications of the ACM*, 53(10), 22–4.
Cusumano, Michael A. and Gawer, Annabelle (2002) 'The Elements of Platform Leadership', *MIT Sloan Management Review*, 43(3), 51–8.
Gawer, A. and Cusumano, M. A. (2002) *Platform Leadership: How Intel, Microsoft, and Cisco Drive Industry Innovation*, Cambridge, MA, Harvard Business School Press.

Portfolio management

Portfolio management (PM) is a dynamic new product development (NPD) decision process (Cooper *et al.*, 1999). Deciding on the right set of NPD projects is a critical role of senior management in **product innovation** (Cooper *et al.*, 2001). In PM, NPD projects are continually evaluated, selected and prioritised (Cooper *et al.*, 1999). Existing projects may be accelerated, killed or reprioritised; and resources may be reallocated to new products. Best-practice companies have a clearly defined and explicit PM. They consistently apply portfolio management to all projects (Cooper *et al.*, 2001). Hewlett-Packard utilises PM to guide all discretionary investments across a broad range of its businesses, strategic initiatives, investment programmes, platforms, products and markets (Menke, 2005).

In the 1960s and 1970s portfolio selection models were highly mathematical. Later methods include:

- Financial models and financial indices based on net present value (NPV), internal rate of return (IRR) and various financial ratios.

- Probabilistic financial models such as Monte Carlo Simulation and decision trees such as the **expected commercial value (ECV)**.
- Options pricing.
- Scoring models and checklists.
- Analytical hierarchy, behavioural and mapping approaches, and bubble diagrams.

The majority of firms use combinations of financial and strategic approaches to PM. Some use a strategic approach combined with bubble diagrams or three port-folio methods in conjunction: a financial method, a strategic approach and a scoring model. Others use strategic buckets to allocate resources across all inno-vation projects in the portfolio. A *strategic bucket* is a collection of NPD programs in alignment with a particular innovation strategy (Chao and Kavadias, 2008). The NPD programs in a strategic bucket may involve four types of development proj-ects. Wheelwright and Clark (1995) identified these as *breakthrough (technical) proj-ects* such as radical next-generation technological research or ground-breaking R&D initiatives; *platform projects* such as new architecture; *derivative (sustaining) projects* such as minor product modifications; and *support projects* such as process improvements and cost reductions. Michael McGrath (2004) refers to the use of on-demand portfolio analysis from *development chain management (DCM) systems* as *dynamic portfolio management*. Uncertainty in dynamic environments influences the strategic balance between incremental and radical innovation projects. Chao and Kavadias (2008) found that environmental complexity shifted the balance towards **radical innovation**, while environmental instability shifted the balance towards **incremental innovation**. **Best-practice** companies do a better job of resource allocation (that is, they match number of projects with available resources) and achieve a better strategic balance than poor-performing companies (Cooper and Edgett, 2008).

See also: **platform**

Further reading

Chao, Raul O. and Kavadias, Stylianos (2008) 'A Theoretical Framework for Managing the New Product Development Portfolio: When and How to Use Strategic Buckets', *Management Science*, 54(5), 907–21.

Cooper, Robert G. and Edgett, Scott J. (2008) 'Maximizing Productivity in Product Innovation', *Research Technology Management*, 51(2), 47–58.

Cooper, Robert G., Edgett, Scott J. and Kleinschmidt, Elko J. (1999). 'New Product Portfolio Management: Practices and Performance', *Journal of Product Innovation Management*, 16, 333–51.

Cooper, Robert, Edgett, Scott, and Kleinschmidt, Elko (2001) 'Portfolio Management for New Product Development: Results of an Industry Practices Study', *R&D Management*, 31(4), 36180.

McGrath, Michael (2004) *Next Generation Product Development: How to Increase Productivity, Cut Costs, and Reduce Cycle Times*, New York, McGraw-Hill.

Menke, Michael (2005) 'Using Portfolio Management to Drive Profitable Growth: Benchmarking the Best-in-Class', Presentation given at the PDMA-IIR Conference on Strategic and Operational Portfolio Management, February 25, Cambridge, MA.

Wheelwright, Steven C. and Clark, Kim B. (1995) *Leading Product Development*, New York: Free Press.

Postponement

Postponement is also known as *delayed differentiation*. It is a supply-chain strategy to delay product differentiation to a point closer to the customer (Van Hoek, 2001). In postponement, a product is customised quickly and inexpensively once the actual demand is known. Postponement is also a specific inventory strategy to deploy inventory further away from the customer while fulfilling service-level objectives, reducing inventory costs and minimising risks (Wadhwa *et al.*, 2008). In **new product development** postponement is used to manage the uncertainty and risk associated with *forecasting demand for new products*.

There are two types of postponement: *time postponement* and *form postponement* (Zinn, 1990). In time postponement the key differential is the timing of inventory deployment to the next location in the distribution process. With form postponement, risk is minimised by delaying assembly, packaging, labelling and/or manufacturing activities until inventory replenishment is required (Bowsersox *et al.*, 1999). Computers are often assembled, packaged and labelled to meet specific configurations during the customer order process. Form postponement also involves forward deployment of materials or components. The shipment of house paint to retailers as a neutral base with subsequent mixing to customer-specific colours is a classic example of postponing until the time of the end-consumer purchase.

With postponement firms avoid anticipatory inventory deployment and reduce the risk associated with forecasting demand variations.

See also: **forecasting**

Further reading

Bowersox, Donald J., Stank, Theodore P. and Daugherty, Patricia J. (1999) 'Lean Launch: Managing Product Introduction Risk', *Journal of Product Innovation Management*, 16. 557–68.

Van Hoek, R. (2001) 'The Rediscovery of Postponement: A Literature Review and Directions for Research', *Journal of Operations Management*, 19, 161–84.

Wadhwa, S., Bhoon, K. S. and Chan, F. (2008) 'Postponement Strategies for Re-engineering of Automotive Manufacturing: Knowledge-Management Implications', *International Journal of Advanced Manufacturing Technology*, 39(3/4), 367–87.

Zinn, W. (1990) 'Should You Assemble Products before an Order Is Received?', *Business Horizons*, 33, 70–73.

Prediction markets

P

A new product firm can use prediction markets (PM) to select the most promising new product ideas for **development** and to forecast **demand** before they are launched (Ho and Chen, 2007). It is a method to aggregate a large amount of information from various individuals to create a forecast (Ivanov, 2009). PM outperforms other forecasting models, including surveys, for four main reasons: the wisdom of crowds, incentives, performance-based weighting and real-time dynamic (*ibid.*).

The principle behind prediction markets is to create a **market** where a number of individuals can place bets on a specific outcome of interest (Borison and Hamm, 2010). The market is closed when the event occurs, the outcome is

revealed, and the participants win or lose based on that outcome. The first proto-
type of the current prediction market was developed in 1982 by Charles Plott and
Shyam Sunder and was further enhanced in 1988 (Ho and Chen, 2007). The Iowa
Electronic Market (IEM) was the first field application (*ibid.*).

Hewlett-Packard uses PM to forecast sales, General Electric uses it in **new
product development** to filter out the best ideas, and Siemens and Microsoft use
PM to forecast product **launch** and completion dates (Ivanov, 2009).

See also: **Delphi method, forecasting**

Further reading
Borison, Adam and Hamm, Gregory (2010) 'Prediction Markets: A New Tool for Strategic
    Decision Making', *California Management Review*, 52(4), 125–41.
Ho, Teck-Hua and Chen, Kay-Yut (2007) 'New Product Blockbusters: The Magic and
    Science of Prediction Markets', *California Management Review*, 50(1), 144–58.
Ivanov, Aleksandar (2009) 'Using Prediction Markets to Harness Collective Wisdom for
    Forecasting', *Journal of Business Forecasting*, 28(3), 9–14.

Process innovation

See **product and process innovation**

PRO INNO Europe

PRO INNO Europe is an innovation policy initiative of the European Union. It is
managed by the EU Directorate General Enterprise and Industry and is planned to
become the focal point for innovation policy analysis, learning and development.
It unites the previously separate PAXIS initiative, the TrendChart on Innovation in
Europe, including the European Innovation Scoreboard, and a series of innovation
policy studies in a common framework.

PRO-INNO Europe is based on three pillars, accommodating a total of eight
modules. Each of these follows a specific key aim of the initiative and collectively
forms an integrated policy approach intended to develop new and better innovation policies based on policy analysis and reliable statistics.

http://www.proinno-europe.eu/ (accessed on December 9, 2010).

Product and process innovation

The technology life cycle describes the dynamics of product and process innovations (Utterback and Abernathy, 1975; Clark, 1985; Klepper, 1996). **Innovation**
can be driven by the external requirements of the market (Schmookler, 1966) or
the activities and internal capabilities of the firm (Dosi, 1982). The drivers of
product innovations are customer demand for new products and the firm's
desire to penetrate new markets. The drivers of **process innovations** are reduction in delivery lead time, lowering of operational costs and increase in flexibility.
The Oslo Manual (2005) defines process innovation as 'the implementation of a
new or significantly improved production or delivery method'. Process innovation
involves significant changes in techniques, equipment, software or other components of the system (Bloch, 2007).

Utterback and Abernathy (1975) developed a dynamic model of product and process innovation. The central theme of this model is that the technology employed to support a product changes over time with a consistent pattern. Utterback (1994) described how product and process innovations change during three phases of the industry life cycle: *fluid, transitional* and *specific*. In the fluid phase, product design is fluid and product innovation is focused on improving product performance. The rate of product innovation is highest in the fluid phase. Thus leads to a 'transitional phase' in which the rate of major product innovations slows down and the rate of process innovations speeds up. In the transitional phase, the optimal product configuration is reached and the **dominant design** emerges. As a product becomes more standardised and is produced in a systematic process (that is, the dominant design takes root), the innovation emphasis shifts to process innovations to lower costs. During the transitional phase of the automotive industry, public drivers and car companies defined the modern automobile. At the same time, increasing process innovations enabled automobile companies to lower manufacturing costs.

In the specific phase, the rate of major innovations decelerates for both product and process as some industries focus primarily on cost, volume and capacity. The airline industry offers an example of decelerating product and process innovations. In digital and information-based technologies, the life cycle can be extended by increasing performance at a stable price (Adner and Levinthal, 2001). A culture of cost reductions and continuous improvements supports **incremental innovations** while becoming an impediment to **radical innovations.**

A key observation drawn from a 100-year perspective is that industries experience waves of product and process innovation interspersed with periods of stability and consolidation.

Firms invest in innovative activities differently under different market environments. Lin and Saggi (2002) found that firms do more product innovation under price competition while investing more in process innovation under quantity competition. Dawid *et al.* (2009) found that competitors invest more in process innovation in order to push the potential innovator to introduce the new product, since this reduces competition for the existing product.

Further reading
Adner, Ron and Levinthal, Daniel (2001) 'Demand Heterogeneity and Technology Evolution: Implications for Product and Process Innovation', *Management Science,* 47(5), 611–28.
Bloch, Carter (2007) 'Assessing Recent Developments in Innovation Measurement: The Third Edition of the Oslo Manual', *Science & Public Policy,* 34(1), 23–34.
Clark, K. B. (1985) 'The Interaction of Design Hierarchies and Market Concepts in Technological Evolution', *Research Policy,* 14(5), 235–51.
Dawid, Herbert, Kopel, Michael and Dangl, Thomas (2009) 'Trash It or Sell It? A Strategic Analysis of the Market Introduction of Product Innovations', *International Game Theory Review,* 11(3), 321–45.
Dosi, G. (1982) 'Technological Paradigms and Technological Trajectories', *Research Policy,* 11, 147–62.
Klepper, S. (1996) 'Entry, Exit, Growth and Innovation over the Product Life Cycle', American Economic Review, 86(3), 562–83.
Lin, P. and Saggi, K. (2002) 'Product Differentiation, Process R&D, and the Nature of Market Competition', *European Economic Review,* 46, 201–11.

P

OECD/Eurostat (2005) *Guidelines for Collecting and Interpreting Innovation Data – The Oslo Manual*, 3rd edn, Paris, OECD.

Phillips, Wendy, Noke, Hannah, Bessant, John and Lamming, Richard (2006) 'Beyond the Steady State: Managing Discontinuous Product and Process Innovation', *International Journal of Innovation Management*, 10(2), 175–96.

Schmookler, J. (1966) *Invention and Economic Growth*, Cambridge, MA, Harvard University Press.

Utterback, J. and Abernathy, W. (1975) 'A Dynamic Model of Process and Product Innovation', *Omega*, 3, 639–56.

Utterback, James M. (1994) *Mastering the Dynamics of Innovation: How Companies Can Seize Opportunities in the Face of Technological Change,* Boston, MA, Harvard Business School Press.

http://www.zanthus.com/databank/innovation/market_dynamics.php (accessed on November 13, 2010).

Product and technology adoption curves

In 1957, Iowa State University researchers developed the technology-adoption life-cycle model (Cravotta, 2003). The model classified technology adopters into early adopters, early majority, late majority and laggards. Early adopters work with incomplete and pre-certified products and development tools, while the early majority requires the product and tools to support the necessary functions (*ibid.*). There is an institutional, financial and skills gap during the transition from an emerging technology favoured by early adopters to the creation of a compelling new market-driven business embraced by the early majority (Barr *et al.*, 2009). This gap is referred to as the *valley of death* in the **commercialisation** of technology (Markham, 2002).

The *technology S-curve* theory suggests that as a technology becomes more mature, maintaining the ability to improve performance takes an ever-increasing amount of effort (Rifkin, 1994). The S-curve framework suggests that technological progress may begin more slowly, speed up and then slow down after maturity (Christensen, 1992).

The *new product adoption curve* has a similar shape to the technology S-curve. For some end-user segments, the adoption curve can actually lead the S-curve because of self-developed applications by **lead users,** as discussed by Eric von Hippel (1986). In other cases the S-curve leads the adoption curve. The new technology adoption gap represents huge challenges to managers in technology-intensive businesses.

Further reading

Barr, Steve H., Baker, Ted., Markham, Stephen K. and Kingon, Angusi (2009) 'Bridging the Valley of Death: Lessons Learned from 14 Years of Commercialization of Technology Education', *Academy of Management Learning & Education*, 8(3), 370–88.

Christensen, C. (1992) 'Exploring the Limits of the Technology S-Curve. Part I: Component technologies', *Production and Operations Management*, 1(4), 334–57.

Cravotta, Robert (2003) 'Welcome to the Jungle', *EDN Europe*, 48(11), 57–61.

Markham, S. K. (2002) 'Moving Technologies from Lab to Market', *Research-Technology Management*, 45(6), 31–42.

Rifkin, Glenn (1994) 'Wrestling with the S-Curve', *Harvard Business Review*, 72(1), 10.

Von Hippel, E. (1986). 'Lead Users: A Source of Novel Product Concepts', *Management Science*, 32(7), 791–806.

Product architecture

Product architecture is defined as 'the specification of the interfaces among interacting physical components' (Ulrich, 1995). The term refers to the arrangement of functional elements of a product into physical blocks. Each physical block or chunk is made up of a collection of components that implement the function of the product. Product architecture emerges when a design is decomposed into its component parts and the ways in which the component parts will interact in the design are determined (Sanchez, 2001). Product architecture decisions, made in the early phases of the **product development process**, exert a significant influence on subsequent product development and **commercialisation** activities (Ulrich, 1995).

Ulrich and Eppinger (1995) recommend a four-step process to establish product architecture:

1. Creating a schematic of the product.
2. Clustering the elements of the schematic.
3. Creating a rough geometrical layout.
4. Identifying the fundamental and incidental interactions.

There are two types of product architecture: **modular architecture** and integral architecture (Shibata *et al.*, 2005). **Architectural innovation** focuses on cost reduction. Wang (2008) suggests that a low-cost and low-price strategy allows indigenous manufacturers to compete with multinational corporations. For example, the Geely Group in China successfully implemented a quasi-open architecture for its low-cost, competitive strategy that led to the car maker's success (Wang, 2008).

Further reading

Sanchez, R. (2001) 'Product, Process, and Knowledge Architectures in Organizational Competence', in R. Sanchez (ed.), *Knowledge Management and Organizational Competence*, New York, Oxford University Press, pp. 227–50.

Shibata, T., Yano, M. and Kodama, F. (2005) 'Empirical Analysis of Evolution of Product Architecture. Fanuc Numerical Controllers from 1962 to 1997', *Research Policy*, 34, 13–31.

Ulrich, K. T. (1995) 'The Role of Product Architecture in the Manufacturing Firm, *Research Policy*, 24, 419–40.

Ulrich, Karl T. and Eppinger, Steven D. (1995) *Product Design and Development*, New York, McGraw-Hill.

Wang, Hua (2008) 'Innovation in Product Architecture – A Study of the Chinese Automobile Industry', *Asia Pacific Journal of Management*, 25(3), 509–35.

Product autonomy

The term 'product autonomy' refers to the phenomenon whereby a product can initiate and complete certain tasks on its own without human intervention (Rijsdijk and Hultnik, 2003). *Autonomous products* show 'proactive and self-starting behavior, work together with human beings, and take over some of the user's normal decision-making functions' (*ibid.*). The automower by the Swedish firm Electrolux is an example of an autonomous product (Rijsdijk and Hultnik, 2009).

Rijsdijk and Hultnik (2003) illustrate four levels of product autonomy (manual, bounded, supervised and symbiosis) with a washing machine as an example of an autonomous product. In the *manual level of autonomy*, a user puts laundry in the washing machine, selects the operating parameters, closes the door and starts the machine. In *bounded autonomy*, the washing machine starts itself after the user has put the laundry in the machine, selects the operating parameters and closes the door. In *supervised autonomy*, while the user issues commands, the washing machine gives suggestions about operating parameters (programme settings). In *symbiosis autonomy*, ongoing communication occurs between the user and the washing machine. For example, the machine may open the door when the user approaches it with laundry in his or her hands, and close the door after the laundry is loaded into the machine. Also, the machine would wash clothes with appropriate programme settings without input from the user.

Autonomy is the first of seven dimensions of *smart products* or *product smartness* (Rijsdijk and Hultnik, 2009): autonomy, adaptability, reactivity, multifunctionality, ability to cooperate, humanlike interaction and personality.

Further reading
Rijsdijk, Serge A. and Hultnik, Erik Jan (2003) 'Honey, Have You Seen Our Hamster? Consumer Evaluations of Autonomous Domestic Products', *Journal of Product Innovation Management*, 20(3), 204–16.
Rijsdijk, Serge A. and Hultink, Erik Jan (2009) 'How Today's Consumers Perceive Tomorrow's Smart Products', *Journal of Product Innovation Management*, 26(1), 24–42.

Product champion

The Materials Advisory Board of the National Academy of Engineering defines a **champion** as 'an individual who is intensely interested and involved with the overall objectives and goals of the project and who plays a dominant role in many of the research-engineering interaction events through some of the stages, overcoming technical and organizational obstacles and pulling the effort through its final achievement by the sheer force of his will and energy' (Chakrabarti, 1974). Schon (1967) argues that a product champion is essential for successful **product innovation**. From his personal experience as an executive champion, Frey (1991) emphasises that a champion must be prepared to pull people together from many different functions and suppliers and be persistent. Howell (2005) found that 'effective champions convey confidence and enthusiasm about the innovation, enlist the support and involvement of key stakeholders, and persist in the face of adversity'.

Markham and Aiman-Smith (2001) outline ten truths about champions, suggesting that championing is a sociopolitical process. Champions associated with **new product development** projects get resources and keep projects alive; are passionate about their activities; are persuasive and take risks; their existence is independent of having a formal NPD process with NPD teams; and they provoke antagonists and respond to high-level strategy (*ibid.*).

Champions pose a challenge to managers. Managers must accurately understand what can be expected from champions and what motivates them in order to utilise the energies of champions in product innovation (*ibid.*).

Further reading
Chakrabarti, Alok K. (1974) 'The Role of Champion in Product Innovation', *California Management Review*, 17(2), 58–62.

Frey, Don (1991) 'Learning the Ropes: My Life As a Product Champion', *Harvard Business Review*, 69(5), 46–52.

Howell, Jane M. (2005) 'The Right Stuff: Identifying and Developing Effective Champions of Innovation', *Academy of Management Executive*, 19(2), 108–19.

Markham, Stephen K. and Aiman-Smith, Lynda (2001) 'Product Champions: Truths, Myths, and Management', *Research Technology Management*, 44(3), 44.

Schon, Donald A. (1967) *Technology and Change*, New York, Dell Publishing.

Product data management (PDM)

Product data management (PDM) is the essential technology of **concurrent engineering** (Sung and Park, 2007). It is the management of all product data and life-cycle information. PDM is commonly used to manage and control engineering information and as a tool to integrate product-related information from multiple sources. It allows authorised users to access relevant product data that is always the latest version (*ibid.*). PDM gives multiple advantages such as reducing time to market, improving design and manufacturing accuracy, and improving project management (*ibid.*). A PDM system captures and consistently enforces a specific **product development process** in support of other organisational processes (Huang *et al.*, 2004). However, PDM systems have access limits for users at different locations, a requirement for global manufacturing. This shortfall can be overcome with a *component-based product data management (CPDM)* system (Sung and Park, 2007) or *web-based product data management (WPDM)* system (Huang *et al.*, 2004).

Further reading

Huang, M.Y., Lin, Y.J. and Xu, H. (2004) 'A Framework for Web-Based Product Data Management Using J2EE', *International Journal of Advanced Manufacturing Technology*, 24(11/12), 847–52.

Sung, C. S. and Park, Sam Joon (2007) 'A Component-Based Product Data Management System', *International Journal of Advanced Manufacturing Technology*, 33(5/6), 614–26.

Product definition

The PDMA Handbook of New Product Development defines product as describing all goods, services and knowledge that are sold (Rosenau *et al.*, 1996). The PDMA considers products as bundles of attributes. Armstrong and Kotler (2000) define a product as an offering 'to a market for attention, acquisition, use, or consumption, and that might satisfy a want or a need'.

Product definition is the functional or technical specifications for a product. During the definition phase of the **new product development process**, input data about customer preferences and competitive products are transferred into design specifications (Pugh and Gardiner, 1991). In addition, customers, marketers and engineers work together to translate the **voice of customers** into product specifications (that is, specific functional requirements) (Yu *et al.*, 2008). **Quality function deployment (QFD)** and kansei engineering are examples of methods used for making rational decisions in product design (*ibid.*). Cooper and Kleinschmidt (1987) found that the **product definition** phase is a critical success factor for new products. According to Cooper (1993), early and sharp definition

P

imposes discipline on the product development process and eliminates costly and time-consuming changes in specifications.

Further reading
Armstrong, G. and Kotler, P. (2000) *Marketing: An Introduction*, 5th edn. Englewood Cliffs, NJ, Prentice-Hall.
Cooper, Robert G. (1993) *Winning at New Products*, Reading, MA: Addison-Wesley.
Cooper, Robert G. and Kleinschmidt, Elko J. (1987) 'New Products: What Separates Winners and Losers?', *Journal of Product Innovation Management*, 4(3), 169–84.
Pugh, S. and Gardiner, K. M. (1991) *Total Design: Integrated Methods for Successful Product Engineering*, Wokingham, Addison Wesley.
Rosenau et. al. (Eds). (1996). *The PDMA Handbook of New Product Development.* New York: John Wiley and Sons.
Yu, Li, Wang, Liya and Yu, Jianbo (2008) 'Identification of Product Definition Patterns in Mass Customization Using a Learning-Based Hybrid Approach', *International Journal of Advanced Manufacturing Technology*, 38(11/12), 1061–74.

Product development

See **new product development**

Product development process

See **new product development process**

Product development strategies

Firms use two generic product development strategies: **vertical** and **horizontal innovation**. They may differentiate their products by 'developing the current product technology or its characteristics (i.e. produce vertical innovation) or horizontally by introducing additional product characteristics (i.e. produce horizontal innovation)' (Koski and Kretschmer, 2007). **Vertical innovation** takes place when all customers consider a product has been improved at the same price, while **horizontal innovation** occurs when only some customers regard a product as improved (*ibid.*). When horizontal and vertical innovations are imitated widely by competitors, a **dominant design** emerges (*ibid.*).

Commonly practised **new product development (NPD)** strategies include pioneer, imitator, rapid innovation, disruptive technology introduction, preannouncement, partnering, standard setting and the use of platforms (Nadeau and Casselman, 2008). In the *pioneer strategy*, companies introduce new products where no other comparable product or approach exists to fulfil a market need. For instance, pharmaceutical company Biovail introduced Wellbutrin, a product with a time-release formulation enabling a convenient once-daily dosage, which is protected by a patent (*ibid.*). *Imitators* follow a pioneer, modifying products to improve on any initial mistakes that a competitor may have made. They can also diminish the value of the pioneering product by turning it into a commodity during the growth and maturity phases of its life cycle (*ibid.*). *Rapid innovation strategy* centres on a shortened planning time in the product development cycle, and implementing smaller incremental improvements more frequently (Stalk, 1988). In *disruptive strategy*, firms sacrifice product performance in some areas compared to existing products. Disruptive products are often not perceived as valuable by mainstream customers (Christensen, 2000). For example, voice over internet products from Skype and NetZero Voice are disruptive products. Pre-announcement strategies involve making public disclosures of intended product releases (Nadeau and Casselman, 2008). Airbus is an example where several European aircraft manufacturers partnered and formed a consortium to launch new aeroplanes. Intel's microprocessors, Microsoft's xBox or the Sony PlayStation are examples of *platform strategy* in which future new products are built on a proprietary technology that also acted as a base (*ibid.*).

Millson and Wilemon (2008) found that there is no difference in the quality or the risk associated with the development of new products across **new product development (NPD)** entry strategies (e.g. in-house developments and joint ventures). NPD organisations also have a choice between innovation and imitation entry strategies. The choice is guided by three factors: cost, incumbent reaction and profitability (Ofek and Turut, 2008). Innovative companies such as Intel and EMC commit to innovation strategies when threatened by innovative and imitative entrants, respectively (*ibid.*).

Zhang *et al.* (2009) found that subsidiaries of multinational companies can have product development strategies with breakthrough focus, platform focus and incremental focus. Breakthroughs focus calls for intensive knowledge utilisation, while the incremental focus involves minor modifications requiring only the extension of prior knowledge (*ibid.*).

See also: **competitive strategy**

Further reading
Christensen, C. M. (2000) *The Innovators Dilemma: When New Technologies Cause Great Firms to Fail*, New York, Harper Collins.
Cusumano, M. (2002) 'The Elements of Platform Leadership', *MIT Sloan Management Review*, 43(3), 51–8.
Koski, Heli and Kretschmer, Tobias (2007) 'Innovation and Dominant Design', *Mobile Telephony. Industry & Innovation*, 14(3), 305–24.
Millson, Murray R. and Wilemon, David (2008) 'Impact of New Product Development (NPD) Proficiency and NPD Entry Strategies on Product Quality and Risk', *R&D Management*, 38(5), 491–509.
Nadeau, John and Casselman, R. Mitch (2008) 'Competitive Advantage with New Product Development: Implications for Life Cycle Theory', *Journal of Strategic Marketing*, 16(5), 401–11.
Ofek, Elie and Turut, Ozge (2008) 'To Innovate or Imitate? Entry Strategy and the Role of Market Research', *Journal of Marketing Research*, 45(5), 575–92.
Stalk, G. (1988) 'Time – The Next Source of Competitive Advantage', *Harvard Business Review*, 66(4), 41–51.
Zhang, Junfeng, Di Benedetto, C. Anthony and Hoenig, Scott (2009) 'Product Development Strategy, Product Innovation Performance, and the Mediating Role of Knowledge Utilization: Evidence from Subsidiaries in China', *Journal of International Marketing*, 17(2), 42–58.

P

Product development team

See **new product development team**

Product family

A product family is a group of similar products that are derived from one or more product **platform**(s), but possess specific features/functionalities to satisfy different customer needs (Meyer and Lehnerd 1997) while achieving economies of scale during production (Ye *et al.*, 2009). Commonality in the cockpit has helped fuel demand for Airbus aircraft (Aboulafia, 2000). However, commonality also can lead to a lack of product distinctiveness (Robertson and Ulrich, 1998) and cannibalisation (Fruchter *et al.*, 2006).

To minimise the risk of internal competition among similar products within the product family, Alizon *et al.* (2009) introduced the commonality versus diversity index (CDI), commonality being how well the components and functions can be shared across a product family. The CDI provides a rationale for making commonality/diversity decisions within a family of products or across families by relating to a potentially ideal design (*ibid.*). To balance the trade-off between product commonality and variety (the range of different products in a product family), Ye *et al.* (2009) introduced the product family evaluation graph (PFEG) to quantify the trade-off at the product family planning stage relative to the desired competitive focus. Kumar *et al.* (2009) proposed a market-driven product family design (MPFD) methodology to examine the variety in product offerings and the cost savings associated with commonality.

A single-platform product family design has the advantage of simplicity, but may make some low-end products overdesigned and some high-end products underdesigned. This can result in losing market share in high-end market niches or wasting capital investment in low-end niches (Dai and Scott, 2004). A multiple-platform product design offers opportunities to generate more efficient and effective product families (de Weck *et al.*, 2003) by allowing the platform variable to take more than one common value.

See also: **platform**

Further reading
Aboulafia, R. (2000) 'Airbus Pulls Closer to Boeing', *Aerospace America*, 38, 16–18.
Alizon, Fabrice, Shooter, Steven and Simpson, Timothy (2009) 'Assessing and Improving Commonality and Diversity within a Product Family', *Research in Engineering Design*, 20(4), 241–53.
Dai, Z. and Scott, M. J. (2004) 'Product Platform Design through Sensitivity Analysis and Cluster Analysis', *Proceedings of the ASME Design Engineering Technical Conference*, Salt Lake City, UT, ASME, pp. 893–905.
De Weck, O. L., Suh, E. S. and Chang, D. (2003) 'Product Family and Platform Portfolio Optimization', *Proceedings of the ASME Design Engineering Technical Conferences*, Sept. 2–6, Chicago, IL.
Fruchter, G. E., Fligler, A. and Winer, R. S. (2006) ,Optimal Product Line Design: Genetic Algorithm Approach to Mitigate Cannibalization', *Journal of Optimization Theory Applications*,131, 227–44.
Kumar, Deepak, Chen, Wei and Simpson, Timothy W. (2009) 'A Market-Driven Approach to Product Family Design', *International Journal of Production Research*, 47(1), 71–104.
Meyer, M. H. and Lehnerd, A. P. (1997) 'The Power of Product Platforms: Building Value and Cost Leadership', New York, Free Press.
Robertson, D. and Ulrich, K. (1998) 'Planning for Product Platforms', *Sloan Management Review.*, 39, 19–31.

P

Simpson, T. W., Maier, J. R. A. and Mistree, F. (2001) 'Product Platform Design: Method and Application', *Research Engineering Design,* 13(1), 2–22.

Ye, Xiaoli, Thevenot, Henri J., Alizon, Fabrice, Gershenson, John K., Khadke, Kiran, Simpson, Timothy W. and Shooter, Steven B. (2009) 'Using Product Family Evaluation Graphs in Product Family Design', *International Journal of Production Research,* 47(13), 3559–85.

Product innovation

The Oslo Manual (2005) defines product innovation as 'the introduction of a good or service that is new or significantly improved with respect to its characteristics or intended uses'. Story *et al.* (2009) describe product innovation as the innovation 'in offering to market, the object of market exchange, and the basis upon which money from a customer flows in into the firm'. Market knowledge (Atuahene-Gima, 1995) and cross-functional collaboration are two fundamental resources for successful product innovation. *Cross-functional collaboration* refers to 'the degree of cooperation and the extent of representation by marketing, research and development (R&D), and other functional units in the product innovation process' (Song *et al.,* 1997).

Product innovation performance is the market reward for new products measured in the products' contributions to sales or profits (Zhang *et al.,* 2009). New products include totally new products, new product lines, modifications and derivatives (Li and Atuahene-Gima, 2001).

Zhang *et al.* (2009) found a U-shaped relationship between resources and the product innovation performance of highly innovative products (breakthrough products), a positive relationship between resources and moderately innovative products (platform products) and no correlation between resources and *incremental innovation* performance. In addition, knowledge utilisation is a key predictor of the performance of highly and moderately innovative products (Zhang *et al.,* 2009).

See also: **incremental innovation, product and process innovation, radical innovation**

Further reading

Atuahene-Gima, Kwaku (1995) 'An Exploratory Analysis of the Impact of Market Orientation on New Product Performance: A Contingency Approach', *Journal of Product Innovation Management,* 12(4), 275–93.

Bloch, Carter (2007) 'Assessing Recent Developments in Innovation Measurement: The Third Edition of the Oslo Manual', *Science & Public Policy,* 34(1), 23–34.

Li, Haiyang and Atuahene-Gima, Kwaku (2001) 'Product Innovation Strategy and the Performance of New Ventures in China', *Academy of Management Journal,* 44(6), 1123–35.

OECD/Eurostat (2005) *Guidelines for Collecting and Interpreting Innovation Data – The Oslo Manual,* 3rd edn, Paris, OECD.

Song, Michael, Montoya-Weiss, Mitzi M. and Schmidt, Jeffrey B. (1997) 'Antecedents and Consequences of Cross-Functional Cooperation: A Comparison of R&D, Manufacturing, and Marketing Perspectives', *Journal of Product Innovation Management,* 14(1), 35–47.

Story, Vicky, Hart, Susan and O'Malley, Lisa (2009) 'Relational Resources and Competences for Radical Product Innovation', *Journal of Marketing Management,* 25(5/6). 461–81.

P

Zhang, Junfeng, Di Benedetto, C. Anthony and Hoenig, Scott (2009) 'Product Development Strategy, Product Innovation Performance, and the Mediating Role of Knowledge Utilization: Evidence from Subsidiaries in China', *Journal of International Marketing*, 17(2), 42–58.

Product innovation charter (PIC)

There is no uniformity concerning the concept and definition of the product innovation charter (PIC) (Bart and Pujari, 2007). Merle Crawford (1980) originally defined it as a 'set of policies and objectives designed to guide new product development'. Other researchers such as Cooper (1987), Crawford (1997), Crawford and di Benedetto (2008) and Kuczmarski (1994) built on Crawford's definition. Cooper viewed PIC more as a strategic planning tool influencing every stage of the product development process. Cooper's PIC also focused more on entry method (that is, internal development, licensing, joint venture or acquisition) (Bart, 2002). However, all PICs indicate the existence of goals, a target market or product arena and a set of new product policies specifying how objectives should be achieved (Bart and Pujari, 2007). PIC unifies the elements of the NPD strategy, guides new product development activities and provides clear direction (Durmu o lu *et al.*, 2008).

In practice, the product innovation charter is a document prepared by senior management. It emphasises that the strategy is for products and not processes, that it is for innovation and that it is a charter that outlines the conditions under which an organisation operates (Crawford and di Benedetto, 2008). A PIC can be thought of as a *mission statement* adapted for new products (Bart, 2002). It sets out a strategic plan for the **opportune identification** of new products (di Benedetto, 2004). The contents of a PIC include the target business arenas, the goals of product innovation (including quantitative metrics), the activities to achieve the goals, the strengths to exploit and the weaknesses to avoid (Durmu o lu *et al.*, 2008).

Further reading
Bart, C. K. (2002) 'Product Innovation Charters: Mission Statements for New Products', *R&D Management*, 32(1), 23–34.
Bart, Chris and Pujari, Ashish (2007) 'The Performance Impact of Content and Process in Product Innovation Charters', *Journal of Product Innovation Management*, 24(1), 3–19.
Cooper, R. G. (1987) 'Defining the New Product Strategy', *IEEE Transactions on Engineering Management*, 34(3), 184–94.
Crawford, C. Merle (1980) 'Defining the Charter for Product Innovation', *Sloan Management Review*, Fall, 3–12.
Crawford, C. M. (1997) *New Products Management*, Homewood, IL, Richard D. Irwin.
Crawford, Merle and di Benedetto, Anthony (2008) *New Products Management*, 9th edn, New York, McGraw-Hill.
di Benedetto, C. Anthony (2004) 'From the Editor', *Journal of Product Innovation Management*, 21(3), 153.
Durmu o lu, Serdar S., McNally, Regina C., Calantone, Roger J. and Harmancioglu, Nukhet (2008) 'How Elephants Learn the New Dance When Headquarters Changes the Music: Three Case Studies on Innovation Strategy Change', *Journal of Product Innovation Management*, 25(4), 386–403.
Kuczmarski, T. D. (1994) 'Inspiring and Implementing the Innovation Mind Set', *Planning Review*, September–October, 37–48.

Product life cycle (PLC)

The product life cycle is a theoretical tool for understanding corporate financial behaviour, including growth, risk and return (Gup and Agrrawal, 1996). Steiner (1969) postulated that every product goes through a life cycle. For Porter (1980), the product life cycle is a strong predictor of the probable course of industry growth. According to Gup and Agrrawal (1996), the sales volume and profits of a product follow an *S-shaped curve*, while the product evolves through four phases of pioneering, expansion, stabilisation and decline, with each phase defined by inflections in growth rate of sales (Gup and Agrrawal, 1996).

- The pioneering phase begins when a new product is introduced in the market.
- The expansion phase is characterised by increasing sales, increasing competition and falling prices. Total industry profits rise and peak during this phase. Large numbers of competitors also fail during the expansion phase.
- In the stabilisation phase sales peak. This phase is characterised by a small number of surviving firms that try to maintain or expand their market shares.
- The declining phase of the life cycle is characterised by diminishing sales and profits and fewer firms.

John Stark (2004) defines **product life-cycle management (PLM)** as 'a new paradigm for product manufacturing, enables a company to manage its products all the way across their life cycles in the most effective way. It helps companies get products to market faster, provide better support for their use, and manage end-of-life better.' Practical PLM is configuration management (CM) with the addition of higher-level management of the product's progress along its life cycle (Rosen, 2010). CM comprises two main processes: documentation release and documentation change control. The eight stages of the product life cycle in practical PLM are planning, concept development, system-level design, detailed design, testing and refinement, production ramp-up, on-going production and phase-out (*ibid.*).

The International Organization for Standards (ISO) developed a framework for *life-cycle assessment* in LCA protocol ISO14040. A life-cycle assessment is the full assimilation of the environmental impact of a product, process or service throughout its life span (also known as 'environmental footprint'). LCA aids in the implementation of process or product development in the context of environmental sustainability (Khan *et al.*, 2002). Bevilacqua *et al.* (2007) propose a method to integrate *design for environment (DfE)* and LCA into the determination of product function and goals, conceptual design, prototype assembly testing and detailed final design activities of the **new product development process**.

P

See also: **dynamics of innovation, ecodesign**

Further reading
Bevilacqua, M., Ciarapica, F. E. and Giacchetta, G. (2007) 'Development of a Sustainable Product Lifecycle in Manufacturing Firms: A Case Study', *International Journal of Production Research*, 45(18/19), 4073–98.
Gup, Benton E. and Agrrawal, Pankaj (1996) 'The Product Life Cycle: A Paradigm for Understanding Financial Management', *Financial Practice & Education*, 6(2), 41–8.

Khan, F. I., Sadiq, R. and Husain, T. (2002) 'GreenPro-I: A Methodology for Risk-Based Process Plant Design Considering Life Cycle Assessment', *Journal of Environmental Modeling & Software*, 17, 669–92.

Porter, Michael F. (1980) *Competitive Strategy: Techniques for Analyzing Industries and Competitors*, New York, Free Press.

Rosen, Jason (2010) 'Product Lifecycle Management and You', *Industrial Engineer: IE*, 42(1), 44–9.

Stark, John (2004) 'Product Lifecycle Management: 21st Century Paradigm for Product Realization', http://www.johnstark.com/PLM_Paradigm.html (accessed on December 19, 2010).

Steiner, George A. (1969) *Top Management Planning*, Toronto, Collier-Macmillan.

Product life cycle management

See **product life cycle**

Product management

Product management refers to the holistic management of a product from the time it is conceived as an idea to the time it is discontinued and withdrawn from the market (Haines, 2009). It is business management at the **product,** product line or *product portfolio* level. In **product management**, companies face the challenge of balancing the needs of the customers with the needs of production.

In *The Product Manager's Desk Reference*, Haines (2009) outlines the following activities for successful product management:

- Creation of a master plan of record for your product.
- Formulation of actionable strategies.
- Incorporation of market data into important decisions.
- Setting the stage for creating innovative products.
- Optimising existing products and product portfolios.
- Applying financial techniques to manage product profitability.

While the traditional **product life-cycle** theory teaches that the products go through introduction (launch to take-off), growth (take-off to slowdown) and maturity (slowdown to decline), *product managers* can achieve long-term, profitable growth by leveraging interdependence in the evolution of products and markets (Agnihotri and Hu, 2009). Christiansen *et al.* (2010) show how product value is co-created through qualification and requalification processes. They use the 50-year sales history of the Egg Lounge Chair, designed by Arne Jacobsen, launched in the 1950s and currently sold by the Danish Company Fritz Hansen, to illustrate how markets, value, products, services and brands are constructed and co-constructed and are stable (for a time) only when fragile networks are created and established. The Egg Lounge Chair gained a 73 per cent increase in the manufacturer's sales volume during 2005–2008 (*ibid.*).

Further reading
Agnihotri, Raj and Hu, Michael Y. (2009) 'The Changing Landscape of Product Management, *Marketing Review*, 9(4), 275–88.

Christiansen, john, K., Claus J. Varnes, Marta Gasparin, Diana Storm-Nielsen and Erik Johnsen Vinthe. (2010). Living Twice: How a Product Goes through Multiple Life Cycles. Journal of Product Innovation Management, 27 (6): 797–827.
Haines, Steven (2009) *The Product Manager's Desk Reference*, New York, McGraw-Hill.

Product platform strategy

A product platform is 'a set of subsystems and interfaces that form a common structure from which a series of derivative products can be efficiently developed and produced' (Meyer, 1996). Elements of a platform strategy are:

1. The underlying elements of the product platform are clearly understood.
2. The defining technology of the product platform is clearly distinguishable from other platforms.
3. The unique differentiation of the product platform provides a sustainable competitive advantage.
4. One product platform should serve one market (*ibid.*).

Product platforms are useful in defining product platform strategy, determining the core platform building blocks, designing the platform and developing a road map for derivative products. Product platforms are also useful for organising the implementation team (Stake, 2003).

Product platform strategy focuses management to make key decisions at the right time (McGrath, 2001). It enables rapid and consistent deployment of products and encourages the long-term view of **product strategy** while leveraging significant operational efficiencies (*ibid.*). Platform strategy also helps management to anticipate when to replace a major product platform (*ibid.*).

Further reading
McGrath, Michael (2001) *Product Strategy for High-Technology Companies*, New York, McGraw-Hill.
Meyer, Marc (1996) 'Product Platform Strategy', Final report, August 1996.
Stake, Roger (2003) 'Product Platforms and Families', www.iip.kth.se/documents/courses/4K1112/ProductPlatforms1.pdf (accessed on January 6, 2011).

Product strategy

Strategy begins with a clear understanding of how **product innovation** goals tie into the broader **business** goals (Cooper and Edgett, 2010). Management defines five elements of strategy (*ibid.*): goals and roles for NPD, select strategic arenas, attack strategy and entry strategy, and resource commitment (strategic and tactical) (*ibid.*). According to McGrath (2001), product strategy can be viewed as a structure consisting of core elements such as **core strategic vision (CSV)**; **product platform strategy**; product line strategy; **new product development (NPD)**; and the marketing platform plan (MPP) framework.

Wheelwright and Clark (1992) suggest that product strategy addresses four key questions: Who will be the target customer? What products will be offered? How products will reach the customer? Why will customers prefer the firm's products to those of competitors?

P

Further reading
Cooper, Robert G. and Edgett, Scott J. (2010) 'Developing a Product Innovation and Technology Strategy for your Business', *Research Technology Management*, 53(3), 33–40.
McGrath, Michael (2001) *Product Strategy for High-Technology Companies*, New York, McGraw-Hill.
Wheelwright, Steven and Clark, Kim (1992) *Revolutionizing Product Development*, New York, Free Press.

Product visualisation

Visualisation is a cognitive tool used during product evaluations and product-adoption decisions. It aids in the understanding of data by leveraging the highly tuned ability of human visual systems to see patterns, spot trends and identify outliers (Heer *et al.*, 2010). A variety of visual encoding and interaction techniques are available for visualisation. Emerging domains such as bioinformatics and text visualisation will find new and creative representations of data. However, all visualisation techniques do the same; that is, 'the principled mapping of data variables to visual features such as position, size, shape, and color' (*ibid.*).

Self-visualisation enables a consumer to imagine a product purchase, to simulate a product experience and to assess the consequences of product usage (Dahl and Hoeffler, 2004). It works for incremental product extensions because the consumer is familiar with the product category from which an incremental product extension is generated. Familiarity enables individuals to produce images where they can see themselves using the new product. However, for really new products (RNP), consumers experience difficulty in visualising the full application of an RNP to their current consumption behaviour. So they underestimate the usefulness and adoption of radically new features, and the advantage of self-visualisation over other visualisations is lost for really new products (*ibid.*).

Zhao *et al.* (2009) compared imagination-focused visualisation with memory-focused (self)-visualisation for really new products and found that imagination-focused visualisation gave higher evaluations of an RNP, but had no effect on incremental new products.

Further reading
Dahl, Darren W. and Hoeffler, Steve (2004) 'Visualizing the Self: Exploring the Potential Benefits and Drawbacks for New Product Evaluation', *Journal of Product Innovation Management*, 21(4), 259–67.
Heer, Jeffrey, Bostock, Michael and Ogievetsky, Vadim (2010) 'A Tour through the Visualization Zoo', *Communications of the ACM*, 53(6), 59–67.
Zhao, Min, Hoeffler, Steve and Dahl, Darren W. (2009) 'The Role of Imagination-Focused Visualization on New Product Evaluation', *Journal of Marketing Research*, 46(1), 46–55.

Productivity index

The productivity index is a tool used for reviewing and ranking projects. It is useful to prioritise projects during portfolio reviews, especially when resources are constrained (Cooper, 2009). Cooper (2009) described the productivity index as a powerful extension of the net present value (NPV) method. It uses a quantitative, financial approach based on the theory of constraints for maximising the value of a portfolio subject to a constraining resource. It is calculated by dividing the factor that is to be maximised (e.g. the NPV) by the constraining resource (e.g. person-days or costs required to complete the project).

Productivity Index (PI) = Forecasted NPV/Person-Days to Complete Project

or

PI = Forecasted NPV/Cost to Complete Project

Cooper (2009) recommends that the projects be ranked according to this index. Top-ranked projects are 'go' projects. They are resourced and accelerated to market. Once the resources run out, projects beyond the resource limit go on hold or are killed.

The productivity index method is designed to maximise the value of the development portfolio while staying within the resource limit.

Further reading
Cooper, Robert G. (2009) 'How Companies Are Reinventing Their Idea-to-Launch Methodologies', *Research Technology Management*, 52(2), 47–57.

Product-service system (PSS)

There are several interpretations of the product-service system (PSS). Most consider PSS as 'product(s) and service(s), combined in a system, to deliver required user functionality in a way that reduces the impact on the environment' (Baines *et al.*, 2007). Baines *et al.* (2007) interpret the PSS as an integrated product and service offering that delivers value in use. It utilises the knowledge of the designer-manufacturer both to increase value (increase output) and to decrease material and other costs (decrease input) to a system (*ibid.*).

Goedkoop *et al.* (1999) defines the key elements of a PSS as:

- Product: a tangible commodity manufactured to be sold and capable of fulfilling a user's needs.
- Service: an activity (work) done for others with an economic value, on a commercial basis.
- System: a collection of elements, including their relations.

There are two evolving trends in support of PSS: servicisation and productisation. Morelli (2003) saw *servicisation* as the evolution of product identity where the material component is inseparable from the service system. Similarly, *productisation* is the evolution of the service component to include a product or a new service component marketed like a product. Convergence of these two trends leads to a single offering; that is, a PSS (Baines *et al.*, 2007). Morelli (2002) recommends the involvement of designers in the development of the PSS and prescribes tools to support the design process.

P

Further reading
Baines, T. S., Lightfoot, H. W.. Evans, S., Neely, A., Greenough, R., Peppard, J., Roy, R., Shehab, E., Braganza, A., Tiwari, A., Alcock, J. R., Angus, J. P., Bastl, M., Cousens, A., Irving, P., Johnson, M., Kingston, J., Lockett, H., Martinez, V. and Michele, P. (2007)

Proceedings of the Institution of Mechanical Engineers – Part B – J. Engineering Manufacture, 221(10), 1543–52.

Goedkoop, M., van Haler, C., te Riele, H. and Rommers, P. (1999) 'Product Service-Systems, Ecological and Economic Basics', Report for Dutch Ministries of Environment (VROM) and Economic Affairs (EZ).

Morelli, N. (2003) 'Product Service-Systems, a Perspective Shift for Designers: A Case Study – The Design of a Telecentre', Design Studies, 24(1), 73–99.

Morelli, Nicola (2002) 'Designing Product/Service Systems: A Methodological Exploration', *Design Issues*, 18(3), 3–17.

Product-servicisation

Product-servicisation is also referred as the **product-service system (PSS)** in manufacturing and product design. Other similar terms include *product service combinations, product-to-service, integrated product service offerings* and *integrated product and service engineering* (Jung and Nam, 2008).

Product-servicisation refers to a marketable (bundled) set of products and services capable of fulfilling a user's need (Mont, 2002). Service in product-servicisation refers to intangible benefits or values added on to existing products (Jung and Nam, 2008). Williams (2007) classified service factors into 'Product-oriented services (e.g. maintenance contract, advice and consultancy), Use-oriented services (e.g. leasing or sharing) and Result-oriented services, (e.g. mobility)'.

Further reading

Jung, Mi-Jun and Nam, Ki-Young (2008) 'Design Opportunities in Service-Product Combined Systems', *Proceedings of Design Research Society Biennial Conference*, Sheffield, July, http://www3.shu.ac.uk/ (accessed on 17 September, 2009).

Mont, O. K. (2002) 'Clarifying the Concept of Product-Service System', *Journal of Cleaner Production*, 10, 237–45.

Williams, Andrew (2007) 'Product Service Systems in the Automobile Industry: Contribution to System Innovation?', *Journal of Cleaner Production*, 15, 1093–103.

Product-variety paradigm

See **growth theory**

Profit

The traditional concept of profit is related to achieving a positive net financial gain. In his book *The Ultimate Question*, Fred Reichheld (2006) challenged the idea that all profits are good and suggested that satisfied customers promoting a company's products and services are necessary for *sustained profitable growth*. In some firms more than 30 per cent customers come under the heading of bad profits (Reichheld and Markey, 2006). Reichheld (2006) describes bad profits and good profits as follows:

- *Bad profits*: Bad profits extract value from customers, rather than creating value. Some companies save money by delivering a poor **customer experience** and gain bad profits. Customers who are badly treated by a company become its *detractors*.

- *Good profits*: Companies earn good profits when customers are enthusiastic about them and their products. A company earns good profits when it delights its customers. Satisfied customers become its *promoters*. One technique to build promoters is to create a *community of customers*. Card company Hallmark's Idea Exchange builds customer communities by including each of the company's target customer groups: mothers with children at home, grandparents, Hispanic mothers and so on. These *customer communities* deliver real-time insights to Hallmark that are deeper and richer than those provided by traditional customer feedback and market research tools.

Reichheld (2006) introduced the concept of *net promoter score (NPS)*. NPS is calculated by the equation NPS = P − D, where P = promoters and D = detractors. Intuit was one of the first firms to adopt NPS, while 20 per cent of executive bonuses at General Electric are driven by the score (Reichheld, 2006).

See also: **customer emotional clusters**

Further reading
Reichheld, Fred (2006) *The Ultimate Question: Driving Good Profits and Growth*, Boston, MA, Harvard Business School Publishing.
Reichheld, Fred and Markey, Rob (2006) 'Blowing the Whistle on Bad Profits', *Strategic Finance*, 88(2), 8–10.

Profiting from innovation

The profiting from innovation (PFI) framework explains why some innovators profit from innovation while others do not (Pisano and Teece, 2007). The three building blocks of the PFI framework developed by Teece (1986) are:

1. Appropriability regime: The appropriability regime can be classified as tight if the technology is relatively easy to protect (e.g. the formula for Coca-Cola syrup) and weak if the technology is almost impossible to protect (e.g. the Simplex algorithm in linear programming). Companies such as Merck keep the upstream appropriability regime weak in order to protect their ability to continue to leverage their downstream assets in development and commercialisation (Pisano and Teece, 2007). Merck created the Merck Gene Index of the expressed human gene sequence and put gene data into the public domain (*ibid.*). This is a property-preempting investment preventing the privatisation of genes that could block Merck's future research objectives (*ibid.*).

2. Dominant design paradigm: According to Abernathy and Utterback (1978) and Dosi (1982), technological evolution and competition among firms manifests themselves in competition among designs. After trial and error, one design or a narrow class of designs emerges as the **dominant design**. The Model T Ford, the IBM 360 and the Douglas DC-3 are examples of dominant designs in the automobile, computer and aircraft industries, respectively. Competition shifts to price and away from design after the dominant design emerges.

P

3. Complementary assets: The **commercialisation** of innovation requires generic, specialised and co-specialised complementary assets. Marketing, competitive manufacturing and after-sales support are specialised assets. Deficient marketing can lead to improper positioning of a new product or process and market failure. In addition, a lack of required manufacturing and supply capacities may lead to a loss of market leadership to suppliers or competition, and the profits from innovations may go to suppliers or competitors rather than to innovators and owners of the intellectual property.

Business strategy as it relates to the decision of the innovating firm whether to integrate and collaborate is also a key factor in realising profits from innovation (Teece, 1986).

See also: **strategic management**

Further reading
Abernathy, W. J. and Utterback, J. M. (1978) 'Patterns of Industrial Innovation', *Technology Review,* 80(7), 40–47.
Dosi, G. (1982) 'Technological Paradigms and Technological Trajectories', *Research Policy*, 11, 147–62.
Pisano, Gary P. and Teece, David J. (2007) 'How to Capture Value from Innovation: Shaping Intellectual Property and Industrial Architecture', *California Management Review,* 50(1), 278–96.
Teece, David J. (1986) 'Profiting from Technological Innovation: Implications for Integration, Collaboration, Licensing and Public Policy', *Research Policy,* 15(6), 285–305.

Programme management

The Project Management Institute (PMI) states that a *programme* is 'a group of related projects managed in a coordinated way to obtain benefits and control not available from managing them individually' (Shaltry, 2007). Programme management (PgM) is 'the centralized, coordinated management of a program to achieve the program's strategic objectives and benefits' (*ibid.*). For Martinelli and Waddell (2004), PgM is 'the coordinated management of interdependent projects over a finite period of time in order to achieve a common set of business goals'. It is widely practised in the US Department of Defense, although it is an emerging practice in the industry (Hodgkinson, 2010).

PgM is different from *project management*. Programme management is strategic, with a focus on business success, while project management is tactical and stays focused at task accomplishment (Riznic, 2007). Project management manages vertically within a single project; programme management manages horizontally across the functional projects involved (Milosevic *et al.*, 2007). Milosevic *et al.* (2007) view project management as a subset of PgM. PgM also differs from **portfolio management**. While PgM coordinates projects and allocates resources over them, it also stimulates processes of development by managing horizontally across line organisations (Buijs, 2010). A focus on interrelations and coherence between projects would not be possible for project managers working separately in line organisations (Partington *et al.*, 2005).

Programmes are action systems in complex governance processes such as PgM. Each action system consists of subsystems and is embedded in larger systems (Teisman, 2008). Three lines of connectivity are essential for the successful governance of PgM (Buijs, 2010):

1. Programme and projects: There is a development of separate projects and their interconnectedness at programme level.
2. Programme and line organisation: The connection between the programme and the line organisation is about the boundary management and autonomy of the programme in relation to adaptation in the line organisation and the embedding of programme results.
3. Programme and stakeholders in its environment: Stakeholders or other programmes in its environment might hinder or contribute to the programme during its development.

According to Martinelli and Waddell (2004), two high-tech companies, Intel and Tektronix, successfully utilised programme management to:

- Link business strategy to project output.
- Integrate the efforts of multiple project teams to deliver the 'whole product' and achieve a common set of business goals.
- Structure highly matrix organisations into cross-functional programme core teams to align the work and deliverables of multiple project teams.

Further reading
Buijs, Jean-Marie (2010) 'Understanding Connective Capacity of Program Management from a Self-Organization Perspective', *Emergence: Complexity & Organization*, 12(1), 29–38.
Hodgkinson, Jeff (2010) 'Navigating the New PgMP® Credential', http://www.pmosig.org/?page=progprof (accessed on December 20, 2010).
Martinelli, Russ and Waddell, Jim (2004) 'Demystifying Program Management. Linking Business Strategy to Product Development', *PDMA Visions*, 28(1), 20.
Milosevic, Dragan Z.. Martinelli, Russ J. and Waddell, James M. (2007) *Program Management for Improved Business Results*, New York, John Wiley & Sons.
Partington, D., Pellegrinelli, S. and Young, M. (2005) 'Attributes and Levels of Program Management Competence: An Interpretive Study', *International Journal of Project Management*, 23(2), 87–95.
Riznic, Jovica R. (2007) 'Program Management for Improved Business Results', *Engineering Management Journal*, 19(2), 51.
Shaltry, Paul E. (2007) 'Program Management for Improved Business Results', *Project Management Journal*, 38(3), 92.
Teisman, G. R. (2008) 'Infrastructure Investments on the Edge of Public and Private Domains', in G. Arts, W. Dicke and L. Hancher (eds), *New Perspectives on Investment in Infrastructures*, Amsterdam, Amsterdam University Press, pp. 319–47.

P

Prototyping

Prototyping is the process of building a model of a system. It is an essential tool for both **innovation** and **design**, as it helps to visualise novel concepts that may be difficult to communicate otherwise. Creating a tangible prototype is an important step in the proof-of-concept step of a **new product** or **innovation process**. Brodersen *et al.* (2008) recommend staging imaginative places for participatory prototyping.

Prototyping comes in many forms, from low-technology paper sketches to high-technology operational systems using CASE (computer-aided software engineering) or 3D physical modelling.

Many types of prototypes can be developed based on key factors, such as the problem to be solved or the mandate of the customer. For example:

- *Concept prototypes* are developed during the early stages of a project to illustrate the overall vision of a system with respect to its functionality, design, structure and operational characteristics.
- *Technical prototypes* are developed to verify that critical components of the technical architecture of a system integrate properly and are capable of meeting the business needs of a project.
- *Horizontal or user interface prototypes* are developed in the early stages of system analysis to model the outer shell of an entire system for clarifying the scope and requirements of a project.
- *Vertical prototypes* are developed during the later stages of analysis to demonstrate a working, stripped-down version of a system with its core features.
- *Functional storyboarding* is sometimes used to model a business function by describing its user interface and the application flow.

See also: **rapid prototyping**

Further reading
Brodersen, Christina; Dindler, Christian and Iversen, Ole Sejer (2008) 'Staging Imaginative Places for Participatory Prototyping', *CoDesign*, 4(1), 19–30.

Pseudo innovations

Mensch (1979) defines pseudo innovations as 'changes that have benefits neither to the supply nor to the demand side'. According to Mensch (1979), pseudo innovations do not reduce production costs, nor do they increase productivity on the supply side. They also do not improve the quality or usefulness of the product for the final user. They are typically changes in the appearance of a commodity (*ibid.*).

Examining economic cycles of bust and boom, Mensch (1979) suggested that a series of basic or strategic innovations (innovation clusters around 1825, 1886 and 1935) began a new phase of economic prosperity. These innovation clusters generated periods of economic growth at an increasing rate, followed by periods of stagnation. Mensch saw a pattern in that basic innovation, in an industry, is followed by improvement innovations. At first these improvements radically enhanced the basic innovation. Improvement innovations, however, became progressively less significant and lapsed into pseudo innovations. When pseudo innovations dominate (that is, there is a lack of sufficient basic and improvement innovation), it amounts to a technological stalemate (*ibid.*).

If governments pursue growth policies in a technological stalemate, they will stimulate pseudo innovation in existing industries and in a declining life cycle, protect inefficient industries from foreign competition and expand surplus productive capacities (*ibid.*).

Further reading
Mensch, G. (1979) *Stalement in Technology: Innovations Overcome the Depression*, Cambridge, MA, Ballinger.

QualiQuant research

QualiQuant research (also called *quali-quant, qualiquanti, fusion* or *hybrid research*) is a fast-growing component of marketing research with multiple applications to **new product development** and management (Bortner, 2008). The internet and Web 2.0 revolutions, the need for more frequent and holistic understanding of consumers and the need for cheaper and faster research are driving the adoption of QualiQuant research.

Market research is traditionally broken down into qualitative and quantitative research. *Qualitative research* is focused on understanding emotions, social and cultural dimensions and needs that typically cannot be expressed using small samples. It uses techniques such as one-on-one interviews, **focus groups** or **ethnography**. *Quantitative research* is focused on generating hard data for decision making using large representative samples and statistical analysis (surveys, panels etc.).

Merging both qualitative and quantitative approaches, QualiQuant allows for qualitative research with large samples that can be selected to be representative of a broader population (*online focus groups, online communities, netnography* or online behaviour) and for quantitative research generating qualitative insights (online projective techniques, visual associations, recreation of real-life situations) (*ibid.*).

QualiQuant research can be controversial, however. Quantitative research often includes open-ended questions to obtain qualitative insights, and even numerical questions may conceal various meanings. Qualitative research and focus groups in particular are often used for decision making, and even verbal responses can be counted.

Further reading
Bortner, Brad (2008) 'Fused Research Modes Will Save You Money, How to Master the Faster and Cheaper Imperative In Stark Economic Times', Forrester Research, February 24.
Pawle, John and Cooper, Peter (2001) 'New QualiQuant™ Technology: Accelerating Innovation on the Web', *Proceedings of ESOMAR Worldwide Internet Conference*, February, Barcelona.

Quality

Joseph Juran defined quality as fitness for use, but that is not a sufficient definition. So he provided two additional definitions of features (design quality) and freedom from deficiencies (delivery quality) (Bisgaard, 2008b). *Design quality* relates to the features and grade of the product, process or service and expresses intentions (Bisgaard, 2008a). *Delivery quality* is about how well the intentions are actually executed. It is a measure of how many deficiencies there were in delivering the product or service (*ibid.*).

When a firm improves the quality of design or delivery, it engages in innovative efforts. So from an economic perspective, innovation can be considered an umbrella concept that includes quality improvement, and **Six Sigma** as its subsidiary function (*ibid.*). In quality terms, Juran refers to **product innovation** as *quality planning* (Juran, 1989) and *Design for Six Sigma (DFSS)* in Six Sigma terminology (Bisgaard, 2008a). However, the relationship between *total quality management (TQM)* and **technological innovation** is not linear. It depends on the organizational culture, **competitive strategy** and competitive forces (Perdomo-Ortiz *et al.*, 2009).

Dervitsiotis (2010) considers quality the current value-generating capacity of firms and **innovation** as their future value-generating capability. As firms engage in parallel quests for *effectiveness* and *efficiency* (Hill, 1988), Dervitsiotis (2010) proposes that, under a quality initiative, effectiveness is 'doing the right things' and efficiency is 'doing the right things right'; while for innovation, effectiveness is concerned with 'doing the right kind of innovation' and efficiency is 'doing the right kind of innovation right'.

Further reading

Bisgaard, Soren (2008a) 'Geared toward innovation', http://pramleeelvis.wordpress.com/2010/04/11/geared-toward-innovation-an-article-extracted-from-asq-org/ (accessed on December 21, 2010).

Bisgaard, Soren (2008b) 'Quality Management and Juran's Legacy', *Quality Engineering*, 20(4), 390–401.

Dervitsiotis, Kostas N. (2010) 'A Framework for the Assessment of an Organization's Innovation Excellence', *Total Quality Management & Business Excellence*, 21(9), 903–18.

Hill, C. W. L. (1988) 'Differentiation versus Low Cost or Differentiation and Low Cost: A Contingency Framework, *Academy of Management Review*, 13(3), 401–12.

Juran, Joseph M. (1989) *Juran on Leadership for Quality*, New York, Free Press.

Perdomo-Ortiz, Jesus, Gonzalez-Benito, Javier and Galende, Jesus (2009) 'The Intervening Effect of Business Innovation Capability on the Relationship between Total Quality Management and Technological Innovation', *International Journal of Production Research*, 47(18), 5087–107.

Q

Radical innovation

Radical innovation involves 'the commercialization of products based on significant leaps in technology development, with the potential for entirely new features or order of magnitude improvements in performance or cost compared with existing substitutes' (Leifer *et al.*, 2000). Support for radical innovation often takes a cyclical path. Firms may choose to invest in new business opportunities (Burgelman and Valikangas, 2005) in boom times but shift resources to short-term efforts during an economic downturn (Sykes and Block, 1989). Mary Brenner of the Wharton School reports that Wall Street analysts talk glowingly about **sustaining innovations** extending the current technology of incumbents, while downplaying incumbents' initiatives to capitalise on the next wave of technology (Kirby, 2010). For example, from 1990–2001 there were 2821 mentions of Kodak film-based hybrid products and only 158 mentions of Kodak digital products (*ibid.*). *Sustaining technologies* improve the performance of existing products or services and produce short-term, positive financial performance, which is rewarded by Wall Street.

Entrenched market players (that is, large, established, incumbent companies) need to be innovative. However, most are poorly equipped to implement a growth strategy based on radical innovation, because they do not have the right organisation, culture, leadership practices or employees to collect and successfully commercialise radical new ideas (Stringer, 2000). In companies focused on sustaining innovations, organisational structures and cultures discourage bringing big ideas to market; they rely too heavily on internal R&D and do not attract or retain radical innovators (*ibid.*). To remedy this problem, O'Connor *et al.* (2009) suggest that firms must understand the discovery, incubation and acceleration phases of radical innovations:

- *Discovery*: The mission in the discovery phase is to create and identify opportunities in the marketplace and explore the fit between technological capabilities and marketplace needs.
- *Incubation:* This phase involves experimenting in order to create a new business that delivers breakthrough value to customers and to the firm.
- *Acceleration*: The mission is to nurture the business until it can stand on its own.

O'Connor *et al.* (2009) suggest that, during discovery, employees should do bench science or technological experimentation to satisfy a marketplace need; during incubation, employees need to experiment and anticipate the impact of breakthrough business on corporate strategy; and during acceleration, established busi-

ness capabilities such as ramp-up processes, discipline in execution and specialisation help achieve the mission.

Kelley (2009) found that established companies can promote technology-based
innovation by:

- Balancing strategic clarity with the need to allow for creativity and exploration.
- Structuring programmes so that innovations benefit from the organisation's
resources while minimising constraints that impede creative activities.
- Providing accountability and effective resource allocation, while not restrict-
ing the flexibility required for successful innovation.

Kelley (2009) also reported that radical innovation programmes introduced into
established corporate environments are successful when they are highly flexible
systems that maintain organisational connectedness as they evolve. Song and
Thieme (2009) found the greatest positive impact of suppliers in radical innovation
projects when they were engaged in **commercialisation** activities rather than
market information-gathering activities during the pre-design phase of the NPD
process.

Koen *et al.* (2010) diagnose the *breakthrough innovation* problem of large
companies differently. They acknowledge that many large incumbents succeed in
sustaining innovations, but that they fail to implement breakthrough **business-
model innovations** that are 'outside the box' (e.g. in new market spaces). They
cite Microsoft and Sony as examples of companies with a proven ability to
succeed in breakthrough technology innovations but that have failed at business-
model innovation, which opened up new market entry opportunities to others.
Sony changed the way people listened to music with portable music players (e.g.
the Sony Walkman), but yielded the MP3 market to Apple with the iPod and
iTunes. Similarly, Microsoft made a computer in every home a reality, but ceded
the search market to Google (*ibid.*). Koen *et al.* (2010) suggest that breakthrough
business model innovations require new *value networks*. The value network
dimension includes how a firm identifies, interacts with and reacts to its
customers, suppliers and competitors (*ibid.*).

See also: **diffusion of innovations, game-changing innovations**

Further reading
Burgelman, R. A. and Valikangas, L. (2005) 'Managing Internal Corporate Venturing
Cycles', *MIT Sloan Management Review*, 46(4), 26–34.
Kelley, Donna (2009) 'Adaptation and Organizational Connectedness in Corporate
Radical Innovation Programs', *Journal of Product Innovation Management*, 26(5), 487–
501.
Kirby, Julia (2010) 'Wall Street Is No Friend to Radical Innovation', *Harvard Business
Review*, 88(7/8), 28.
Koen, Peter A., Bertels, Heidi, Elsum, Ian R., Orroth, Mike and Tollett, Brenda L. (2010)
'Breakthrough Innovation Dilemmas', *Research Technology Management*, 53(6), 48–51.
Leifer, R., McDermott, C., O'Connor, G., Peters, L., Rice, M. and Veryzer, R. (2000)
Radical Innovation: How Mature Companies Can Outsmart Upstarts, Boston, MA, Harvard
Business School Press.
O'Connor, Gina Colarelli, Corbett, Andrew and Pierantozzi, Ron (2009) 'Create Three
Distinct Career Paths for Innovators', *Harvard Business Review*, 87(12). 78.

R

Song, Michael and Thieme, Jeff (2009) 'The Role of Suppliers in Market Intelligence Gathering for Radical and Incremental Innovation', *Journal of Product Innovation Management*, 26(1), 43–57.

Stringer, Robert (2000) 'How to Manage Radical Innovation', *California Management Review*, 42(4), 70–88.

Sykes, H. B. and Block, Z. (1989) 'Corporate Venturing: Obstacles, Sources, and Solutions', *Journal of Business Venturing*, 4, 159–67.

Rapid prototyping

Rapid prototyping (RP) refers to the fabrication of a three-dimensional (3D) physical model from **computer-aided design (CAD)** data using an additive process (that is, by layer-by layer deposition without using tools) (Byun and Lee, 2005). RP processes begin with a stereolithography (STL) file that describes a model created by a CAD surface or a solid modeller (*ibid.*). The RP models can be used to visualise or verify designs; to check for form, fit and function; or to produce a tooling (or master) pattern for casting or moulding (Williams *et al.*, 1996). In addition to STL, selective laser sintering (SLS), fused deposition modelling (FDM), laminated object manufacturing (LOM), electron beam melting (EBM) and 3D printing (3DP) are some of the technologies used for rapid prototyping. RP systems are used in manufacturing industries such as automobiles, electric home appliances and aerospace industries (Byun and Lee, 2005).

In the early, planning stages of **new product development**, physical prototypes allow the evaluation of subassembly fit, system aesthetics and overall quality before **development** and **launch** (Wright, 2005). The physical prototype guides the consumer product design process, enabling ideation and **focus groups** to respond to the real product (*ibid.*). In the later stages of NPD, the prototype can be used to coordinate suppliers (*ibid.*). Rapid prototyping (RP) reduces product development time and cost (Byun and Lee, 2005).

Besides using additive fabrication for rapid prototyping, some companies are also using it to manufacture parts for use, directly, in their end products (Engineer, 2010). This application of additive fabrication technology is also called *direct digital manufacturing* (DDM), *rapid manufacturing, additive manufacturing* or *free-form fabrication* (*ibid.*).

Further reading

Byun, H. S. and Lee, K. H. (2005) 'A Decision Support System for the Selection of a Rapid Prototyping Process Using the Modified TOPSIS Method', *International Journal of Advanced Manufacturing Technology*, 26(11/12), 1338–47.

'What's New Rapid Prototyping', *Engineer*, 295(7803), 40–44.

Williams, R. E., Komaragiri, S. N., Melton, V. L. and Bishu, R. R. (1996) 'Investigation of the Effect of Various Build Methods on the Performance of Rapid Prototyping (Stereolithography)', *Journal of Materials Processing Technology*, 61, 173–8.

Wright, Paul K. (2005) 'Rapid Prototyping in Consumer Product Design', *Communications of the ACM*, 48(6), 36–41.

Real options analysis

See **risk-quantification techniques**

Regional innovation systems

A regional innovation system is 'the institutional infrastructure supporting innovation within the production structure of a region' (Pekkarinen and Harmaakorpi, 2006). Its primary task is to increase innovative capability in the region.

According to Cooke and Morgan (1998), the systematic dimension of the regional innovation system comes from innovation networks. *Innovation networks* are often formed from a heterogeneous group of firms, universities, technology centres and development organisations, reflecting a top-down model of innovation. The ability of multiple actors to interact in these networks is a decisive success factor in improving innovative capability in the region (Pekkarinen and Harmaakorpi, 2006). This requires integration and exploitation of several assets, such as knowledge, economic strength, creativity and proper governance. Science and technology parks can be considered a special example of a regionalised national innovation system.

Howells (1999) describe five key processes as a requirement for the operation and viability of a regional system of innovation:

- Localised communication patterns at an individual, firm and group level.
- Localised search and scanning procedures for collaborative partners or equipment within the region.
- Localised invention and learning patterns to support user-producer interactions, skilled labour and investment capital.
- Localised knowledge sharing through exposure to solutions and personal interaction.
- Localised innovation performance supported by high levels of trust and reciprocity.

The biotechnology hub around Boston is an example of a regional system of innovation (Voort and De Jong, 2004).

See also: **clusters**

Further reading
Asheim, B. (2001) 'Learning Regions as Development Coalitions: Partnership as Governance in European Workfare States?', *Concepts and Transformation,* 6(1), 73–101.
Cooke, P. and Morgan, K. (1998) *The Associational Economy: Firms, Regions, and Innovation,* Oxford, Oxford University Press.
Howells, J. (1999) 'Regional Systems of Innovation,' in D. Archibugi, J. Howells and J. Michie (eds), *Innovation Policy in a Global Economy,* Cambridge, Cambridge University Press.
Pekkarinen, Satu and Harmaakorpi, Vesa (2006) 'Building Regional Innovation Networks: The Definition of an Age Business Core Process in a Regional Innovation System', *Regional Studies,* 40(4), 401–13.
Rutten, Roel and Boekema, Frans (2007) 'Spatial Innovation Systems: Theory and Cases – an Introduction', *European Planning Studies,* 15(2), 171–7.
Voort, Haiko Van der and De Jong, Martin (2004) 'The Boston Bio-Bang: The Emergence of a Regional System of Innovation', *Knowledge Technology & Policy,* 16(4), 46–60.

R

Requirements engineering (RE)

Zave (1997) defines requirements engineering (RE) as 'the branch of software engineering concerned with the real-world goals for, functions of, and constraints on software systems'. It is also concerned with the relationship of these factors to precise specifications of software behaviour, and to their evolution over time and across software families (*ibid.*). Software cannot function in isolation from the system in which it is embedded. Stevens *et al.* (1998) characterise RE as a branch of *systems engineering* whose goal is to deliver desired systems behaviour to its stakeholders (*ibid.*).

Requirements are often expressed in terms of conflicting characteristics that a system should meet (Hal *et al.*, 2005). For example, regulation and legislation constrain systems, while they are also required to provide security, dependability, safety, reliability, maintainability, portability and so on, which are often in conflict (*ibid.*). The trade-offs between conflicting requirements are often expressed through the choice of high-level *software architecture* (*ibid.*). While RE is concerned with the purpose of a software system and the context of its use (Nuseibeh and Easterbrook, 2000), software architecture is about the study of the structure of software (Perry and Wolf, 1992).

The software industry continues to struggle with the implementation of RE, the most critical process within the development of embedded systems (Parviainen and Tihinen, 2007). Parviainen and Tihinen (2007) observe that the greatest obstacle for its application 'is the availability of information and evidence of the applicability of the methods to different kinds of products, projects and environments' (*ibid.*). In response, Gorschek and Wohlin (2006) propose a market-driven, product-centred requirements engineering model, called the requirements abstraction model (RAM). RAM design orients towards a product perspective and supports continuous requirement engineering effort (*ibid.*). Gorschek *et al.* (2008) evaluated the implementation of RAM at Danaher Motion (DHR) and ABB. They found that both DHR and ABB achieved significant increases in the accuracy of actions performed in requirements engineering/**product management**, and in requirements quality.

Further reading
Gorschek, T. and Wohlin, C. (2006) 'Requirements Abstraction Model', *Requirements Engineering*, 11(1), 79–101.
Gorschek, Tony, Garre, Per, Larsson, Stig and Wohlin, Claes (2008) 'Industry Evaluation of the Requirements Abstraction Model', *Requirements Engineering*, 13(3), 163–90.
Hal, Jon G., Mistrík, Ivan, Nuseibeh, Bashar and Silva, Andrés (2005) 'Relating Software Requirements and Architectures', *IEE Proceedings – Software*, 152(4), 141–2.
Nuseibeh, B. and Easterbrook, S. (2000) 'Requirements Engineering: A Roadmap', *Proceedings of the International Conference on Software Engineering (ICSE-2000)*, pp. 35–46.
Parviainen, Päivi and Tihinen, Maarit (2007) 'A Survey of Requirements Engineering Technologies and Their Coverage', *International Journal of Software Engineering & Knowledge Engineering*, 17(6), 827–50.
Perry, D. E. and Wolf, A. L. (1992) 'Foundations for the Study of Software Architecture', *SIGSOFT Software Engineering Notes*, 17(4), 40–52.
Stevens, R., Brook, P., Jackson, K. & Arnold, S. (1998) *Systems Engineering: Coping with Complexity*, London, Prentice Hall Europe.
Young, Ralph R. (2004) *The Requirements Engineering Handbook*, Norwood, MA, Artechhouse.
Zave, P. (1997) 'Classification of Research Efforts in Requirements Engineering', *ACM Computing Surveys*, 29(4), 315–21.

R

Research and development (R&D)

Research and development (R&D) involves creative work undertaken systematically to increase the stock of knowledge and its application in scientific and engineering arenas. Common types of R&D efforts include:

- *Exploratory research* in the early stages of research in areas not yet well understood before getting full programmatic support.
- *Basic research,* where the objective is to gain more complete knowledge or understanding of the subject under study, without specific applications in mind.
- *Applied research*, which is aimed at discovering new scientific knowledge that has specific commercial objectives. It attempts to determine the means by which specific, recognised needs may be met.
- *Incremental R&D*, which involves small advances in technology to exploit existing knowledge in new ways.
- *Radical R&D*, which focuses on the creation of radically new knowledge with the explicit aim of applying that knowledge to a useful purpose.
- *Fundamental R&D*, which focuses on reaching out into the unknown for developing depth of research competence in specific scientific or engineering areas in fields of potential future technology, and preparing for future commercial exploitation of these fields.
- *Industrial R&D*, which aims at broadening and deepening a company's technological capabilities to support and expand its existing business or drive new business.
- *Interdisciplinary R&D*, which involves practitioners of a number of relevant scientific disciplines.

David *et al.* (2000) define *R&D investment* as 'the validated way to encourage an organization's operations by conducting R&D activities for commercially technological innovations, process improvements, and productive capacities under special government science and technology plans'. R&D investment can be classified as government-funded R&D and R&D funded by private enterprise (Gonzalez and Pazo, 2008). Whether public support complements or substitutes for private R&D is a fundamental issue in the design of technology policy (García-Quevedo, 2004). García-Quevedo (2004) reports that the relationship between public funding of business R&D and private R&D expenditure is ambiguous. However, Wang *et al.* (2009) found that the public impact on industry research incentives was strong in the life sciences (both agriculture and medicine) because public finance generated technological opportunities, which industry exploited profitably (*ibid.*). Wang *et al.* (2009) argue that much of the complementarity of the government's downstream research comes from its breadth of subject matter and orientation towards social and long-term rather than immediate commercial value. Losing sight of that niche would undermine a major part of the justification for public life science effort (*ibid.*).

In *research management,* universities are increasingly held responsible for research output in a collaborative, contractual, legal, financial and ethical environment (Kirkland, 2009). Kirkland (2009) recommends that all research universi-

R

ties must find new ways to maximise the use of resources, and how the results are accounted for and applied. In the USA, Atkinson and Pelfrey (2010) see a trend towards closer relations between universities and industry and claim that 'this trend is encouraging new ways of conducting scientific research and new forms of organization within the research university' (*ibid*.).

There is a concern among scholars of R&D, beyond manufacturing, that *services R&D* is underappreciated by innovation policy makers and that even some service managers view problems of innovation in services as problems of project management rather than as problems of R&D (Miles, 2007). Services managers with such a view may miss out on opportunities for collaborative research and innovation development (*ibid*.). To overcome this problem, Miles (2007) encourages **service innovation** managers to interact with R&D policy makers and network with professional communities engaged in new technologies that are relevant to services.

Further reading
Atkinson, Richard C. and Pelfrey, Patricia A. (2010) 'Science and the Entrepreneurial University', Research & Occasional Paper Series, CSHE.9.10. Berkeley, CA, Center for Studies in Higher Education, July.
David, P. A., Hall, B. H. and Toole, A. A. (2000) 'Is Public R&D a Complement or Substitute for Private R&D? A Review of the Econometric Evidence', *Research Policy*, 29(4–5), 497–529.
Garcia-Quevedo, José Kyklos (2004) 'Do Public Subsidies Complement Business R&D? A Meta-Analysis of the Econometric Evidence', 57(1), 87–102.
Gonzalez, X. & Pazo, C. (2008) 'Do Public Subsidies Stimulate Private R&D Spending?', *Research Policy*, 37(3), 371–89.
Kirkland, John (2009) 'Research Management', *Perspectives: Policy & Practice in Higher Education*, 13(2), 33–6.
Miles, Ian (2007) 'Research and Development (R&D) beyond Manufacturing: The Strange Case of Services', *R&D Management*, 37(3), 249–68.
Wang, Chenggang, Xia, Yin and Buccola, Steven (2009) 'Public Investment and Industry Incentives in Life-Science Research', *American Journal of Agricultural Economics*, 91(2), 374–88.

Research and development (R&D) management

See **research and development**

Research journals on innovation

Scholarly research on innovation has expanded significantly since the 1980s and continues to grow at an unprecedented pace. Version 4 of the Association of Business Schools (ABS) Academic Journal Quality Guide published in March 2010 rates the relative quality of research journals in the field of innovation as shown in Table 1.

Grade Four are the top journals with the highest citation impact factor in the field. They publish the most original and best-executed research. Grade Two are well regarded in the field and publish original research of an acceptable standard. Grade One are modest standard journals that publish research of a recognised standard.

Table 1 Relative quality of academic journals on innovation

ISSN	Journal Title	Impact Factor 2008
	Grade Four	
0737-6782	Journal of Product Innovation Management	1.1
	Grade Two	
1363-9196	International Journal of Innovation Management	
1366-2716	Industry and Innovation	
1465-7503	International Journal of Entrepreneurship and Innovation	
	Grade One	
0963-1690	Creativity and Innovation Management	
1368-275X	International Journal of Entrepreneurship and Innovation Management	
1460-1060	European Journal of Innovation Management	
0892-9912	Journal of Technology Transfer	

Source: Based on Harvey, Charles, Kelly, Aidan, Morris, Huw and Rowlinson, Michael (eds) (2010) *The Association of Business Schools – Academic Journal Quality Guide,* Version 4

Resource-based view of the firm

The resource-based view of the firm (RBV) is a theory about the nature of firms, a core theory in **strategy** and management (Lockett *et al.*, 2009). In the resource-based view, the firm is seen as a pool of tangible and intangible resources. Tangible resources include financial and physical assets; intangible resources include intellectual property assets, organisational assets and reputational assets; and intangible resources that are capabilities (Galbreath and Galvin, 2004). Chisholm and Nielsen (2009) include **social capital** in the intangible assets, which can create **competitive advantage** and superior profits. Under the RBV, managerial responsibilities include the need to reposition the firm as opportunities change and its resource set evolves (Lockett, 2009). Managers are both adaptive and proactive; that is, they are 'enactors' (Lado and Wilson, 1994) engaged in the search for the most profitable use of the resources at their disposal (Lockett, 2009).

See also: **knowledge-based view of the firm**

Further reading
Barney, J. B. (ed.) (1991) 'The Resource Based View of Strategy: Origins, Implications, and Prospects', Special Theory Forum in *Journal of Management*, 17, 97–211.
Chisholm, Andrew M. and Nielsen, Klaus (2009) 'Social Capital and the Resource-Based View of the Firm', *International Studies of Management & Organization*, 39(2), 7–32.
Galbreath, Jeremy and Galvin, Peter (2004) 'Which Resources Matter? A Fine-Grained Test of the Resource-Based View of the Firm', *Academy of Management Proceedings*, L1–L6.
Lado, A. and Wilson, M. (1994) 'Human Resource Systems and Sustained Competitive Advantage: A Competency-Based Perspective', *Academy of Management Review*, 19, 699–727.
Lockett, Andy, Thompson, Steve and Morgenstern, Uta (2009) 'The Development of the Resource-Based View of the Firm: A Critical Appraisal', *International Journal of Management Reviews*, 11(1), 9–28.

R

Return on innovation investment (ROI2)

The term 'return of innovation investment (ROI2)' refers to a performance measure that is used for evaluating the effectiveness of a company's investment in new products or services. It looks at the company's cumulative net profits from new products launched during a 3–5-year period divided by total expenditure on new products (Kuczmarski, 2000). It is a single, standard measure for comparing performance between divisions, over time and within an industry (*ibid*.).

Further reading
Kuczmarski, Thomas D. (2000) 'Measuring Your Return on Innovation', *Marketing Management*, 9(1), 24–32.

Reverse engineering

Companies routinely disassemble competing products to see how they work. A key principle for reverse engineering is that it is not legal if a product is reproduced without authorisation, but it is legal if it enables learning about how it works (Hiawatha, 2000). In reverse engineering, a 3D virtual model of an existing physical part is created (Thilmany, 2009). That model is used in 3D CAD, **computer-aided manufacturing** or other computer-aided engineering applications (*ibid*.). In 1993, Sega Enterprises, a maker of home computer game consoles, sued a company called Accolade for producing illegal games for use on Sega equipment (Hiawatha, 2000). Accolade had used reverse engineering to learn how to make Sega-compatible games. Sega claimed that Accolade had no right to do this, but a US federal appeals court held that Accolade had a right to study Sega's source code to create a compatible product (*ibid*.).

China advanced its innovative problem-solving design and manufacturing capabilities by learning about advanced technologies with reverse engineering. A recent example is the successful reverse engineering for the variability design of offset spoilers in the high-speed train in China (Lu and Xu, 2009). Reverse engineering is a commonly practised method in biomedical and other bioscience research and applications (Thilmany, 2008) and a topic of study for biomedical engineering students in the USA (*ibid*.).

Further reading
Hiawatha, Bray (2000) 'Software Firm's Suit Tests Reverse Engineering', *Boston Globe*, 28 March..
Lu, Bihong and Xu, Jixiang (2009) 'Reverse Engineering for the Variability Design of Offset Spoilers in High-speed Train' in Hu Shuhua and Hamsa Thota (eds), *Proceedings of the 4th International Conference on Product Innovation Management*, Wuhan, China, Hubei People's Press, Part 1, 23–6.
Thilmany, Jean (2008) 'Working Backward', *Mechanical Engineering*, 130(7), 30–34.
Thilmany, Jean (2009) 'Outside In', *ME Magazine*, August, http://memagazine.asme.org/Articles/2009/August/Outside.cfm (accessed on December 22, 2010).

Revolutionary innovation

See **radical innovation**

Risk analysis

Risk analysis (or *risk assessment*) is 'the process of quantifying the probabilities of potential consequences in various hazardous scenarios (*risk determination*) and of evaluating that information to determine whether and how to act, under conditions of uncertainty (*risk evaluation or risk management*)' (Haimes, 2004).

Firms accept the prospect of investing in the costly and risky attempt to **launch** a really new product and be a pioneer in new markets. Tellis and Golder (1996) predicted higher failure rates for market pioneers, while other researchers found that *first-mover advantage* protects the market pioneer from outright failure (Robinson and Min, 2002). For example, Robinson and Min (2002) reported a ten-year survival rate of 66 per cent for market pioneers versus 48 per cent for early followers. Additional research by Min *et al.* (2006) clarified this confusion. These authors reported that, in markets started by a really new product, the pioneer (that is, the first to market) is often the first to fail, while in markets started by an incremental innovation, the first-mover advantages protected the pioneer from outright failure.

Day (2007) developed a *risk matrix* to assess risk across an innovation portfolio. In his risk matrix, each innovation can be positioned by determining its score on two dimensions: familiarity of the company with the intended market (that is, customers and not geography) on the x-axis and the familiarity of the product or technology on the y-axis. The size of the dot is proportional to the estimated revenue (*ibid.*). Another way to conduct risk analysis is to focus on risk events before market launch, analysis of expert assessments and preventive actions taken by the company (Gasparini *et al.*, 2004). New products and services have different types of risk, such as commercial, manufacturing, environmental or financial risks (*ibid.*). Assessment of risk events associated with suppliers, vendors and external partners is necessary in **open innovation** projects.

Further reading
Day, George S. (2007) 'Is It Real? Can We Win? Is It Worth Doing?', *Harvard Business Review*, 85(12), 110–20.
Gasparini, Mauro, Margaria, Gabriella and Wynn, Henry P. (2004) 'Dynamic Risk Control for Project Development, *Statistical Methods & Applications*, 13(1), 73–88.
Haimes, Yacov Y. (2004) *Risk Modeling, Assessment and Management*, 2nd edn, New York, John Wiley & Sons.
Min, Sungwook, Kalwani, Manohar U. and Robinson, William T. (2006) 'Market Pioneer and Early Follower Survival Risks: A Contingency Analysis of Really New versus Incrementally New Product-Markets', *Journal of Marketing*, 70(1), 15–33.
Robinson, William T. and Min, Sungwook (2002) 'Is the First to Market the First to Fail? Empirical Evidence for Industrial Goods Businesses', *Journal of Marketing Research*, 34(February), 120–28.
Tellis, Gerard J. and Golder, Peter (1996) 'First to Market, First to Fail? Real Causes of Enduring Market Leadership', *Sloan Management Review*, 2(Winter), 65–75.

R

Risk-quantification techniques

Eight risk-quantification techniques commonly used in innovation project management are described below.

1. **Decision tree analysis**: The decision tree method comes from operations research and games theory. Decision tree analysis (DTA) supports investment decisions in well-defined alternatives. It structures the decision problem by mapping out all project options contingent on the actual market and environment responses at the time of making initial project decisions (Junkui Yao and Jaafari, 2003). DTA is useful for analysing sequential investment decisions when uncertainty can be resolved at distinct, discrete points in time (*ibid.*). Decision trees help to structure choices in the face of uncertainty and to quantify outcomes.

2. **Expected monetary value (EMV)**: Expected monetary value is one of the earliest methods for analysing decisionmaking under conditions of uncertainty (Piney, 2003). It takes each cost and revenue item, multiplies each risk of gain and loss by the probability of its occurrence, and sums all these values algebraically (*ibid.*). Expected monetary value does not account for the organisation's risk tolerance. Piney (2003) recommends that project managers consider the utility value of the impact. This will take into account the risk tolerance of the organisation and is known as the *expected utility value* (*ibid.*). EMV is applicable in the selection of product configuration options.

3. **Failure mode and effects analysis (FMEA)**: FMEA provides a framework for cause-and-effect analysis of potential product failures. It is used, in the early stages of product design, to meet product quality requirements (Chin *et al.*, 2008). In **new product development**, cross-functional teams (that is, team members from design, process, production and quality) conduct FMEA analysis. Each failure mode is assessed for severity, probability of occurrence and difficulty of detection. While FMEA is used in the early stages of product design to meet product quality requirements, uncertainty and imprecise information exist in the conceptual design stage (*ibid.*). Chin *et al.* (2008) propose a fuzzy FMEA approach to correct the problems of using FMEA for evaluating new product concepts. In fuzzy FMEA fuzzy logic and a knowledge-based approach are combined with the FMEA methodology to achieve competitive product design and development (*ibid.*).

4. **Monte Carlo simulation**: This refers to a computerised simulation of the project model for assessing the probability of meeting time and cost targets (Githens, 2001). Monte Carlo simulation is used to develop estimates and confidence intervals for completion dates and expenses. For instance, a Monte Carlo simulation provides more realistic results than 'What if?' scenarios in petroleum risk assessment to determine the economic feasibility of oil and gas exploration and development projects (Kok *et al.*, 2006). It is 'more commonly used than PERT to estimate the criticality of different activities and paths, and the probability distribution of the project duration' (Pollack-Johnson and Liberatore, 2005).

5. **Program evaluation and review technique (PERT)**: PERT addresses project schedule uncertainty by considering the uncertainty related to activity duration (*ibid.*). The primary tool used in the PERT approach is the network or flow plan (Roman, 1962). The network is a series of related events and activ-

ities. Events are required sequential accomplishment points and are represented by circle or square symbols. Activities are the time-consuming element of the programme and are used to connect the various elements. PERT models uncertain activity durations by collecting optimistic, most likely and pessimistic duration estimates of all activities (Pollack-Johnson and Liberatore, 2005). The number of events shown on any flow plan should reflect significant programme accomplishments (Roman, 1962). The elements of planning information, cost estimates, estimated completion dates and manpower requirements are then related to the elements of the network (Hitchcock and Bliss, 1964). The PERT system uses this information in preparing reports for integrating the plan of action and establishing schedules, budgets and manpower assignments (*ibid.*).

6. **Real options analysis**: The real options extend *financial option theory* to options on real assets. *Option pricing theory* quantifies flexibility in strategic investment projects. Real options analysis allows managers to identify risk components and make investments under uncertainty. R&D projects face technological and product market uncertainties at the **front end** of product development. For example, if pharmaceutical research is successful in the *discovery* phase, the company concerned makes further investments in *development* (that is, clinical testing). If the drug proves safe and effective, the company will make investments, produce and market it (Copeland and Keenan, 1998). This decision process preserves flexibility and allows for frequent adjustment. Real options analysis is useful for making investment decisions as projects progress through the **development funnel**. Options are also useful for assessing the value associated with technology adoption, such as IT platform adoption (Fichman, 2004).

7. **Scenario analysis**: Scenarios are coherent and credible alternative stories about the future designed to help companies challenge their assumptions, develop their strategies and assess their plans (Cornelius *et al.*, 2005). For instance, Royal Dutch Shell uses scenarios at three levels: global, focused and project specific. It designed *global scenarios* to formulate its overall tactical and strategic policies, *focused scenarios* with country-specific issues or business strategies and *project scenarios* to make project-specific investment decisions (*ibid.*).

8. **Sensitivity analysis**: Sensitivity analysis examines how an output changes in response to a change in inputs. Each input is changed one at a time or in combination, while observing the effect of the change on **cash flow** or profit-and-loss (P&L) projection (Olson, 2010). Sensitivity analysis is commonly used for testing pricing and profitability assumptions during **business case** development. Price sensitivity refers to how individuals perceive and respond to changes or differences in prices for products or services (Wakefield and Inman, 2003).

Further reading
Chin, Kwai-Sang, Chan, Allen and Jian-Bo Yang (2008) 'Development of a Fuzzy FMEA Based Product Design System, *International Journal of Advanced Manufacturing Technology*, 36(7/8), 633–49.
Copeland, Thomas E. and Keenan, Philip T. (1998) 'Making Real Options Real', *McKinsey Quarterly*, 3, 128–41.

Cornelius, Peter, Van de Putte, Alexander and Romani, Mattia (2005) 'Three Decades of Scenario Planning in Shell', *California Management Review*, 48(1), 92–109.

Fichman, Robert G. (2004) 'Real Options and IT Platform Adoption: Implications for Theory and Practice', *Information Systems Research*, 15(2), 132–54.

Githens, Gregg (2001) 'Risk Management Practices for NPD Programs', http://www.cata-lystpm.com/Risk%20ToolBook%20Chapter.htm (accessed on December 23, 2010).

Hitchcock, R. P. and Bliss, Shirley (1964) 'Introduction to Critical Path Scheduling', Olympia, DC, Washington State Board for Vocational Education.

Junkui Yao, Frank J. and Jaafari, Ali (2003) 'Combining Real Options and Decision Tree: An Integrative Approach for Project Investment Decisions and Risk Management', *Journal of Structured & Project Finance*, 9(3), 53–70.

Kok, Mustafa Versan, Kaya, Egemen and Akin, Serhat (2006) 'Estimation of Expected Monetary Values of Selected Oil Fields', *Energy Sources Part B: Economics, Planning & Policy*, 1(2), 213–21.

Olson, Linden (2010) 'Sensitivity Analysis Identifies Risk, Benefit', *National Hog Farmer*, 54(1), 12–14.

Piney, Crispin (2003) 'Applying Utility Theory to Risk Management', *Project Management Journal*, 34(3), 26–31.

Pollack-Johnson, Bruce and Liberatore, Matthew J. (2005) 'Project Planning Under Uncertainty Using Scenario Analysis', *Project Management Journal*, 36(1), 15–26.

Roman, Daniel D. (1962) 'The PERT System: An Appraisal of Program Evaluation Review Technique', *Journal of the Academy of Management*, 5(1), 57–65.

Wakefield, K. and Inman, J. (2003) 'Situational Price Sensitivity: The Role of Consumption Occasion, Social Context and Income', *Journal of Retailing*, 79(4), 199–212.

Roadmapping

Roadmapping is a widely used technique for supporting **innovation strategy** (Phaal *et al.*, 2007). Roadmaps attempt to answer three basic questions considering a variety of perspectives, including markets, products and technology (*ibid.*):

1. Where are we going?
2. Where are we now?
3. How can we get there?

Kostoff and Schaller (2001) describe four types of roadmaps: science and technology; industry technology; corporate or product technology; and product/portfolio management. Of these, product technology roadmaps have become a major planning tool in many technology-oriented companies (Lichtenthaler, 2008). Technology roadmaps identify technology needs for future success (future **competitive advantage**) that involve a consensus understanding of the risks associated with the plan in terms of timing, costs and the capabilities of the organisation (Compton, 1997). Roadmaps also serve as decision aids to improve the coordination of activities and resources (Wells *et al.*, 2004). However, product technology roadmaps are insufficient to support the external technology utilisation needed in **open innovation**. To remedy this shortfall, Lichtenthaler (2008) proposed an *integrated technology commercialisation roadmap* showing the links at three levels: links in the technologies, the internal commercialisation projects and the external commercialisation projects. Lichtenthaler's (2008) roadmap also shows the connections between these three levels.

While company-level roadmaps are generally highly confidential, sector-level roadmaps are actively promoted and disseminated. The semiconductor industry embraced the use of the technology roadmap to identify the technology and product development required to meet *Moore's Law* (Kumar and Krenner, 2002), which captures the pace of continuous improvement in the semiconductor industry (that is, that the number of transistors on a chip doubles every 18 months).

Further reading

Compton, W. D. (1997) *The Management of World Class Manufacturing Enterprises*, Upper Saddle River, NJ, Prentice Hall.

Kostoff, R. N. and Schaller, R. R. (2001) ,Science and Technology Roadmaps', *IEEE Transactions on Engineering Management*, 48(2), 132–43.

Kumar, Sameer and Krenner, Nicole (2002) 'Review of the Semiconductor Industry and Technology Roadmap', *Journal of Science Education and Technology*, 11(3), 229–36.

Lichtenthaler, Ulrich (2008) 'Integrated Roadmaps for Open Innovation', *Research Technology Management*, 51(3), 45–9.

Phaal, R., Farrukh, C. J. P. and Probert, D. R. (2007) 'Strategic Roadmapping: A Workshop-based Approach for Identifying and Exploring Strategic Issues and Opportunities', *Engineering Management Journal*, 19(1), 3–14.

Wells, R., Phaal. R., Farrukh. C. and Probert. D. (2004) 'Technology Roadmapping for a Service Organization', *Research-Technology Management*, 47(2), 46–51.

Robust design

The robust design method, also called the Taguchi method, was pioneered by Dr Genichi Taguchi. Its goal is to create products that are insensitive to noise factors and thus perform consistently on target (Hasenkamp *et al.*, 2007). Identification of potential noise factors and an assessment of their impact on product performance are necessary early in the design process (*ibid.*).

Robust parameter design calls for the dual objective of optimisation of the signal (usually the mean) and minimisation of the variability transmitted to the response via noise (uncontrollable) factors (Shore and Arad, 2004). The problem of robust parameter design is to find settings for the design parameters in support of the dual objective. Design parameters are decision variables incorporated into the design of a product or a process once their values are determined (*ibid.*).

One example of the use of robust design capabilities is Honda's development of a world car on a standardised platform incorporating adjustable brackets (Swan *et al.*, 2005). This innovation extended the static view of a standardised component core to include a dynamic and robust core that broadened the ability to provide a larger product family (*ibid.*). This global platform allowed the company to provide different Honda Accords to different countries and to make a $1200 saving per car (*ibid.*).

Further reading

Hasenkamp, Torben, Adler, Tommy, Carlsson, Anders and Arvidsson, Martin (2007) 'Robust Design Methodology in a Generic Product Design Process', *Total Quality Management & Business Excellence*, 18(4), 351–62.

Shore, Haim and Arad, Ram (2004) 'Product Robust Design and Process Robust Design: Are They the Same? (No)', *Quality Engineering*, 16(2), 193–207.

Swan, K.. Masaaki Kotabe, Scott and Allred, Brent B. (2005) 'Exploring Robust Design Capabilities, Their Role in Creating Global Products, and Their Relationship to Firm Performance', *Journal of Product Innovation Management*, 22(2), 144–64.

Sales forecast

In manufacturing, sales forecasts are available at the product level for a short time horizon. *Demand planning systems* provide a forecast for medium-term and long-term horizons. They forecast demand at the product, product group, customer group and regional levels. In demand planning, market research is used as an input to determine the phase-in and phase-out periods of a new product. Demand forecasting is also an input to a 'one number' forecast for *sales and operations planning (SOP)* (Urs, 2008).

Errors in the *demand forecast* for new products are a concern. Kahn (2009) identified 10 biases in forecasting demand for new products: accountability/management commitment, advocacy, anchoring, confirmation, 'hard' data, optimism, overconfidence, planning fallacy, post-decision audit and sunk cost fallacy. Kahn (2009) also prescribed ways for managing their impact on the new product forecast.

Further reading
Kahn, Kenneth B. (2009) 'Identifying the Biases in New Product Forecasting', *Journal of Business Forecasting*, 28(1), 34–7.
Shu, Zhou, Jackson, Peter, Roundy, Robin O. and Zhang, Rachel Q. (2007) 'The Evolution of Family Level Sales Forecasts into Product Level Forecasts: Modeling and Estimation', *IIE Transactions*, 39(9), 831–43.
Urs, Rajiv (2008) 'How to Use Demand Planning System for Best Forecasting and Planning Results', *Journal of Business Forecasting*, 27(2), 22–9

Scenario analysis

See **risk-quantification techniques**

Scenario boards

Scenario boards are a flexible concept-development tool for defining user needs and opportunities for product and feature innovations with minimal investment in time and resources (Jordan and Geiselhart, 2006). They consist of a series of sketches that loosely illustrate how the new product or service concept is integrated into the user's workflow and environment (*ibid.*). They elicit early-stage user input to guide the introduction of new-to-the-world technologies, and aid in the selection of a concept direction from a range of possible solutions (*ibid.*).

Applications of scenario boards include the introduction of new technologies, directional testing of concepts, definition of a new workflow or **business model** and the communication of development decisions to internal stakeholders (*ibid.*). Scenario boards focus on process rather than on product form or features.

See also: **concept testing**

Further reading
Jordan, Matthew and Geiselhart, Ed (2006) 'Scenario Boards Can Be Used to Improve Validation of Early Product and Service Concepts', *PDMA Visions*, 30(1), 16.

Scenarios

Scenarios are 'carefully crafted tales that link certainties and uncertainties about the future to the decisions that must be made today' (Warshall, 1999), of which Pierre Wack was a revered teacher (*ibid.*). For him, scenario work is an iterative process in which intuitive exploration through scenario building alternates with significant and deep research and analysis of the questions raised by the scenarios (*ibid.*). The challenge in scenario planning is to build an optimal strategy for the plausible future conditions within which an organisation might have to operate (Muneer and Sharma, 2008). Scenario planning helps by answering key questions such as: 'Should we make an acquisition?' 'Should we expand into global markets?' 'Does the alliance with the no. 2 player in the market make sense?' (*ibid.*).

The term 'scenarios' also describes narratives of **personas** in action under specific situations (Beale and Sutton, 2008). Such situations include use of a product, buying a product or thinking of buying a product, including product research. Scenarios are useful for gaining insight into the different ways customers use products and for identifying opportunities to improve products or introduce new features (*ibid.*). They also function as a standalone tool or can be used with other visual tools, such as mapping and **storyboards**.

See also: **visualisation**

Further reading
Beale, Claire-Juliette and Sutton, Tricia (2008) 'Personas and Scenarios Workshop Launches PDMA Chapter in South Carolina', *PDMA Visions*, 32(3), 26–7.
Muneer, Sami and Sharma, Chetan (2008) 'Enterprise Mobile Product Strategy Using Scenario Planning', *Information Knowledge Systems Management*, 7(1/2), 211–24.
Pruitt, John and Adlin, Tamara (2006) *The Persona Lifecycle: Keeping People in Mind Throughout Product Design,* San Francisco, CA, Morgan Kaufmann.
Warshall, Peter (1999) 'Pierre Wack', *Whole Earth*, 96, 89–91.

Schumpeterian framework

See **growth theory**

Schumpeterian growth theory

See **growth theory**

Scrum

Scrum is a project management technique for managing software projects, based on the principles of **agile development** (Olson, 2010). Agile means responding to changes in technology, customer demand and market opportunities (*ibid.*).

S

Scrum differs from traditional project management in that it accepts that things will change during development and that project managers' role is to influence and facilitate the development. Scrum project managers are called 'scrum masters' (*ibid.*).

According to Olson (2010), scrum work is delivered in six-week 'sprints'. Each sprint delivers a single, usable piece of work called the 'product increment' (*ibid.*).

In scrum, a product owner collects and maintains all requests for feature additions. The list is called the 'product backlog'. This list is prioritised by stakeholders to reduce the overall project risk. Sprint planning meetings determine the backlog items to be delivered by the next sprint, the development team breaks work into a series of tasks, and the work begins (*ibid.*). Sprints are monitored for work remaining with 'burn-down' charts (*ibid.*).

At the end of each sprint, the new product increment is demonstrated and made available for evaluation and use by the customer (*ibid.*). A retrospective meeting is held to capture learning in order to improve in the next sprint (*ibid.*).

Honda, Toyota and Google have adopted the scrum approach and achieved greater quality and value for customers (*ibid.*).

Further reading
Brandl, Dennis (2009) 'Agile Software Development', *Control Engineering*, 56(7), 18.
Olson, Kirk B. (2010) 'HEI Embraces Scrum Project Management', *SMT: Surface Mount Technology*, 25(5), 44–52.
Salo, O. and Abrahamsson, P. (2008) 'Agile Methods in European Embedded Software Development Organizations: A Survey on the Actual Use and Usefulness of Extreme Programming and Scrum', *IET Software*, 2(1), 58–64.

Second-mover advantage

A great deal of research has been published about the advantages of being first to market (**first-mover advantage**). However, being first to market is not always a successful strategy for some firms (Tellis and Golder, 2001). BetaMax was the first mover in home video systems, but JVC won the Betamax vs VHS standard war with its Video Home System (VHS) (Kopel and Löffler, 2008). Microsoft Word superseded first-mover WordStar in the software market for word processing (*ibid.*).

In the automotive industry, General Motors (GM) and Volkswagen were the early leaders in the Chinese market, but were leapfrogged by late entrants Hyundai and Chery. GM and Volkswagen's market share, sales and profits fell due to increased competition (Roberts *et al.*, 2005). As another example, Cho *et al.* (1998) describe how latecomers to the semiconductor industry in Japan and Korea achieved market dominance.

See also: **competitive advantage**

Further reading
Cho D. S., Kim, D. J. and Rhee, D. K. (1998) 'Latecomer Strategies: Evidence from the Semiconductor Industry in Japan and Korea', *Organization Science*, 9(4), 489–505.
Kopel, Michael and Löffler, Clemens (2008) 'Commitment, First-Mover, and Second-Mover Advantage', *Journal of Economics*, 94(2), 143–66.
Roberts, D., Ihlwan, M., Rowley, I. and Edmondson, G. (2005) 'GM and VW: How not to Succeed in China', *Business Week*, May, 22–3.

Tellis, G. J. and Golder, P. N. (2001) *Will and Vision: How Latecomers Grow to Dominate Markets*, New York, McGraw-Hill.

Service

Service is a substantial part of the global economy. Services differ in degree from manufactured goods along four dimensions: intangibility, simultaneity, heterogeneity and perishability (Easingwood, 1986).

- Intangibility means that service cannot be seen, touched or examined prior to purchase (Rushton and Carson, 1986). It also means that it is easy to design new products (Easingwood, 1986).
- Simultaneity means that operations are responsible for producing and simultaneously delivering service to the customer (*ibid.*). The service is consumed at the same time as it is produced and the customer is in direct contact with the service delivery system (Rushton and Carson, 1986). The customer's perception of product quality may be influenced by the quality of the service itself (Sasser *et al.*, 1978).
- Heterogeneous means that the quality depends on the quality of the service provider's performance (Gronroos, 1982). This may vary from provider to provider, as well as over time.
- Perishable means that services cannot be stored (Lovelock, 1984).

However, not all services have these four dimensions and some manufactured goods may possess one or more of these characteristics (Easingwood, 1986). In addition, in some services such as air traffic control, there is a close interaction between production and consumption (Tether and Hipp, 2002). Air traffic control services are 'co-produced by the users (i.e. airline pilots) and the providers (i.e. air traffic controllers) acting together' (*ibid.*).

Further reading
Easingwood, Christopher J. (1986) 'New Product Development for Service Companies', *Journal of Product innovation Management*, 4, 264–75.
Gronroos, Christian (1982) 'An Applied Service Marketing Theory', *European Journal of Marketing*, 16, 30–41.
Lovelock, Christopher H. (1984) 'Strategies for Managing Demand in Capacity Constrained Service Organizations', *Service Industries Journal*, 4, 12–30.
Rushton, Angela M. and Carson, David J. (1986) 'The Marketing of Services: Managing the Intangibles', *European Journal of Marketing*, 19, 19–40.
Sasser, Earl W., Jr., Olsen, R. Paul and Wyckoff, D. Daryl (1978) *Management of Service Orientations*, Boston, MA, Allyn and Bacon.
Tether, Bruce S. and Hipp, Christiane (2002) 'Knowledge Intensive, Technical and Other Services: Patterns of Competitiveness and Innovation Compared', *Technology Analysis & Strategic Management*, 14(2), 163–82.

Service innovation

Innovation scholars hold an ongoing debate about the similarities and differences between services and the product **innovation process** (Gallouj and Windrum, 2009). Toivonen and Tuominen (2009) define service innovation as 'a new service or such a renewal of an existing service which is put into practice and which provides benefit to the organization that has developed it; the benefit usually derives from the added value that the renewal provides the customers. In addition, to be an innovation the renewal must be new not only to its developer, but in a broader context, and it must involve some element that can be repeated in new situations, i.e. it must show some generalizable feature(s).' A service innovation process 'is the process through which the renewals described are achieved' (Sundbo, 1997).Three approaches to studying service innovation are the demarcation approach, assimilation approach and synthesis approach (Gallouj and Windrum, 2009). In the demarcation approach, service innovation is seen as different from product and manufacturing innovation (Coombs and Miles, 2000). Scholars in support of this approach argue that some service-specific forms of innovation do exist (Gallouj and Windrum, 2009). The assimilation approach treats services as similar to manufacturing (Gallouj, 2002). The synthesis approach integrates insights gained from demarcation and manufacturing studies (Windrum, 2007) with innovation, which includes organisational, product, market, process and input innovation (Gallouj and Windrum, 2009). Gallouj and Savona (2009) argue that the integrative approach is the most promising and constitutes a balanced theory for studying service innovation.

In knowledge-intensive business services, service innovation is the creation of new combinations of knowledge (Amara *et al.*, 2009). Knowledge-intensive and technical service organisations function as knowledge integrators or knowledge-transfer organisations and provide customised solutions (Tether and Hipp, 2002). Their innovation activities are more like product innovation than are those of non-knowledge-intensive services (*ibid.*). Amara *et al.* (2009) identify six types of innovation in knowledge-intensive business services: **product innovation**, **process innovation**, delivery innovation, **strategic innovation**, managerial innovation (including **business model innovation**) and **marketing innovation**:

- Service product innovations consist in the creation of new service products.
- Process innovations renew procedures for developing and delivering the service and are innovations in operational processes (back office) or delivery processes (front office).
- Delivery innovations are in how the enterprise delivers its products (goods or services) to its customers, examples of which are just-in-time delivery, consumer e-commerce and significantly improved home shopping services (*ibid.*).
- Marketing innovations constitute the creation of innovations in marketing and **commercialisation**. Finding a new market opportunity or entering a complementary market in another industry (such as when a large retailing chain begins the sale of financial services) is an example of marketing innovation (*ibid.*).

Song *et al.* (2009) propose a five-stage service innovation model to assess service innovation ideas, evaluate the performance of ongoing service innovations, allocate resources and improve the success rate of service innovations. Their process incorporates all of the familiar stages of the **new product development (NPD)**

process, plus service quality training as a pre-launch training stage to improve service quality.

Further reading
Amara, Nabil, Landry, Rejean and Doloreux, David (2009) 'Patterns of Innovation in Knowledge-Intensive Business Services', *Service Industries Journal*, 29(4), 407–30.
Coombs, R. and Miles, I. (2000) 'Innovation, Measurement and Services: The New Problematic', in J. S. Metcalfe and I. Miles (eds), *Innovation Systems in the Service Economy: Measurement and Case Study Analysis*, Boston, MA, Kluwer Academic, pp. 85–103.
Gallouj, F. (2002) 'Innovation in Services and the Attendant Old and New Myths, *Journal of Socio-Economics*, 31, 137–54.
Gallouj, Faïz and Windrum, Paul (2009) 'Services and Services Innovation', *Journal of Evolutionary Economics*, 19(2), 141–8.
Gallouj and Lavona (2009) details to come.
Love, James H. and Mansury, Mica Ariana (2007) 'External Linkages, R&D and Innovation Performance in US Business Services', *Industry and Innovation*, 14(5), 477–96.
Song, Lisa Z.; Song, Michael and Di Benedetto, C. Anthony (2009) 'A Staged Service Innovation Model', *Decision Sciences*, 40(3), 571–99.
Sundbo, J. (1997) 'Management of Innovation in Services', *Service Industries Journal*, 17(3), 432–55.
Tether, Bruce S. and Hipp, Christiane (2002) 'Knowledge Intensive, Technical and Other Services: Patterns of Competitiveness and Innovation Compared', *Technology Analysis & Strategic Management*, 14(2), 163–82.
Toivonen, Marja and Tuominen, Tiina (2009) 'Emergence of Innovations in Services', *Service Industries Journal*, 29(7), 887–902.
Windrum, P. (2007) 'Innovation in Services', In H. Hanusch and A. Pyka (eds), *The Edward Elgar Companion to Neo-Schumpeterian Economics*, Cheltenham, Edward Elgar.

Service life cycle

Verganti and Buganza (2005) provide a pragmatic definition of the service life cycle: the service cycle starts with the first service design and it never ends. The first service design is the initial radical and systemic design process that leads to the generation of a service new to the company (*ibid.*). According to Verganti and Buganza (2005), the service life cycle is perpetuated unless the company exits that particular business. Even if a radical change modifies the service's core concept, part of the design solution developed in the first service design, or at least the knowledge associated with the first service design, remains in the long term. This definition of the service life cycle is consistent with the evolutionary theories of the firm and the concept of path dependency (*ibid.*).

Further reading
Verganti, Roberto and Buganza, Tommaso (2005) 'Design Inertia: Designing for Life-Cycle Flexibility in Internet-Based Services', *Journal of Product Innovation Management*, 22, 223–37.

Service life-cycle flexibility

Verganti and Buganza (2005) define service life-cycle flexibility as the ability of a firm to launch **innovations** throughout the service life cycle at low cost and within a short interval. They recommend that life-cycle flexibility can be designed in during the *first service design* to reduce customer inertia – customers who are not ready for the innovation or not willing to learn how to take advantage of it – and inertia related to the back-end design (technology and organisation).

See also: **service innovation, service life cycle**

Further reading
Verganti, Roberto and Buganza, Tommaso (2005) 'Design Inertia: Designing for Life-Cycle Flexibility in Internet-Based Services', *Journal of Product Innovation Management,* 22, 223–37.

Service management maturity

Successful companies such as Hewlett-Packard, Apple, Rolls-Royce and General Electric (GE) complement their **product innovations** with **service innovations** (that is, they integrate products and services) to sustain **competitive advantage** (Shelton, 2009). They achieve integration by advancing through four stages of solution management maturity (*ibid.*):

- Stage 1, Product-Centric Manufacturer, limits service innovation to keeping the product in use and ensuring customer satisfaction with product support.
- Stage 2, As Needed Service Provider, provides targeted functional services to meet stated customer needs. Service revenues account for 15–20 per cent of total revenues in this stage.
- Stage 3, Full-Line Service Expert, provides comprehensive and differentiated services and products as solutions to customers' life-cycle problems. Service revenues account for 30–40 per cent of total revenues in this stage.
- Stage 4, Integrated Solutions Provider (ISP), provides integrated solutions 'from cradle to grave' to deliver outstanding customer economic benefits. Service revenues account for more than 50 per cent of total revenues in this stage.

Further reading
Shelton, Robert (2009) 'Integrating Product and Service Innovation Research', *Technology Management,* 52(3), 38–44.

Service-productisation

The term 'service-productisation' refers to making the service offering 'product like' (as in the definition of core process and outcomes) (Jung and Nam, 2008). However, unlike in products, intangible value in services cannot be perceived by sensory means. For this reason, the productisation of professional services is accomplished by associating tangible features with intangible service offerings (*ibid.*). The technical information system, financial services, advice and consulting services are suitable for service-productisation (*ibid.*).

Further reading
Jung, Mi Jun and Nam, Ki Young (2008) 'Design Opportunities in Service-Product Combined Systems', Undisciplined! Design Research Society Conference 2008, Sheffield, Sheffield Hallam University, 16–19 July.
http://shura.shu.ac.uk/535/1/fulltext.pdf (accessed on December 25, 2010).

Services innovation management model

Tidd and Hull (2006) propose a four-cell management typology for managing services innovation. The four organisational types are *simple craft-batch*, *mechanistic bureaucracy*, *hybrid mechanistic–organic* and *organic technical-batch*. Each type is associated with a different service performance. The traditional organic–mechanistic continuum differentiates into a number of almost independent dimensions, including project organisation, knowledge sharing and rewarding collaboration, and cross-functional teams. Similarly, performance is multidimensional. The simple *efficiency–innovation trade-off* is replaced by elements of cost, time, delivery and innovation. The implication for managers is that adhering to industry 'best practices' in new service development may give inferior results compared to tailored practices for improving cost, time and delivery. Innovative service practices tailored to local needs contribute to superior service performance.

Further reading
Tidd, Joe and Hull, Frank M. (2006) 'Managing Service Innovation: The Need for Selectivity rather than 'Best Practice: New Technology', *Work & Employment*, 21(2), 139–61.

Services innovation theory

Barras (1986) constructed a theory of services innovation that is a reversal of Schumpeterian product cycle theory. In Barras's product cycle, **process innovation** precedes **product innovation**. The cycle consists of three phases: improved efficiency, improved quality and new product. In the improved efficiency phase, established firms make an initial investment in new information technology, which leads to *incremental process innovations* such as improving the efficiency of delivery of existing products, increasing labour productivity and reducing cost. In the improved quality phase, the organisation invests in radical process innovation to improve existing products. In the new products phase, competitive strategy shifts to product differentiation, new products, new markets and new business opportunities.

Barras (1990) argues that there is an interaction between new technologies in capital goods industries such as computer manufacturing and innovation in adopting service industries such as financial services. In this interaction, the two product cycles work in opposite directions, so that manufacturing innovation moves from an emphasis on product to one on process, while services innovation does the reverse. This means that as product innovation declines in a particular industry, services innovation accelerates.

Barras's argument may have its roots in history. According to Cain and Hopkins (1993), the service sector was the 'most successful and dynamic element' in the British economy after 1850. In the second half of the nineteenth century, growth in the service sector made up for the relative decline of the manufacturing sector.

Further reading
Barras, R. (1986) 'Towards a Theory of Innovation in Services', *Research Policy*, 15, 161–73.

S

Barras, R. (1990) 'Interactive Innovation in Financial and Business Services: The Vanguard of the Service Revolution', *Research Policy*, 19, 215–37.

Cain, P. J. and Hopkins, A. G. (1993) *British Imperialism: Innovation and Expansion, 1688–1914*. London, Longman.

Shared risk

See **alliances**

Simulation

The use of simulation in product development is increasing. Simulation reduces the need for physical prototypes. It lowers development costs and speeds up innovation cycle time. It also contributes to the robustness and quality of a new product (Jusko, 2009).

Becton, Dickinson and Company (BD) utilises the *Abaqus finite element analysis* for structural analysis to optimise part performance and structural integrity (*ibid.*). It also uses *Moldflow* and *Fluent* and *CFX software* in injection-moulding simulations, to predict manufacturability and optimise fluid flow. The simulation group is involved early in the **new product development process** (*ibid.*). At BD, it participates in the determination of project scope, concept selection and initial design. Input from the simulation group at the **front end of innovation** adds greater value to the project (*ibid.*).

Feller *et al.* (2005) introduced the use of business process simulation as a tool for process learning and innovation in collaborative R&D. Process learning is the creation and internalisation of new process knowledge towards **process innovation** (Nonaka *et al.*, 2000). It also creates a virtual community of practice (Wenger, 1998), where the visualised process flow is simulated through joint discussion based on the personnel's experience and joint imagination.

Further reading

Feller, J., Hirvensalo, A. and Smeds, R. (2005) 'Inter-partner Process Learning in Collaborative R&D – A Case Study from the Telecommunications Industry', *Production Planning & Control*, 16(4), 388–95.

Jusko, Jill (2009) 'The Simulation Solution', *Industry Week*, 258(8), 53.

Nonaka, I., Toyama, R. and Konno, N. (2000) 'SECI, Ba and Leadership: A Unified Model of Dynamic Knowledge Creation', *Long Range Planning*, 33, 5–34.

Wenger, E. (1998) *Communities of Practice: Learning, Meaning and Identity*, Cambridge, Cambridge University Press.

Six Sigma for Product Development (SSPD)

An engineer named Bill Smith at Motorola invented Six Sigma for Product Development (SSPD) in 1986 (Tegel and Kriva, 2004). In the 1980s the standard measurement of quality was 'defects in thousands of opportunities', whereas the Six Sigma methodology measures quality in 'defects per million opportunities' (*ibid.*). Its fundamental method is DMAIC (Define, Measure, Analyse, Improve, Control) (Johnson, 2006). The DMAIC method emphasises the relatively short term, cost reduction, production and administration (*ibid.*).

In 2003, *Design for Six Sigma in Technology and Product Development* by Creveling *et al.* accelerated Motorola's development of the Six Sigma programme. The company integrated Six Sigma through all phases of its new product development process: front-end portfolio planning, technology development, software development, marketing and product commercialization. In addition, Motorola integrated SSPD into product review scorecards and the Stage-Gate™ process before deployment in its business units. SSPD brought increased clarity and rigour to product development (Tegel and Kriva, 2004).

A corporate culture dominated by Six Sigma principles works best in **incremental innovation** when there is an expressed commercial purpose (Dodge, 2007). In companies such as 3M, the focus of research is to create commercial products and solve customer problems. 3M utilised Six Sigma to improve the performance of its research scientists (*ibid.*).

In his industry workshops, Johnson (2006) found that *Design for Six Sigma (DFSS)* positioned R&D organisations as key drivers of long-range, sustainable growth. DFSS refers to the design of products that are manufactured and services that are delivered with defect rates at or below Six Sigma (*ibid.*). It helped General Electric (GE) obtain record financial results in 1999 (Carleysmith *et al.*, 2009). Companies that linked Six Sigma and Design for Six Sigma with corporate R&D strategy generated superior products (*ibid.*).

When lean thinking and Six Sigma are used in combination, this is *Lean Six Sigma (LSS)* or *Lean Sigma* (*ibid.*). The *Theory of Constraints (ToC)* is an important partner tool for Lean Sigma. ToC, developed by Goldratt (Goldratt and Cox, 2004), focuses on identifying and understanding the constraint to the flow of product (Carleysmith *et al.*, 2009). GlaxoSmithKline and Bristol-Myers Squibb are examples of pharmaceutical companies deploying LSS in their R&D (*ibid.*).

See also: **ambidextrous organisation**

Further reading
Carleysmith, Stephen W., Dufton, Ann M. and Altria, Kevin D. (2009) 'Implementing Lean Sigma in Pharmaceutical Research and Development: A Review by Practitioners', *R&D Management*, 39(1), 95–106.
Creveling, Clyde M., Slutsky, Jeff and Antis, Dave (2003) *Design for Six Sigma in Technology and Product Development*, Upper Saddle River, NJ, Prentice Hall.
Dodge, John (2007) 'Sun Sets on Six Sigma at 3M R&D', *Design News*, 62(18), 69–72.
Goldratt, E. M. and Cox, J. (2004) *The Goal: A Process of Ongoing Improvement*, 2nd edn, Great Barrington, MA, North River Press.
Johnson, A. (2006) 'Lessons Learned from Six Sigma in R&D', *Research Technology Managment*, 49(2), 15–19.
Tegel, Dan and Kriva, Rick (2004) 'Motorola's Next Generation of Six Sigma – Six Sigma for New Product Development', *PDMA Visions*, 28(4), 14–16.

Smart products

Smart products are differentiated from traditional products because they can process information (Rijsdijk and Hultink, 2003). Three distinguishable characteristics of smart products are that they communicate with other products; they are more flexible; and they become autonomous decision makers (*ibid.*). Digital cameras and wireless communication between palmtops and mobile phones are examples of flexible capability (*ibid*.). Smart products also adapt their actions to different situations. For example, Mercedes-Benz developed a front passenger seat that automatically recognised whether a child was in the safety seat and reduced airbag power when the airbag deployed (*ibid*.). Electrolux's autonomous lawnmower and robotic vacuum cleaner are examples of autonomous decision-making capability (*ibid*.).

Rijsdijk and Hultink (2009) tested consumer responses to smart products and found that autonomy has relatively few disadvantages, while multifunctionality and the ability to cooperate are problematic for consumers. Multifunctionality is the ability of a single product to fulfil multiple functions, as with mobile phones and personal digital assistants (PDAs) (*ibid*.).

Further reading

Rijsdijk, Serge A. and Hultnik, Erik Jan (2003) 'Honey, Have You Seen Our Hamster? Consumer Evaluations of Autonomous Domestic Products', *Journal of Product Innovation Management*, 20(3), 204–16.

Rijsdijk, Serge A. and Hultink, Erik Jan (2009) 'How Today's Consumers Perceive Tomorrow's Smart Products', *Journal of Product Innovation Management*, 26, 24–42.

Social capital

There are many definitions of social capital. OECD defines social capital as 'networks together with shared norms, values and understandings that facilitate cooperation within or among groups' (OECD, 2001). Portes (1998) defines it as 'the ability of actors to secure benefits by virtue of memberships in social networks and other social structures'. All the definitions emphasise a shared value system, norms and institutions (specifically trust and reciprocity) and institutionalised forms of social interaction such as networks (Lizarralde, 2009).

Nahapiet and Ghoshal (1997) present a theoretical model to illustrate how firms create value through social capital. Social capital originally described the relational resources in community social organisations (Tsai and Ghoshal, 1998). The three dimensions of social capital are structural, relational and cognitive (Nabapiet and Ghoshal, 1997). The structural dimension includes social interaction. People use personal contacts to gather information about products and services, learn about culture and social mores and for personal, family and professional advice (*ibid*.). The relational dimension refers to trust and trustworthiness. Trust is a characteristic of a relationship, but trustworthiness is a characteristic of a person involved in the relationship. People collaborate with people they trust and a trustworthy team member is likely to get others' support for achieving agreed goals that would otherwise not be possible without trust-based collaboration (*ibid*.). Tsai and Ghoshal (1998) confirm that social interaction and trust significantly influence resource exchange among intrafirm networks and affect **product innovation**.

The cognitive dimension facilitates the mutual understanding of shared goals and appropriate behaviour in a social system. The cognitive dimension captures the 'the public good aspect of social capital' (Coleman, 1990).

Further reading

Coleman, J. S. (1990) *Foundations of Social Theory*, Cambridge, MA, Harvard University Press.

Lizarralde, Iosu (2009) 'Cooperatism, Social Capital and Regional Development: The Mondragon Experience', *International Journal of Technology Management & Sustainable Development*, 8(1), 27–38.

Nahapiet, J. and Ghoshal, S. (1997) 'Social Capital, Intellectual Capital and the Creation of Value in Firms', *Academy of Management Best Paper Proceedings*, 35–9.

OECD (2001) *The Well-being of Nations: The Role of Human and Social Capital*, Paris, OECD.

Portes, A. (1998) 'Social Capital: Its Origins and Application in Modern Sociology', *Annual Review of Sociology*, 24, 1–24.

Tsai, Wenpin and Ghoshal, Sumantra (1998) 'Social Capital and Value Creation: The Role of Intrafirm Networks', *Academy of Management Journal* 41(4), 464–76.

Social entrepreneurship

A social entrepreneur is an innovator in the social sector. Social entrepreneurs differ from commercial entrepreneurs in the opportunities exploited and the value sought in the entrepreneurial process. Social entrepreneurs seek innovative, sustainable programmes to deliver mass social benefits. They seize opportunities to solve social problems (such as hunger or poverty) and measure performance by the achievement of social value (Smith *et al.*, 2008).

Social entrepreneurship can be considered in Schumpeterian terminology as dynamic behaviour in the non-economic areas of society (Tapsell and Woods, 2008). It is characterised by change and the creation of something new rather than the replication of existing enterprises or processes (Tapsell and Woods, 2008). Contemporary social entrepreneurs include Mohammad Yunus, founder of the Grameen Bank in Bangladesh; Wendy Kopp, founder of Teach for America; and Paul Farmer, founder of Partners in Health (Bloom and Chatterji, 2009).

However, scalability is a significant problem in achieving a wider application of social entrepreneurship (*ibid.*).

Further reading

Bloom, Paul N. and Chatterji, Aaron K. (2009) 'Scaling Social Entrepreneurial Impact', *California Management Review*, 51(3), 114–33.

Smith, Brett R., Barr, Terri Feldman, Barbosa, Saulo D. and Kickul, Jill R. (2008) 'Social Entrepreneurship: A Grounded Learning Approach to Social Value Creation', *Journal of Enterprising Culture*, 16(4), 339–62.

Tapsell, Paul and Woods, Christine (2008) 'A Spiral of Innovation Framework for Social Entrepreneurship: Social Innovation at the Generational Divide in an Indigenous Context', *Emergence: Complexity & Organization*, 10(3), 25–34.

S

Social innovation

Mumford (2002) defines social innovation as the commercialisation of new ideas about people and their interactions within social systems. Social innovation can be viewed broadly as **business model innovation** and **marketing innovation** to deliver sustainable economic, environmental and social prosperity. In India, Aravind Eye Care provides cataract surgery for the poor (Sirkin and Hemerling, 2008) and Narayana Hrudayalaya set up tele-health centres to provide medical help to the poor in rural parts of India (Chauhan, 2008).

Mumford (2002) studied the social innovations attributed to Benjamin Franklin and found that he used *causal analysis* to achieve his social innovations such as

the introduction of paper currency, the establishment of volunteer fire departments and the founding of subscription libraries. Franklin did a detailed analysis
of the operative causes of an observed deficiency and identified a few critical
causes that were subject to manipulation, such as 'want of regulation' for fire
prevention and 'inflation' for paper currency. He argued for the introduction of a
paper currency tied to a tangible asset, specifically land (*ibid.*).

The industrialisation of socially responsible innovation is an emerging field of
study. In *industrialised social innovation*, the *role of industrial designers* is evolving
(Morelli, 2007). Designers become enablers of innovation (*ibid.*). They provide
scenarios, technology platforms and **strategy maps** to enable local networks
of small organisations, not-for-profit groups and individual customers to co-
produce their own solutions (*ibid.*).

Further reading
Chauhan, Chetan (2008) 'Banking Tools for Rural India a Big Hit in West', *Hindustan
    Times*, 29 July.
Marcy, Richard T. and Mumford, Michael D. (2007) 'Social Innovation: Enhancing
    Creative Performance through Causal Analysis', *Creativity Research Journal,* 19(2/3),
    123–40.
Morelli, Nicola (2007) 'Social Innovation and New Industrial Contexts: Can Designers
    "Industrialize" Socially Responsible Solutions?', *Design Issues*, 23(4), 3–21.
Mumford, M. D. (2002) 'Social Innovation: Ten Cases from Benjamin Franklin', *Creativity
    Research Journal*, 14, 253–66.
Sirkin, Harold L. and Hemerling, James W. (2008) 'Rethinking Innovation', *Washington
    Times*, 3 August.

Spiral development

The Stage-Gate™ NPD process is a widely practised method. It is most effective for incremental projects where the impact of changes is known and understood (Oosterwal, 2009). However, as projects become more complex, its
effectiveness diminishes because it promotes point-based linear development.
Redesign loops tend to arise during the 'Test and Develop' phase (*ibid.*). Redesign
loops drive waste throughout the development portfolio as they require the
unplanned reallocation of resources to fix unexpected problems (*ibid.*).

One approach to modifying the strict Stage-Gate® system is spiral development,
which puts a premium on speed but still requires cross-disciplinary input to the
process (Hauser *et al.*, 2005). Spiral development extends the design–test–redesign
loop directly to the customer in order to cope with changing or fluid information
as the project moves through the NPD process (Oosterwal, 2009). Top-performing
NPD teams build a series of 'build–test–feedback–revise' iterations or spiral development into their Stage-Gate® system (Cooper, 2008).

In spiral development, product development teams develop an initial version of
the product (this could be a virtual prototype) and seek immediate and early feedback by testing it with customers (Cooper and Edgett, 2008). The team uses
customer feedback to build the next, more complete version (a working model)
(*ibid.*). Spiral development teams avoid unnecessary work and move quickly
through a series of these iterative steps (build–test–feedback–and–revise loops) to
finalise the product (*ibid.*). They begin using the iterative loops from early-stage

scoping through the development stage and into testing. When this process is sketched on a flow diagram, these loops look like spirals, hence the name 'spiral development' (*ibid.*).

Further reading
Cooper, Robert G. (2008) 'What Leading Companies Are Doing to Reinvent Their NPD Processes?', *Visions*, 32(3), 6–8.
Cooper, Robert G. and Edgett, Scott J. (2008) 'Maximizing Productivity in Product Innovation', *Research Technology Management*, 51(2), 47–58.
Hauser, J., Tellis, G. J. and Griffin, A. (2005) 'Research on Innovation: A Review and Agenda for Marketing Science', Special Report 05-200, Cambridge, MA, Marketing Science Institute.
Oosterwal, Dantar (2009) 'The Truth behind the Ineffectiveness of Phase-Gate Methodology – and Why a Knowledge-based Process May Be Better', *Visions*, 33(2), 18–19.

Stage-Gate™ new product development process
See **new product development process**

Strategic alliances

Alliances are strategic choices that firms make for growth and access to technology. An alliance is a 'deal between two companies where neither intends to buy anything from the other' (McCamey, 2003). Another definition is 'a relationship among two or more entities intended to achieve agreed objectives through a combination of resources, with mutual interdependence defined by a limited long term agreement' (*ibid.*).

In practice more than 30 per cent of alliances fail, and an additional 23 per cent achieve limited objectives (*ibid.*). Common failure factors for alliances include poor strategy; poor business planning; win–lose legal and financial terms and conditions; lack of trust; or damaged relationships between firms (*ibid.*). Successful alliance partners invest in the alliance and work together to create alliance capability. They cooperate to find real customer problems and generate solutions (*ibid.*).

The alliance framework process consists of seven steps (Sagal, 2003): appointing the planning and negotiating team; achieving internal consensus; approaching potential partners; strategic fit analysis; resource fit analysis; selecting the partner; and negotiating an agreement. Key success factors for alliances are senior management commitment; the right people and resources; aligned direction and plans; clear responsibilities and expectations; robust communication; effective decision making; a disciplined improvement approach; aligned work systems; and constructive conflict resolution (*ibid.*).

GlaxoSmithKline (GSK) wanted to enter the fast-growing teeth-whitening market (Andersen *et al.*, 2007). Oratech wanted a partner to develop and launch a disposable teeth-whitening system (*ibid.*): 'Oratech had the product idea, patents, and manufacturing capabilities while GSK had the product development as well as consumer and professional marketing, sales, and distribution capabilities' (*ibid.*). Their strategic alliance launched a successful whitening product under the Aquafresh® brand name.

The key to GSK and Oratech's success was to find the right technology and the right partner to address consumer needs, while making an effort to match the philosophies of both companies and their people (*ibid*.). The two companies offered following the following advice to companies planning to implement strategic alliances (*ibid*.):

1. Focus on unmet needs in the marketplace.
2. Have early discussions about the value each partner brings to the alliance and the role for each side.
3. Recognise and leverage differences in culture and capabilities between the two companies.
4. Have a senior **champion** on each side.
5. Recognise that the process takes longer than anticipated, and plan accordingly.

Further reading
Andersen, Scott, Foley, Kevin and Shorter, Lee (2007) 'A Story of What Happens When Opposites Attract', *Visions*, 31(4), 16–17.
McCamey, David (2003) 'Creating High Performance Strategic Alliances: P&G Pharmaceuticals Alliance Capability', *Proceedings of Managing the Front End of Innovation PDMA & IIR Conference*, Boston, MA, May 29–30.
Sagal, Mathew W. (2003) 'Planning and Negotiating Strategic Alliances', *Proceedings of Managing the Front End of Innovation PDMA & IIR Conference*, Boston, MA, May 29-30.

Strategic innovation

According to Constantinos Markides, strategic innovation 'is a fundamental reconceptualization of what the **business** is all about that, in turn, leads to a dramatically different way of playing the game in an existing business' (Markides, 1998). Strategic innovation takes place 'when a company identifies gaps in industry positioning, goes after them, and the gaps grow to become the new mass market' (*ibid*.).

Strategic innovation begins with questioning the status quo and coming up with new ideas about how to compete in new ways (Markides, 1997). Markides (1997) prescribes five general approaches to strategic innovation:

1. Redefine the business.
2. Redefine the who. Ask: Who is the customer? Identify new customers or new customer segments and develop a plan to serve them better.
3. Redefine the what. Ask: What products or services are we offering these customers? Think of new customer needs or wants and develop a plan to satisfy them better.
4. Redefine the how. Build new products or a better way of doing business utilising core competences and then find the right customers.
5. Start the thinking process at different points. Start by asking: What are our unique capabilities? What specific needs can we satisfy? Who will be the right customer?

Evidence suggests that most strategic innovations come from outsiders, and rarely from established players (Markides, 1998). Markides suggests that established firms adopt a **fast-follower strategy** in the face of market-disrupting innovation, adopt it and then scale it up (Yang, 2000). For instance, the disposable razor was a disruptive strategic innovation in the razor market. It targeted price and ease of use and quickly gained market share. Gillette was an established player in the razor market with a focus on 'closeness of shave'. In response to the disruptive disposable razor, it produced its own disposable razors, but also innovated by creating two new products, the Sensor and the Mach3 (Charitou and Markides, 2003). Gillette played the disposable razor game, which is in conflict with its established business, while also innovating in that established business.

When the current **business model** is not disrupted, a tight alignment of innovation strategy with business strategy is appropriate (Govindarajan and Trimble, 2005). Cisco Systems leveraged its capability in networking technology to transform its internal processes and achieved significant productivity improvements.

Further reading
Charitou, Constantinos D. and Markides, Constantinos C. (2003) 'Responses to Disruptive Strategic Innovation', *MIT Sloan Management Review*, 44(2), 55–63.
Govindarajan, Vijay and Trimble, Chris (2005) 'Organizational DNA', *California Management Review*, 47(3), 47–76.
Markides, Constantinos (1997) 'Strategic Innovation', *Sloan Management Review*, 38(3), 9–23.
Markides, Constantinos (1998) 'Strategic Innovation in Established Companies', *Sloan Management Review*, Spring, 31–42.
Yang, Paul (2000) 'Strategic Innovation: Constantinos Markides on Strategy and Management', *Academy of Management Executive*, 14(3), 43–5.

Strategic management

Strategic management began in the 1960s with the concept of strategic adaptation to vague environmental factors (Herrmann, 2005). Strategic management addresses the fundamental question of how firms achieve sustainable competitive advantage.

Michael Porter (1980) provided the first 'dominant design' in strategic management with his classic book *Competitive Strategy*, considered the most influential book in the field. This was the first era. The second era was created with the resource-based view of the firm. This attributes sustainable **competitive advantage** to the development and use of valuable resources. In the early twenty-first century, a new era of knowledge-driven innovation is marked by rapid advances in technology. In this new era, boundaries among firms are blurred and a new **dominant design** is needed (Herrmann, 2005).

Strategic niche management (SNM) focuses on socially beneficial innovations that serve long-term goals such as sustainability and on radical novelties (Schot and Geels, 2008). More specifically, it focuses on the early adoption of new technologies in sustainable development (*ibid.*).

Management of technological innovation includes uncertain activities, cumulative knowledge, differentiated skills, sustained and extensive collaboration, interaction and implementation (Pavitt, 1990).

According to Pavitt (1990), the *strategic management of technology* and the implementation of innovation require:

S

- Cross-functional integration of specialists.
- Continuous questioning of the **business model** (customers, markets, strategies and skills for exploitation of technological opportunities).
- A long-term view of technological accumulation within the firm.

There is a great deal of concern among innovation professionals that traditional financial measures deter future-oriented innovation investments. *Strategic management accounting (SMA)* allows management accounting to focus on the consumer value generated relative to competitors (Bromwich and Bhimani, 1994). It also considers the firm's long-term performance in the marketplace (*ibid.*). SMA is concerned with strategically oriented information for decision making and control (Yi and Tayles, 2009).

See also: **balanced scorecard**

Further reading
Bromwich, M. and Bhimani, A. (1994) *Management Accounting: Pathways to Progress*, London, Chartered Institute of Management Accountants.
Herrmann, Pol (2005) 'Evolution of Strategic Management: The Need for New Dominant Designs', *International Journal of Management Reviews*, 7(2), 111–30.
Pavitt, Keith (1990) 'What We Know about the Strategic Management of Technology', *California Management Review*, 32(3), 17–26.
Porter, M. E. (1980) *Competitive Strategy: Techniques for Analyzing Industries and Competitors*, New York, Free Press.
Schot, Johan and Geels, Frank W. (2008) 'Strategic Niche Management and Sustainable Innovation Journeys: Theory, Findings, Research Agenda, and Policy', *Technology Analysis & Strategic Management*, 20(5), 537–54.
Yi, Ma and Tayles, Mike (2009) 'On the Emergence of Strategic Management Accounting: An Institutional Perspective', *Accounting & Business Research*, 39(5), 473–95.

Strategy

Corporate strategy is about value creation and capture for an organisation in a specific product market. According to Raynor (1997), the essence of strategy is clearly to define the organisation and the product market, as well as the role of new products in achieving business goals. However, there is a strategy paradox: 'precisely the same behaviors that maximize an organization's probability of great success also maximize its chance of complete catastrophe' (Raynor, 2007). For example, the executions of a value-capture strategy led to the failure of Sony Betamax, while the Walkman and PlayStation 2 were successful (Raynor, 1997).

Firms can choose their strategy from four strategic types (Griffin and Page, 1996):

1. [Prospectors: First to market, not all efforts are profitable.
2. Analysers: Seldom first; fast followers with more cost-effective or innovative product.
3. Defenders: Maintain niche in stable product or service area.
4. Reactors: Respond only when forced by strong environmental pressures.

According to PDMA best practices research (Griffin and Page, 1996), the strategic type is a key success measure for product development (PD) projects (Manion and

Cherion, 2009). For example, prospectors should place greater importance on PD success measures (such as sales from new products) when changing product lines and on early market entry. Prospectors should not put any emphasis on measuring success using development programme return on innovation (ROI) and fit to business strategy (*ibid*.). Analysers should measure improving products and being early followers in newer markets. Defenders should place greater importance on PD success measures relating to stable product lines and market penetration (*ibid*.).

Eisenhardt *et al.* (2008) identify the elements of strategy as: Who will be my targeted customers? What products and services should be offered? How should I offer these products and services to my targeted customers in an efficient and innovative approach? This includes identifying strategic assets and capabilities and creating the right organisational environment.

At the business unit level, Kahn and Barczak (2006) identify four levels of strategy sophistication. Level one companies do not have **new product development (NPD)** goals and do not allocate resources for NPD as part of the annual budget process. Level two companies have NPD goals derived from the organisational mission. Level three companies align goals with their organisational mission and strategic plan and utilise the mission and plan to identify areas of NPD opportunity. Level four companies use the mission and strategic plan to aid in opportunity identification. In addition, they allocate resources to pursue key innovations and fund NPD as a long-term strategic effort.

Further reading
Cooper, Robert G., Edgett, Scott J. and Kleinschmidt, Elko J. (2002) 'Improving New Product Development Performance and Practices', Benchmarking Study. Houston, TX, American Productivity & Quality Center.
Eisenhardt, Kathleen, Grove, Andrew, Markides, Costas, Kay, John and Nascimento Rodrigues, Jorge (2008) 'Special Report: Strategy Classics', *Business Strategy Review*, 19(3), 50–91.
Griffin, Abbie and Page, Albert L. (1996) 'PDMA Success Measurement Project: Recommended Measures for Product Development Success and Failure', *Journal of Product Innovation Management*, 13(6), 478–96.
Kahn, Kenneth B. and Barczak, Gloria (2006) 'Perspective: Establishing an NPD Best Practices Framework', *Journal of Product Innovation Management*, 23(2), 106–16.
Manion, Michael T. and Cherion, Joseph (2009) 'Impact of Strategic Type on Success Measures for Product
Development Projects', *Journal of Product Innovation Management*, 26, 71–85.
Raynor 1997 details to come
Raynor, Michael E. (2007) 'What Is Corporate Strategy Really?', *Ivey Business Journal*, 71(8), 1–8.

S

Strategy canvas

The strategy canvas is an action framework for formulating **blue ocean strategy** (Kim and Mauborgne, 2005). It graphically captures the current state of play in the known market space. It allows firms to understand the factors on which an industry competes, where the competition is currently investing and what solutions customers are receiving from competitive products. Embedded in the strategy canvas is strategic knowledge about the current status and future prospects of a business (*ibid*.).

The strategy canvas is a two-dimensional graph (*ibid.*). The horizontal axis captures factors in which the industry invests and competes. The vertical axis captures the level of offering that buyers receive across key competing factors. Value curves drawn on the strategy canvas graphically represent a firm's relative performance across the competitive factors of its industry (*ibid.*). These are drawn by plotting the performance of a company's offering relative to other alternatives along the key customer value elements.

The 'as is' strategy canvas captures a company's current strategic profile in the known market space, while the 'to be' strategy canvas captures a company's future strategic profile in a blue ocean (*ibid.*).

Further reading
Kim, Chan and Mauborgne, Renee (2005) *Blue Ocean Strategy*, Boston, MA, Harvard Business School Press.

Strategy map

The strategy map is an adaptation of Kaplan and Norton's **balanced scorecard (BSC)** (Kaplan and Norton, 2001). It helps an organisation to visualise its strategies and provides a framework to link *intangible assets* to shareholder value creation through four perspectives: financial, customer, internal process, and learning and growth (Kaplan and Norton, 2004). In the *financial perspective,* outcomes of the strategy are return on investment, shareholder value, profitability, revenue growth and lower unit costs. The *customer perspective* defines the **value proposition** to create a context in which intangible assets create value. The *customer value proposition* includes product/service attributes (such as price, quality, availability, selection and functionality), relationship with the customer (such as a service and partnership) and brand image (*ibid.*). The *internal process perspective* identifies key processes that create and deliver a differentiating customer value proposition. This includes perspectives on operations management, customer management, innovation, the regulatory environment and social improvement (*ibid.*). The *learning and growth perspective* identifies intangible assets that must be aligned with the critical internal processes such as human capital, information capital and organisational capital (*ibid.*). The intangible assets fit into the strategy map by creating alignment and readiness between strategic job families, the strategic IT portfolio and the organisation's change agenda (*ibid.*).

Further reading
Kaplan, Robert S. and Norton, David P. (2001) *The Strategy-Focused Organization: How Balanced Scorecard Companies Thrive in the New Business Environment.* Boston, MA, Harvard Business School Press.
Kaplan, Robert S. and Norton, David P. (2004) 'Measuring the Strategic Readiness of Intangible Assets', *Harvard Business Review,* 82(2), 52–63.

Sustainability

Some companies erroneously believe that a trade-off between competitiveness and the implementation of environmentally sustainable business practices is inevitable (Nidumolu *et al.*, 2009). Similarly, some CEOs believe that sustainable practices and green product development will handicap their companies versus manufacturers in China and other developing countries. Nidumolu *et al.* (2009) have dispelled those beliefs. Their studies reveal that sustainable development can be a source of organisational and technological innovations. Sustainable business practices produce both bottom-line and top-line returns in innovative companies. In the companies they studied, the drive to become environmentally friendly led to a reduction in inputs and lowered costs.

Organisations go through five stages of change towards the implementation of sustainability (*ibid.*). They consider compliance as opportunity; make value chains sustainable; design sustainable products and services; develop new **business models**; and create **next-practice platforms** (*ibid.*).

Lubin and Esty (2010) studied game-changing historical megatrends and identified that sustainability is an emerging business megatrend. According to them, firms seeking to achieve leadership in sustainability must articulate a **vision** for value creation and then execute it. They identify the four stages of value creation as:

1. Do old things in new ways. For example, 3M's Pollution Prevention Pays developed a business case for the value of eco-efficiency.
2. Do new things in new ways. Du Pont's zero-waste commitment drove business operations.
3. Transform the core business. Dow Chemical's 2015 sustainability goals resulted in technology breakthroughs, from solar roof shingles to hybrid batteries.
4. New business model creation and differentiation. General Electric's ecoimagination initiative built a green aura around the GE brand and was expected to bring in $25 billion revenues in 2010.

Further reading
Lubin, David A. and Esty, Daniel C. (2010) ,The Sustainability Imperative', *Harvard Business Review*, 88(5), 42–50.
Nidumolu, Ram, Prahalad, C. K. and Rangaswami, M. R. (2009) 'Why Sustainability Is Now The Key Driver of Innovation', *Harvard Business Review*, 87(9), 5.

Sustainable development

During his opening statement at the *World Summit on Sustainable Development* held in Johannesburg in 2002, Kofi Annan, Secretary-General of the United Nations, declared that the current paradigm for development benefits few and is flawed for many. A paradigm that harms the environment and leaves the vast majority in poverty is not sustainable. A new paradigm is needed at local, national, regional and global levels.

The *Johannesburg Declaration on Sustainable Development* recognises economic development, social development and environmental protection as pillars of sustainable development. Implementation of the Johannesburg Declaration requires a long-term strategic perspective from policy makers, strategy planners and implementers at all levels.

Good governance, access to key technology and knowledge, and science-based decision making are essential elements in converting cooperation in key industries into tangible sustainable development projects (Annan, 2002).

The National Economic Council of the United States (2009) published The Obama Innovation Strategy, 'American Innovation: Driving Towards Sustainable Growth and Quality Jobs'. The Obama strategy has three parts:

1. Invest in the building blocks of American innovation.
2. Promote competitive markets that spur productive entrepreneurship.
3. Catalyse breakthroughs for national priorities.

President Obama has identified innovation as the key to sustainable development and quality jobs for Americans in the twenty-first century (www.whitehose.gov).

Further reading
Annan, Kofi (2002) 'Report of the World Summit on Sustainable Development, Johannesburg, South Africa, 26 August–4 September', United Nations, New York. http://www.internationalchamberofcommerce.org/uploadedfiles/WBCSD/wssd.pdf?ter ms=sustainable+development+charter (accessed on November 9, 2009).
National Economic Council (2009) 'Strategy for American Innovation: Driving Towards Sustainable Growth and Quality Jobs', http://www.whitehouse.gov/assets/documents/ innovation_three-pager_9-20-09.pdf (accessed on December 27, 2010).

Sustaining innovation

Sustaining innovation is a problem for mature firms. They encounter several inhibitors such as the pursuit of stability, risk avoidance, being constrained by experience, a lack of options, legacy systems, complex power structures and a desire for operational excellence (Braganza *et al.*, 2009). Stability comes from internal systems and structures such as standardisation and **lean product development**. In risk avoidance, organisations avoid the risks associated with implementing **radical innovations**. When they are constrained by experience, organisations do not let go of outdated competencies and strategies. Given a lack of options, organisations get locked in by the resources they do have. Faced with legacy systems, investments in information systems (IS) are made to extend old information systems rather than create new capabilities to support innovation. In complex power structures, **breakthrough innovations** shift power and myopic managers focus internally rather than externally. Operational excellence criteria include excellence in customer intimacy, responsiveness to market needs and a focus on cost reductions. A system geared towards operational excellence may inhibit breakthrough innovations.

O'Connor (2009) defines breakthrough innovation capability as 'an embedded system for initiating, supporting and sustaining breakthrough innovation (BI) activities'. To be sustainable, a BI system must be designed to work with the operational excellence system. Elements of a BI system include a clearly defined mandate and scope, a supportive leadership culture, organisational structures to manage interfaces, governance and decision-making mechanisms, developing talent and learning-based process tools, and innovation metrics and rewards (*ibid.*).

Organisations can overcome innovation inhibitors by using **portfolio management** to allocate resources, train managers to be innovative and communicate the value of innovation. They can also align the organisational mission with inno-

vation policies, and change the governance of innovation with dual power structures to allow the movement of ideas from the bottom up and their implementation from the top down.

See also: **innovation killers, middle managers as innovators, strategic innovation**

Further reading
Braganza, Ashley, Awazu, Yukika and Desouza, Kevin C. (2009) 'Sustaining Innovation Is
   a Challenge for Incumbents', Research Technology Management, 52(4), 46–56.
O'Connor, Gina Colarelli (2009) 'Sustaining Breakthrough Innovation', Research
   Technology Management, 52(3), 12–14.

SWOT analysis

SWOT analysis is a framework to assess a firm's competitive strategy, in which the
firm evaluates strategies to exploit its **competitive advantages** or defend against
its weaknesses. Strengths (S) and weaknesses (W) involve identifying the firm's
internal abilities or disadvantages; while opportunities (O) and threats (T) involve
identifying external factors such as competitive forces, new technology developments, governmental intervention or domestic and international economic trends.

SWOT analysis is the first step in analysing a firm's strategic options. The two
sides of the SWOT analysis matrix often point in opposite directions, leaving
strategists with the paradox of creating alignment either from the outside in
(market-driven strategy) or the inside out (resource-driven strategy).

Systemic innovation

Chesbrough and Teece (1996) make a distinction between autonomous and
systemic innovation. An **autonomous innovation** is a standalone innovation,
and it can be incorporated without any additional adjustments to the system to
which it belongs, while **systemic innovation** requires complementary innovations from the system. Chesbrough and Teece (1996) theorise that if a firm sets out
to innovate in a systemic fashion, the integration of all necessary activities into
the corporation is necessary to guarantee success (ibid.). According to
Chesbrough and Teece, the lack of single-integrated-firm position leads to failure
when the innovation involved is systemic in nature (Laat, 1999). Chesbrough and
Teece (1996) use IBM's eclipse in the personal computer (PC) market to make their
point. While IBM was the original architect of open architecture for the PC, its
independent suppliers (Microsoft and Intel) pursued their own business interests,
resulting in the entry of IBM-compatible PC manufacturers within five years of
IBM's PC success (Laat, 1999). The IBM PC was a systemic innovation (Chesbrough
and Teece, 1996).

Laat (1999) identifies three stages in systemic innovation: the creation and
commercialisation of a systemic innovation; the process of setting standards for
it; and its further development in time. He argues that systemic innovation can
also be undertaken by alliance networks, although networks are vulnerable to
opportunism. In the audio CD industry, the change from vinyl records to the music
compact disk (CD) required simultaneous changes in multiple parts of the industry

value chain. In response, consumer electronics companies, record producers, disc equipment manufacturing companies, music retailers and retail manufacturers adjusted their practices (*ibid.*).

To organise a business for innovation, managers need to determine whether the innovation in question is autonomous or systemic and also whether the capabilities needed for innovation can be outsourced (for instance R&D) or created in house (internal R&D) (Bardhan, 2006). In his study of organising for offshoring innovation, Bardhan (2006) found that a centralised structure is more appropriate for a systemic innovation and a diffused, decentralised structure more compatible with an autonomous innovation.

Further reading
Bardhan, Ashok Deo (2006) 'Managing Globalization of R&D: Organizing for Offshoring Innovation', *Human Systems Management*, 25(2), 103–14.
Chesbrough, Henry W. and Teece, David J. (1996) 'When is Virtual Virtuous? Organizing for Innovation', *Harvard Business Review*, 74(1), 65–73.
Laat, Paul B. De (1999) 'Systemic Innovation and the Virtues of Going Virtual: The Case of the Digital Video Disc', *Technology Analysis & Strategic Management*, 11(2), 159–80.

Systems engineering (SE)

The International Council on Systems Engineering (INCOSE, 2004) defines systems engineering (SE) as 'an interdisciplinary approach and means to enable the realization of successful systems'. According to INCOSE (2004), SE defines customer needs and required functionality early in the development cycle, documents requirements, synthesises and validates design while considering the complete problem. It considers the business and the technical needs of all customers to provide a quality product that meets user needs (*ibid.*).

Sage and Rouse (2009) define SE along three dimensions: structure, function and purpose. Structure in SE refers to the management of technology to analyse and interpret the impacts of systems on stakeholder perspectives. Function is the combinations of methods and tools used in systems engineering. The purpose of SE is to maintain overall quality, integrity and integration. SE effort begins with a product focus and ends when the product is launched.

The *System of Systems Engineering (SoSE)* is a high-level viewpoint and explains interactions between independent systems in a class of complex systems (Jamshidi and Jamshidi, 2009). The need for SoSE came to focus in military, security, aerospace, manufacturing and service industries, environmental systems and disaster management (*ibid.*). The SoSE concept is still evolving (*ibid.*).

Further reading
Blanchard, Benjamin S. and Fabrycky, Walter J. (1998) *Systems Engineering and Analysis*, Englewood Cliffs, NJ, Prentice Hall.
INCOSE Communications Committee (2004) 'What Is Systems Engineering?', International Council on Systems Engineering (INCOSE), June 14. http://www.incose.org/practice/whatissystemseng.aspx (accessed on December 28, 2010).
Jamshidi, Mo and Jamshidi, Mohammad (2009) *Systems of Systems Engineering: Principles and Applications*, Boca Raton, FL, CRC Press.
Sage, Andrew B. and Rouse, William B. (2009) *Handbook of Systems Engineering and Management*, 2nd edn, Hoboken, NJ, John Wiley & Sons.

Technological innovation

In the technology literature, the theory of the S-curve is widely accepted (Sood and Tellis, 2005). According to this theory, an initial period of slow growth allows technologies to evolve and is followed by a period of rapid growth, culminating in a plateau (Utterback, 1994). When growth is plotted versus time, the performance of a new technology starts below that of an existing technology, crosses the performance of the established technology and ends at a higher plateau, thus tracing a single S-shaped curve (Christensen, 1997). A central challenge for managers is to know when to switch investments from the current to the future technology. Under the S-curve, the inflection point is the appropriate time. After this point, the performance of existing technology improves at a decreasing rate until maturity.

Sood and Tellis (2005) studied how new technologies evolve, how competing technologies challenge and how firms deal with technological evolution. They found that a single S-curve is not a reliable predictor of the performance of a technology, and in particular that several technologies showed multiple S-curves. Technological evolution follows a step function, with sharp improvements in performance following long periods of no improvement (*ibid.*).

Sood and Tellis (2005) recommend that entrenched players be vigilant to the emergence of new technologies and combine that vigilance with efforts to improve the old technology.

Further reading
Christensen, Clayton M. (1997) *The Innovator's Dilemma: When New Technologies Cause Great Firms to Fail*, Boston, MA, Harvard Business School Press.
Sood, Ashish & Tellis, Gerard J. (2005) 'Technological Evolution and Radical Innovation', *Journal of Marketing*, 69(July), 152–68.
Utterback, James M. (1994) *Mastering the Dynamics of Innovation*, Boston, MA, Harvard Business School Press.

Technological regime

The technological regime is defined as 'the coherent complex of scientific knowledge, engineering practices, production process technologies, product characteristics, skills and procedures, established user needs, regulatory requirements, institutions and infrastructures' (Smith, 2003). A technological regime basically sets or defines the rules of the game that guide the direction of related technological innovations for meeting social needs or economic demands, such as the way in which fossil fuels have dominated the technological regime for energy production (Kemp *et al.*, 1998).

Each technological regime is constituted by an accumulation of knowledge, values and sunken costs and it is articulated by networks of actors and institutions (Smith, 2003). The application of the theory of technological regime shift to **sustainable development** is in its infancy.

Further reading
Kemp, R., Schot, J. and Hoogma, R. (1998) 'Regime Shifts to Sustainability through Processes of Niche Formation: The Approach of Strategic Niche Management', *Technology Analysis and Strategic Management*, 10(2), 175–95.
Smith, Adrian (2003) 'Transforming Technological Regimes for Sustainable Development: A Role for Alternative Technology Niches?', *Science and Public Policy*, 30(2), 127–35

Technology

Science and technology are necessary to innovation. Rosenberg (1972) defines technology as those tools, devices and knowledge that mediate between inputs and outputs to develop new products or services. This includes product technology (to develop new products) as well as process technology (to mediate between inputs and outputs).

Dodgson *et al.* (2005) suggest that three groups of technologies can be deployed in **product and process innovation**:

- Information and communication technology (ICT), for enabling innovation.
- Operations and manufacturing technology (OMT), for implementing innovation.
- Innovation technology (IvT), for creating innovation.

IvT includes model-based design, **simulation**, **virtual customers**, virtual methods and **data-mining** systems, **data visualisation** and **rapid prototyping** (*ibid.*). For example, **computer-aided design**/computer-aided manufacture (CAD/CAM) links design with manufacture and **rapid prototyping** and permits the creation of three-dimensional objects directly from CAD data.

Further reading
Dodgson Mark, Gann, David and Salter, Amon (2005) *Think, Play, Do: Technology, Innovation, and Organization*, New York, Oxford University Press.
Rosenberg, Nathan (1972) *Technology and American Economic Growth*, Armonk, NY, M. E. Sharpe.

Technology discontinuities

Technological discontinuities are sudden and sharp technological shifts that affect consumers, business sectors, nations or the world with a disruptive force (Tushman and Anderson, 1986). Major technological discontinuities can be classified as competence destroying or competence enhancing for existing firms (*ibid.*).

Competence-destroying product discontinuities involve the creation of a new product class (e.g. fax machines or automobiles) or a substitution for an existing product (transistors instead of vacuum tubes) (*ibid.*). Competence-destroying process discontinuities involve new ways of making products (*ibid.*).

Competence-enhancing innovations build on and reinforce existing competencies, skills and know-how in the industry. The development of the World Wide Web has been competence enhancing and has evolved and progressed through two main phases (Web3.0 M2PPressWire, 2008). According to Todd (2009), Web 1.0 was the content Web, Web 2.0 is the social Web and Web 3.0, which the web is now entering, is the location-aware and moment-relevant internet.

Further reading
Technology Discontinuity: Web 3.0 M2PressWIRE, May 27, 2008.
Todd, L. (2009) 'What is Web 3.0 Definition?', http://www.tourismkeys.ca/blog/2009/01/what-is-web-30-definition/ (accessed on December 29, 2010).
Tushman, Michael L. and Anderson, Philip (1986) 'Technological Discontinuities and Organizational Environments', *Administrative Science Quarterly*, 31, 439–65.

Technology fusion

Technology fusion refers to the complementary and cooperative combination and transformation of a number of different core technologies to create new product markets (Kodama, 1992). For example, the combination of optics and electronics resulted in fibre-optic communication systems. Western firms rely too heavily on one technology domain for breakthroughs (*ibid.*). This ignores innovation opportunities through a combination of technologies, another example of which would be fusing mechanical and electronic technologies to create mechatronic revolution (*ibid.*).

Three basic principles essential to technology fusion are (*ibid.*):

1. The market drives the R&D agenda (that is, it articulates demand), not the company.
2. Intelligence capabilities are required to scan technologies inside and outside the industry. In Japan, all employees are active receivers and seek out usable innovations.
3. Technology fusion grows out of substantial and reciprocal investments in research consortia, joint ventures, alliances and partnerships.

In Japan, Honda, NEC, Sharp, Sony and others involve customers in the product-conceptualisation process, articulate demand and participate in large intelligence-gathering networks (*ibid.*). In the aerospace industry, the development of fly-by-wire systems demanded the fusion of electronics and hydraulic technologies (*ibid.*).

Further reading
Kodama, Fumio (1992) 'Technology Fusion and the New R&D', *Harvard Business Review*, 70(4), 70–78.

Technology life cycle

Both products and technologies have different life cycles. The study of the technology life cycle helps companies to focus management's attention on the relationships among its technologies, products, markets and development activities (Ford and Ryan, 1981). For instance, Rolls-Royce manages its technology in five stages that are aligned with the technology life cycle (Foden and Berends, 2010):

identification and monitoring; selection and approval; development research, acquisition and adaptation; exploitation and review; protection.

Ford and Ryan (1981) relate the technology life cycle to the **Abernathy and Utterback model of product and process innovations.** The technology life cycle goes through six stages of development: technology development, technology launch, application launch, application growth, technology maturity and degraded technology (*ibid.*). Early in the life cycle, a company invests heavily in product development to become a dominant technology player in the market, leading to increased market share and a leadership role. Later in the life cycle, as the technology approaches maturity, there is limited, competitive benefit from further development of technology with new products and features. This coincides with the emergence of **dominant design**. Once a dominant design emerges, a company minimises further investment in the mature technology and invests in another technology that is in the early growth stage. For example, a company that owns both colour television and mobile phone technologies will increase investment in cellular technology development as the market for colour televisions enters the mature phase (*ibid.*).

In some product sectors there is a mismatch between the product life cycle and the technology life cycle. High-volume consumer-oriented electronics such as mobile phones and PDAs adapt to the newest technology quickly (Singh and Sandborn, 2006). However, in avionics and military systems, the technology life cycle is shorter than the product life cycle. This is especially true with the obsolescence of electronic parts (*ibid.*).

Further reading
Foden, James and Berends, Hans (2010) 'Technology Management at Rolls-Royce', *Research Technology Management*, 53(2), 33–42.
Ford, David and Ryan, Chris (1981) 'Taking Technology to Market', *Harvard Business Review*, 59(2), 117–26.
Singh, Pameet and Sandborn, Peter (2006) 'Obsolescence Driven Design Refresh Planning for Sustainment-Dominated Systems', *Engineering Economist*, 51(2), 115–39.

Technology platforms

In technology industries, technology platforms are hubs (Economides and Katsamakas, 2006). A technology platform defines a framework for **research and development** and plans for **product platforms** to be built around a specific technology platform (McGrath, 2000). It represents the core competences of a technology company (*ibid.*).

A technology platform may be proprietary or open source. Microsoft Windows, Apple Macintosh and Intel® Centrino™ mobile platform technologies are proprietary platforms, while Linux is an open-source platform (Economides and Katsamakas, 2006).

Apple's platforms are neither fully open such as Linux nor fully closed such as the Windows PC operating system, but are rather 'closed, but not closed' (Cusumano, 2010). While Apple platforms use its proprietary technology, Apple also controls the user experience as well as to what (applications or content) and to whom (service contracts) it opens the platform (*ibid.*). Apple also has content platforms (iTunes and the App store) (*ibid.*) and has created synergies and positive network effects across its products and complementary services. The iPod, iPhone and iPad devices, as well as the iTunes service, work well with the Macintosh computer and have some interoperability with Windows (*ibid.*).

Technology companies face a significant challenge in new platform design relating to how to add new and improved capabilities without increasing the power required for the desired function (Clark *et al.*, 2005). Intel faced this problem in the design of the Intel® Centrino™ mobile platform technology. Intel® High Definition (HD) Audio mitigated the power impact at a system level by enabling the processor to enter power-saving states even while the audio is in use for CD or DVD playback (*ibid.*).

Based on competition between Microsoft Windows and the Linux open-source operating system, it can be said that when an open-source platform competes with a proprietary system, the proprietary system wins in both market share and profitability (Economides and Katsamakas, 2006).

Further reading

Clark, Tom, Feng, Yang, Kaine, Greg, Leete, Brian and Ranganathan, Sriram (2005) 'Low-Power Audio and Storage Input/output Technologies for the Second-Generation Intel® Centrino™ Mobile Technology Platform', *Intel Technology Journal*, 9(1), 49–59.

Cusumano, Michael A. (2010) 'Technology Strategy and Management Platforms and Services: Understanding the Resurgence of Apple', *Communications of the ACM*, 53(10), 22–4.

Economides, Nicholas and Katsamakas, Evangelos (2006) 'Two-Sided Competition of Proprietary vs. Open Source Technology Platforms and the Implications for the Software Industry', *Management Science*, 52(7), 1057–71.

McGrath, Michael E. (2000) *Product Strategy for Technology Companies,* 2nd edn, New York, McGraw-Hill.

Technology-push innovation

Technology-push innovation is the result of discontinuities in technology. Both demand and supply factors play an important role in the adoption of **breakthrough innovation** generated from discontinuous technology innovations (Walsh, 1984).

Kim and Lee (2009) conducted empirical analysis of the global DRAM market (semiconductor industry), specifically consumers' willingness to pay in technology-push markets. They found an L-type curve where technology push is greater than demand pull in the early stages and then decreases as demand pull becomes greater (*ibid.*).

ALSTOM Power Hydro developed a discontinuous innovation called Powerformer (Schwery and Raurich, 2004). In power generators using the classical insulation technology, for power transmission the voltage of the generator has to be transformed into high voltage by a transformer unit. With the Powerformer technology, voltages corresponding to the level of the usual power grid voltages can be generated, making transformer units obsolete (*ibid.*).

The Powerformer technology represents a technology-push situation. ALSTOM had a question about how to build the framework for launching a technology-push innovation successfully into the market (*ibid.*). Initially, the Powerformer team evaluated the new technology for feasibility (ibid.). A SWOT analysis revealed that the Powerformer project had to overcome external **commercialisation** problems and also that the link from disruptive technology to new market should be described properly (*ibid.*). The Powerformer team chose technology roadmapping and the technology commercialisation process from Jolly (1997). Their approach supported 'unconventional' thinking while integrating with **champions** (power promoters), who are crucial for the success of a **disruptive innovation** (Schwery and Raurich, 2004).

Geels (2004) suggests that 'shifts in technological paradigms are often coupled with shifts in socio-cultural regimes'. Such a shift happened with the introduction of quartz watches in the 1970s. This was both a breakthrough change in technologies (the introduction of semiconductors) and meanings (watches moved from being jewels to being instrument) (*ibid.*).

See also: **innovation management**

Further reading
Geels, F. W. (2004) 'From Sectoral Systems of Innovation to Socio-Technical Systems. Insights about Dynamics and Change from Sociology and Institutional Theory', *Research Policy,* 33, 897–920.
Jolly, V. K. (1997) *Commercializing New Technologies,* Boston, MA, Harvard Business School Press.
Kim, Wonjoon and Lee, Jeong-Dong (2009) 'Measuring the Role of Technology-Push and Demand-Pull in the Dynamic Development of the Semiconductor Industry: The Case of the Global DRAM Market', *Journal of Applied Economics,* 12(1), 83–108.
Schwery, Alexander and Raurich, Vicente F. (2004) 'Supporting the Technology-Push of a Discontinuous Innovation in Practice', *R&D Management,* 34(5), 539–52.
Walsh, Vivien (1984) 'Invention and Innovation in the Chemical Industry: Demand Pull or Discovery Push?', *Research Policy,* 13, 211–34.

Technology-Stage Gate™ (TSG)

The Technology-Stage Gate™ (TSG) system is used to manage high-risk projects within and at the transition between technology development and **new product development** (Ajamian and Koen, n.d.). TSG mitigates the limitation of the traditional Stage-Gate™ system where product development begins even when there is uncertainty about the ultimate outcome of technology development efforts (*ibid.*). Ajamian (2003) recommends these steps for transitioning from technology development to product development:

- Identify technology transition team members: The technology development team leader becomes the technology member of the product development team and the product development team leader leads the transition team.
- Identify technology gaps: What is the gap and who owns it? This could be a technology team member or a product development team member.
- Develop a transition plan.
- Approve the transition plan.

TSG may be totally within, partially outside or completely outside the **new concept development (NCD) model** (*ibid.*). It is completed when the technology development risks have been significantly reduced, and a new technology is ready for transition into product development when its technology feasibility is demonstrated (*ibid.*). Ajamian (2003) recommends the rigorous application of TSG for high-risk projects. Premature introduction of high-risk technologies into the development process leads to project delays, project uncertainty and project cancellation (Ajamian and Koen, n.d.).

Further reading

Ajamian, Greg R. (2003) 'After the Front End; Transitioning to Product or Technology Stage-Gate™', *Proceedings of Managing the Front End of Innovation PDMA & IIR Conference*, May 30, Boston, MA.

Ajamian, Greg R. and Koen, Peter A. 'Technology Stage-Gate™: A Structured Process for Managing High-Risk New Technology Projects', http://www.stevens.edu/cce/NEW/PDFs/TechStageGate.pdf (accessed on December 28, 2010).

Technology strategy

Technological innovation, based on technology capabilities, is a source of **competitive advantage** for firms (Bergek *et al.*, 2009). Technological capabilities include two components: technology strategies and technology activities (*ibid.*). Technology strategies guide corporate intentions (such as what technological resources to use) and how they should be implemented (such as how resources should be utilised in the market) (*ibid.*).

Technology strategies could be related to generic competitive strategies such as cost leadership or **differentiation** (Porter, 1985) and may include issues such as 'extent of commitment to R&D, choice of technology leadership or followership, choice of technologies to develop, technological scope and attitudes towards licensing' (Bergek *et al.*, 2009).

In the era of **open innovation**, multinational companies also require *technology intelligence strategy* (TIS) to source ideas from outside. Kodak European Research (KER) implemented a four-step procedure to look outside the company for technologies relevant to its business across Europe, Africa and the Middle East (Mortara *et al.*, 2010):

1. Understand country context. The context allows for a full appreciation of strengths, weaknesses, opportunities and threats within that country.
2. Identify intermediaries and organise country visits. Intermediaries are a valuable resource for scanning foreign countries for new and useful technologies and for recognising and valuing new ideas, knowledge, devices and artifacts new to the organisation.
3. Scouting visits. Capture information and set up social networks and links.
4. Post-visit. Organise information and disseminate findings.

A sound technology strategy enables a company to model its current priorities as well as anticipate future needs (Porter, 1985).

Further reading

Bergek, Anna, Berggren, Christian and Tell, Fredrik (2009) 'Do Technology Strategies Matter? A Comparison of Two Electrical Engineering Corporations, 1988–1998', *Technology Analysis & Strategic Management*, 21(4), 445–70.

Mortara, Letizia, Thomson, Ruth, Moore, Chris, Armara, Kalliopi, Kerr, Clive, Phaal, Robert and Probert, David (2010) 'Developing a Technology Intelligence Strategy at

Kodak European Research: Scan & Target', *Research Technology Management*, 53(4), 27–38.

Porter, Michael, E. (1985) 'Technology Strategy' in Michael E. Porter, *Competitive Advantage: Creating and Sustaining Superior Performance*, New York, Free Press, pp. 176–93.

Technology transfer

Technology transfer (TT) is evolving from a professional practice into an academic discipline (Lane, 2010).

Lane (2003) defines TT as 'a process of transforming an idea for the novel application of a technology into a viable product'. The TT process can begin with technology push (new discoveries offer opportunities to improve product features and functions), market demand pull or corporate collaboration (internal and external ideas refined through iteration with external stakeholders) (Rothwell, 1992).

Lane (1999) provides a framework for the TT process, otherwise known as the idea–prototype–product process, which consists of activities carried out by stakeholders. Lane (2010) presents a research–development–production model of technology transfer elements and combines it with a 20-step development process linking research to production. He drew heavily on the body of knowledge of the Product Development and Management Association (PDMA) to develop his development process.

Bauer and Flagg (2010) define TT as the process of moving an innovation from a source to a destination context via some transfer mechanism, and suggest that TT intermediaries can become facilitators of this transfer mechanism. TT intermediaries (TTIs) are resource providers who offer various forms of assistance, such as university technology transfer offices, federal laboratories and other federally funded brokers (*ibid.*).

Common types of technology transfer include:

- *Technology transition*, the movement of technology along a value chain within an organisation, generally involving the evolution of a technology into a product within a company.
- *Internal technology transfer*, the movement of technology to direct use in-house. This can include the delivery of internally developed equipment to manufacturing or customisation of acquired products for internal use.
- *External technology transfer*, the movement of technology from one organisation to another. This can include the acquisition of technologies from outside sources, licensing of technologies to others or alliances for cooperative development.
- *Division-to-division technology transfer*, the movement of technology into distinctly separate parts in an organisation.
- *Mergers and acquisitions*, the purchase of technologies and technical capabilities by acquiring other companies.
- *Dissemination of technology*, the movement of technology directly to technical communities.

Further reading
Bauer, Stephen M. and Flagg, Jennifer L. (2010) ,Technology Transfer and Technology Transfer Intermediaries', *Assistive Technology Outcomes and Benefits*, 6(1), 129–50.
Lane, J. (1999) 'Understanding Technology Transfer, *Assistive Technology*, 11(1), 5–19.
Lane, J. P. (2003) 'The State of the Science in Technology Transfer: Applications for the Field of Assistive Technology, *Technology Transfer Society*, 28, 333–54.
Lane, Joseph P. (2010) 'State of the Science in Technology Transfer: At the Confluence of Academic Research and Business Development – Merging Technology Transfer with Knowledge Translation to Deliver Value', *Assistive Technology Outcomes and Benefits*, 6(1), 1–38.
Rothwell, R. (1992) 'Successful industrial-innovation – critical factors for the 1990s', *R&D Management*, 22, 221–39.

Three horizons model

See **horizons of growth**

Time to market (TTM)

Time to market (TTM) is the product development cycle time from initial product concept to successful market entry (McGrath, 2004). During the 1980s and 1990s product development efforts were focused on three areas of **new product development (NPD)** to reduce TTM (ibid.):

- Phase-based decision making.
- Cross-functional project teams.
- Structured development process.

The **Stage-Gate**™ NPD process developed by Robert G. Cooper, **the PACE®** process developed by Pittiglio Rabin Todd & McGrath (PRTM) and the project contract system developed by Wheelwright and Clark enabled companies to implement a structured **new product development** process and to incorporate phase-based decision making in their NPD management system (*ibid.*).

Companies implementing TTM-reduction practices reduce their time to market by 40 to 60 per cent across all new products (*ibid.*). The major benefits from decreased time to market are earlier revenue, which improves company **cash flow** and decreases time to profit; an increase in market share and sales volume over the entire product life time; and an increase in unit profit over later competitors (Kumar and Krob, 2007). However, companies implementing automated project planning systems may inadvertently increase time to market because bottom-up architecture is built into project management software (McGrath, 2004).

Further reading
McGrath, Michael E. (2004) *Next Generation Product Development,* New York, McGraw-Hill.
Kumar, Sameer and Krob, William (2007) 'Phase Reviews versus Fast Product Development: A Business Case', *Journal of Engineering Design*, 18(3), 279–91.

Types of development teams, structures and roles

New product development (NPD) teams can be face to face (co-located teams are the most common) or virtual. Four types of face-to-face teams can be designed for new product development projects (Wheelwright and Clark, 1992):

- *Functional teams*, where the work is completed in the function and coordinated by functional managers. A functional team is led by the functional manager, who controls resources and task assignment. Functional teams can effectively solve technical problems.
- *Lightweight teams*, where a team leader works through functional representatives but has little influence over functional resources or the work. The advantages of a lightweight team include improved communication and coordination. Lightweight teams are suited to product improvements.
- *Heavyweight teams*, where a strong leader exerts a direct, integrating influence over resources and the work done across functional silos. The manager of a heavyweight team has direct access to top management and is responsible for the work of everybody involved in the project. Managers have the authority, experience, influence and dedication of core members. Heavyweight teams are especially good at developing next-generation components or products.
- *Autonomous teams*, where a heavyweight team is removed from the function, dedicated to a single project and often co-located. Use autonomous teams when the challenge is to break from the past, enter a new market or do radical technology or commercial projects (new-to-the-world projects).

The *structure of NPD teams* can significantly affect the complex decision processes involved in NPD. *Team stability* is a key driver for maintaining relationships, encouraging discussion, exchanging information and improving the comprehensiveness of decision making by NPD teams (Slotegraaf and Atuahene-Gima, 2011). Slotegraaf and Atuahene-Gima (2011) found that stability in an NPD team has a curvilinear relationship to team-level debate and the comprehensiveness of decision making, and that comprehensiveness has a curvilinear relation to new product advantage.

Team cognition plays an important role in software development (He *et al.*, 2007). Two cognitive elements, mental models and transactive memory, represent the initial building blocks of team cognition during the role-compilation phase of team development (Pearsall *et al.*, 2010). *Mental models* benefit teams by allowing them to synchronise their actions on the basis of a clear understanding of each other's roles and behaviours, and transactive memory enables team members to develop deep, specialised areas of expertise while still maintaining easy access to each other's knowledge (*ibid.*). Mumford *et al.* (2008) developed a situational judgement test, called the *Team Role Test*, to measure knowledge of 10 roles related to the team: contractor, creator, contributor, completer, critic, cooperator, communicator, calibrator, consul and coordinator.

Virtual NPD teams are those whose members are geographically dispersed and cross-functional (Malhotra *et al.*, 2007). *Leadership of virtual teams* creates additional challenges for NPD managers. Malhotra *et al.* (2007) identified six effective leadership practices for virtual teams:

1. Establish and maintain trust through the use of communication technology.
2. Ensure that distributed diversity is understood and appreciated.
3. Manage the virtual work life cycle (meetings).
4. Monitor team progress using technology.
5. Enhance the visibility of virtual members.
6. Enable individual members of the virtual team to benefit from being part of the team.

See also: **new product development team**

Further reading

He, Jun, Butler, Brian S. and King, William R. (2007) 'Team Cognition: Development and Evolution in Software Project Teams', *Journal of Management Information Systems*, 24(2), 261–92.

Malhotra, Arvind, Majchrzak, Ann and Rosen, Benson (2007) 'Leading Virtual Teams', *Academy of Management Perspectives*, 21(1), 60–70.

Mumford, Troy V., Van Iddekinge, Chad H., Morgeson, Frederick P. and Campion, Michael A. (2008) 'The Team Role Test: Development and Validation of a Team Role Knowledge Situational Judgment Test', *Journal of Applied Psychology*, 93(2), 250–67.

Pearsall, Matthew J., Ellis, Aleksander P. J. and Bell, Bradford S. (2010) 'Building the Infrastructure: The Effects of Role Identification Behaviors on Team Cognition Development and Performance', *Journal of Applied Psychology*, 95(1), 192–200.

Slotegraaf, Rebecca J. and Atuahene-Gima, Kwaku (2011) 'Product Development Team Stability and New Product Advantage: The Role of Decision-Making Processes', *Journal of Marketing*, 75(1), 96–108.

Wheelwright, Steven C. and Clark, Kim B. (1992) *Revolutionizing Product Development*, New York, Free Press.

T

Uu

Universal design

Anyone, regardless of age or ability, can use products made with universal design (UD) (Crawford and Di Benedetto, 2008). On December 12, 2007, the Committee of Ministers of the Council of Europe passed a resolution: 'Universal design is a strategy which aims to make the design and composition of different environments, products, communication, information technology and services accessible, usable and understandable to as many as possible in an independent and natural manner, preferably without the need for adaptation or specialized solutions' (Björk, 2009a).

UD considers the abilities of real people in real situations. Examples of products made with universal design are phones with extra-large buttons, closed-captioned television, automatic garage openers and automatic sliding glass doors (Crawford and Di Benedetto, 2008).

The seven principles of UD are equitable use; flexibility in use; simple and intuitive to use; perceptible information; tolerance for error; low physical effort; and size and space for approach and use (Björk, 2009b). Connell *et al.* (1997) indicate that the seven principles can be applied to assess existing designs, guide the design of new products and environments, and educate both designers and consumers about design principles for more usable products and environments.

Several factors inhibit companies from implementing UD principles in the design of products and environments (Björk, 2009b). In Björk's (2009b) opinion, an excessive focus on shortening development time, production efficiency, lack of time, budget limitations and lack of knowledge, tools or a valid **business case** all prevent the wide adoption of UD principles.

Further reading
Björk, Evastina (2009a) 'Why Did It Take Four Times Longer to Create the Universal Design Solution?', *Technology & Disability*, 21(4), 159–70.
Björk, Evastina (2009b) 'Many Become Losers When the Universal Design Perspective Is Neglected: Exploring the True Cost of Ignoring Universal Design Principles', *Technology & Disability*, 21(4), 117–25.
Connell, B., Jones, M., Mace, R., Mueller, J., Mullick, A., Ostroff, E., Sanford, J., Steinfeld, E., Story, M. and Vanderheiden., G. (1997) 'Guidelines for the Use of the Principles of Universal Design, Version 2.0 (Revised September 9, 2002)', Raleigh, NC, Center for Universal Design, North Carolina State University, http://www.ncsu.edu/www/ncsu/design/sod5/cud/about_ud/docs/use_guidelines.pdf (accessed on December 29, 2010).
Crawford, Merle and Di Benedetto, Anthony (2008) *New Products Management,* 9th edn, New York, McGraw-Hill.

Usability

Usability is how a user experiences a product or service. Usability means designing products and systems that are easy to use, and matching features and functions with user needs and requirements (Dumas and Redish, 1999). The international standard ISO 9241-11 defines usability as 'the extent to which a product can be used by specified users to achieve specified goals with effectiveness, efficiency and satisfaction in a specified context of use'. Two primary definitions of usability are 'the efficiency of a user completing a task, for example, on a web site' and 'ease of use of a system and the user's perception of easiness and positive experience' (*ibid.*). Usability definitions rest on four principles: keeping the focus on users; remembering that people use products to be productive; bearing in mind that users are busy people trying to accomplish tasks; and users decide when a product is easy to use (*ibid.*).

There are two types of usability testing methods: with users and without users (Molich and Dumas, 2008). In usability testing with users, researchers observe selected individual participants (representative of the product's user base) as they work through tasks designed to demonstrate the product's functionality (Dumas and Redish, 1999). Usability is not the same as 'functionality', which is concerned only with the functions and features of the product. In addition, increased functionality does not mean improved usability. Non-user testing includes expert review (no users involved). Molich and Dumas (2008) found no practical difference in usability testing results between users and expert reviews.

In web-based e-commerce, usability issues include use of colours, fonts, limited space, labelling and positioning of buttons, as well as the use of progress indicators. Usability is a key factor in the widespread adoption of Apple products (Dougherty, 2004).

Building usability into products is called usability engineering (Dumas and Redish, 1999).

Further reading
Dougherty, Bret (2004) 'iPod Usability Critique', http://www.unc.edu/~bretd/222ipodusecritique.htm (accessed on December 26, 2010).
Dumas, Joseph S. and Redish, Janice (1999) *A Practical Guide to Usability Testing*, revd edn, Exeter, Intellect Books.
ISO 9241-11:1998(E), *Ergonomic Requirements for Office Work with Visual Display Terminals (VDTs) – Part 11: Guidance on Usability*, http://www.it.uu.se/edu/course/homepage/acsd/vt09/ISO9241part11.pdf (accessed on December 26, 2010).
Molich, Rolf and Dumas, Joseph S. (2008) 'Comparative usability evaluation (CUE-4)', *Behaviour & Information Technology*, 27(3), 263–81.

U

User-centred design

User-centred design (UCD), also called **human-centred design**, includes the user in the design process, participating in the definition of product criteria, the testing of design concepts and validation of the final design. ISO 13407: Human Centred Design Process for Interactive Systems describes four key principles of human-centred design: the active involvement of users; an appropriate allocation of function between user and system; the iteration of design solutions; and multidisciplinary design teams. The four essential human-centred design activities are understanding and specification of the context of use; specification of user and organisational requirements; production of design solutions (prototypes); and evaluation of designs with users against requirements (ISO 13407).

Personas and **scenarios** are examples of UCD methods (Putnam *et al.*, 2009). Personas help design teams understand who the users are; and scenarios help them understand what users want to do. In the computer hardware industry, a user-centred approach optimises a computer system's usability (*ibid.*).

The IBM UCD team learned three lessons during the redesign of the ThinkPad product line:

1. A strong emphasis on family design innovation is an essential contributor to mindshare and satisfaction.
2. A truly usable system starts with a focus on the underlying technology architecture, not the user interface.
3. The usage-based segmentation strategy was beneficial in reducing team workload, improving designers' imagination and **innovation**, crystallising the marketing focus and meeting the end-to-end experience objective (Sawin *et al.*, 2002).

Further reading
ISO 13407: *Human Centred Design Process for Interactive Systems*, http://www.ash-consulting.com/ISO13407.pdf (accessed on December 29, 2010).
Putnam, Cynthia, Rose, Emma, Johnson, Erica J. and Kolko, Beth (2009) 'Adapting User-Centered Design Methods to Design for Diverse Populations', *Information Technologies & International Development*, 5(4), 51–73.
Sawin, David A., Yamazaki, Kazuhiko and Kumaki, Atsushi (2002) 'Putting the "D" in UCD: User-Centered Design in the ThinkPad Experience Development', *International Journal of Human–Computer Interaction*, 14(3/4), 307–34.

User-centric innovation

User-centric innovation is an emerging innovation pattern within the global economy (Prahalad and Ramaswamy, 2000). It includes the implementation of user innovations developed outside the firm (e.g. innovations by **lead users**), dedicated **co-creation** between user and manufacturer, as well as users taking over part of the actual manufacturing process (Warnke et al., 2008). Users are firms or individual consumers that benefit from using a product or a service, whereas manufacturers expect to benefit from selling a product or a service (von Hippel, 2005).

A shift of **business model** towards user-centric innovation will be treated as a discontinuity in market conditions (Warnke *et al.*, 2008).

See also: **democratising of innovation**

Further reading
Prahalad, C. K. and Ramaswamy, V. (2000) 'Co-opting Customer Competence', *Harvard Business Review*, 78(1), 79–87.
Von Hippel, Eric (2005) *Democratizing Innovation*, Cambridge, MA, MIT Press.
Warnke, Philine, Weber, Matthias and Leitner, Karl-Heinz (2008) 'Transition Pathways Towards User-Centric Innovation', *International Journal of Innovation Management*, 12(3), 489–510.

User-driven innovation

Users provide valuable insights to **new product development teams** at the
front end of the innovation process. The lead-user approach, participatory
design and design anthropology are three approaches to bring users into the innovation process (Burr and Mathews, 2008). The **lead-user** approach is one way of
outsourcing innovation and bypassing traditional **market research.**
Participatory design projects are engagements in methods development. These
projects give designers remarkable scope to effect a needed change in organisations and to convert insights gathered from users into actions. **Design anthropology** is a recent development in the USA. It is the application of anthropology
to development practices to provide comprehensive understanding of users.

User-driven innovation deals with insights at both an observable and a more
latent level that are quite difficult to grasp (De Moor *et al.*, 2010). **Living labs** are
an example of user-driven innovation environments (Følstad, 2008). They provide
for inventing, prototyping, interactive testing and marketing of (new) mobile technology applications, for instance (Følstad, 2008; Schumacher and Niitamo, 2008).
While user-driven innovation advocates user involvement in the early stages of
the design and development process, there is no mechanism to resolve conflict
between the narrow scope of a project and the broader scope of user-generated
research. There is also a lack of integration of **best practices** into user-driven
innovation research, such as in the living lab environment (Feurstein *et al.*, 2008;
Følstad, 2008).

Further reading
Burr, Jacob and Matthews, Ben (2008) 'Participatory Innovation', *International Journal of
Innovation Management*, 12(3), 255–73.
De Moor, Katrien, Berte, Katrien, De Marez, Lieven, Joseph, Wout, Deryckere, Tom and
Martens, Luc (2010) 'User-Driven Innovation? Challenges of User Involvement in
Future Technology Analysis', *Science & Public Policy*, 37(1), 51–61.
Følstad, A. (2008) 'Towards a Living Lab for the Development of Online Community
Services', *Electronic Journal for Virtual Organizations and Networks*, 10, 47–58.
Feurstein, K., Hesmer, A., Hribernik, K. A., Thoben, K.-D. and Schumacher, J. (2008)
'Living Labs: A New Development Strategy', in J. Schumacher and V.-P. Niitamo (eds),
European Living Labs: A New Approach for Human Centric Regional Innovation, pp. XX–
XX.
Schumacher, J. and Niitamo, V.-P. (eds) (2008) *European Living Labs. A New Approach for
Human Centric Regional Innovation*, Berlin, WVB.

U

User-initiated innovation (UII)

User-initiated innovation (UII) occurs when a firm that has invented a novel device
first uses it in its internal application as a **process innovation** and secondly as a
product innovation (Foxall, 1988). In the first phase of user-initiated process
innovation, the firm gains from the enhancement of its output as a result of the
improvement of its production methods. In the second phase, it gains from the
fees or royalties accruing from leasing or licensing its technological know-how, or
the profits deriving from the sale of the **new product** (*ibid.*).

In international markets, UII can be licensed to overseas manufacturing and
marketing companies. The licensing of *innovation knowledge* can become a prime

source of **competitive advantage** in both domestic and foreign markets (Ryan, 1985).

User-initiated innovation is a feature of firms rather than households or individual consumers.

Further reading
Foxall, Gordon R. (1988) 'The Theory and Practice of User-Initiated Innovation', *Journal of Marketing Management*, 4(2), 230–48.
Ryan, C. G. (1985) 'Innovative Marketing or Selling Seedcorn?', *Electronics and Power*, July, 503–6.

User innovations

User innovation is not a new phenomenon (von Hippel, 1976). According to the *'manufacturer-active paradigm' (MAP)*, the manufacturer is responsible for all activities in the **innovation process** (von Hippel, 1988). In this traditional view of the **new product development (NPD) process**, users are passive acceptors of an innovation. However, **lead users** such as farmers, athletes and physicians are some of the most prolific innovators. In the early 1980s, Eric von Hippel and his colleagues systematically documented how users, rather than manufacturers, are often the main source of innovation (Raasch *et al.*, 2008). During nearly four decades in the rodeo kayaking industry, '100% of innovations in technique came from users, and users also dominated innovation in hardware (62% of major and 83% of minor innovations) over user–manufacturers (2% major and 13% minor innovations) and other manufacturers (25% of major and 15% of minor innovations)' (Baldwin *et al.*, 2006).

Faulkner and Runde (2009) believe that user innovation research has concentrated on changes in the physical form of the objects to the neglect of changes in their function. To remedy this omission, they advance a theory of the technical identity of a technological object that includes both its form and function implying that changes in function are as much due to technological change as changes in form. They used the rodeo kayak as an example to make the point that the emergence of new techniques associated with its use (Baldwin *et al.*, 2006) attracted new users and thereby expanded the market for that object without any concomitant changes in its form (Faulkner and Runde, 2009).

In some mature firms, there is a constant struggle over whether to encourage or discourage user innovation. Microsoft and Hewlett-Packard restrict users from modifying and improving their products by 'building electronic locks and fences around their properties' (Braun and Herstatt, 2008). On the other hand, companies such as Apple, IBM and Lego benefited (with increases in market share and profits) by opening up parts of their **platforms** to user innovations (*ibid.*).

Further reading
Baldwin, C., Hienerth, C. and von Hippel, E. (2006) 'How User Innovations Become Commercial Products: A Theoretical Investigation and Case Study', *Research Policy*, 35, 1291–313.
Braun, Viktor and Herstatt, Cornelius (2008) 'The Freedom Fighters: How Incumbent Corporations are Attempting to Control User-Innovation', *International Journal of Innovation Management*, 12(3), 543–72.

Faulkner, Philip and Runde, Jochen (2009) 'On the Identity of Technological Objects and User Innovations in Function', *Academy of Management Review*, 34(3), 442–62.

Hyysalo, Sampsa (2009) 'User Innovation and Everyday Practices: Micro-Innovation in Sports Industry Development.',*R&D Management*, 39(3), 247–58.

Raasch, Christina, Herstatt, Cornelius and Lock, Phillip (2008) 'The Dynamics of User Innovation: Drivers and Impediments of Innovation Activities', *International Journal of Innovation Management,* 12(3), 377–98.

von Hippel, E. (1976) 'The Dominant Role of Users in the Scientific Instrument Innovation Process', *Research Policy*, 5(4), 212–39.

von Hippel, E. (1988) *The Sources of Innovation*, New York, Oxford University Press.

User testing

See **usability testing**

Utterback and Abernathy model of product and process innovation

See **dynamics of innovation**

U

Vv

Value chain

Value is the amount buyers are willing to pay for what a firm provides them with. Porter (1985) created the concept of the value chain to explain the sources of **competitive advantage.**

A company's value chain consists of nine interrelated generic activities linked to the activities of channels, suppliers and buyer activities (*ibid.*). The nine activities are divided into primary activities (inbound logistics, operations, outbound logistics, marketing and sales and service) or support activities (procurement, technology development, human resource management and firm infrastructure) (*ibid.*). Linkages exist within a firm's value chain, as well as between a firm's value chain and that of suppliers and channels (vertical linkages). When a buyer purchases a product from a firm, a competitive advantage is created for the firm because it has offered the product at a lower price or improved performance (*ibid.*). In addition, a firm can compare its value chain with that of its competitors to highlight its dimensions of **differentiation** from competition.

A 2009 report from Capgemini and the University of Edinburgh introduced the concept of the *value circle* in the evolution of the value chain concept. The value circle involves interaction at all levels to create a continuous cycle of improvement and has the following characteristics:

- A focus on resourcing in product design and **innovation**. New systems are adopted to capture and incorporate new ideas and innovations.
- A shift from manufacturing to management of manufacturing. Collaborative partnerships drive manufacturing, as well as mutual learning and innovation.
- Closer relationships with fewer suppliers. Close working with selected suppliers achieves competitive advantage for both suppliers and manufacturer.
- Use of new information technology (IT) to leverage the value chain.
- A circular approach to managing active partnerships with customers, linking the two ends of the value chain.

Further reading
Porter, M. (1985) *Competitive Advantage: Creating and Sustaining Superior Performance*, New York, Free Press, New York.
'New Report from Capgemini and University of Edinburgh Reveals Changing Nature of Manufacturing Value Chain; Study Introduces New Concept of "value circle" in the Evolution of the End-to-End Value Chain', *M2PressWIRE*, May 15, 2009.

Value innovation

Introduced by INSEAD professors W. Chan Kim and Renee Mauborgne (1997), value innovation is a strategic concept that defies the value–cost trade-off that is generally associated with conventional competition-based strategy. In conventional competition, strategy is seen as making a choice between **differentiation** and low cost. In contrast, value innovation involves 'the simultaneous pursuit of differentiation and low cost' (Kim and Mauborgne, 1997). It focuses on opening up new and uncontested market spaces with value creation for both buyers and the company.

Value innovation 'is the cornerstone of **blue ocean strategy**' and is distinct from either value creation or technology innovation (Kim and Mauborgne, 2005). A company achieves value innovation by aligning its value, profit and people propositions (*ibid.*). The value proposition (utility minus price) raises buyer utility at the right price; the profit proposition (price minus cost) makes a healthy profit for the company; and the people proposition builds execution into **strategy** (www.blueoceanstrategy.com) .

See also: **blue ocean strategy, competitive advantage, strategic innovation**

Further reading

Kim, Chan W. and Mauborgne, Renee (1997) 'Value Innovation: The Strategic Logic of High Growth', *Harvard Business Review,* 82(7/8), 172.

Kim, Chan W. and Mauborgne, Renee (2005) *Blue Ocean Strategy,* Boston, MA, Harvard Business School Press.

www.blueoceanstrategy.com (accessed November 13, 2009).

Value proposition

The value proposition refers to the unique utility that a **business** offers to its customers. The value proposition for a product or service describes 'what is in it for the buyers'. Sawhney (2010) defines the value proposition as 'the potential benefits of an offering for a target customer that outweigh the total customer sacrifice while being differentiated from available alternatives and supported by reasons to believe'. Some companies such as IKEA clearly articulate their value proposition. Sawhney (2010) describes IKEA's value proposition as 'stylish, space efficient and scalable furniture and accessories at very low price points for young, first time, or price-sensitive buyers who are willing to invest some effort in buying and assembling furniture'.

Value propositions are noteworthy because they guide both product development and product marketing. For product development, a value proposition is a statement of priorities and choices that puts engineering, **product management** and product marketing 'on the same page' and provides a clear direction for what needs to be built (*ibid.*). For product marketing, it defines key themes for positioning and messaging in marketing communications. It also guides choices in pricing, channels, services and partners needed to deliver on the promise (*ibid.*).

Positioning is a promise statement that defines how a firm wants prospects to think and feel about its product or brand relative to alternatives, measured in terms of the key customer perceptions it wants to create or change. Sawhney (2010) describes the value proposition of RIM Company's BlackBerry as 'for business e-mail users who want to better manage the increasing number of messages they receive when out of the office, BlackBerry is a mobile e-mail solution that provides a real-time link to their desktop e-mail for sending, reading, and responding to important messages. Unlike other mobile email solutions, BlackBerry is wearable, secure and always connected.'

A customer value proposition is also needed in business-to-business (B2B) markets. Here suppliers use the term in three different ways (Anderson *et al.*, 2006). Some companies simply list all the benefits their offering might deliver to target customers, while others mistakenly assume that favourable points of difference must be valuable for the customer (*ibid.*). However, best-practice suppliers focus their value proposition on a few value elements that most interest target customers, demonstrate the value of this superior performance, and convey how that performance can support the customer's business priorities (*ibid.*).

Further reading
Anderson, James C., Narus, James A. and van Rossum, Wouter (2006) 'Customer Value Propositions in Business Markets', *Harvard Business Review*, 84(3), 90–99.
Sawhney, Mohan (2010) 'Value Proposition Excellence in Product Development and Marketing', PDMA 34th Annual Global Conference on Product Innovation Management, October 19, Orlando, Florida.

Venture café

A venture café combines a nurturing environment for **entrepreneurs** with networking opportunities among one another and with venture capitalists. It offers 'an informal environment where engineers, business people and potential investors can get together and talk about starting new high-technology companies' (Esser, n.d.). It provides entrepreneurs with an opportunity to discuss, assess and receive feedback on their plans in the informal setting of a bar or café before they are ready to make a formal presentation to venture capitalists (*ibid.*).

Further reading
Esser, Teresa (n.d.) 'The Venture Café, Frequently Asked Questions', http://www.teresa.org/faq.html (accessed on December 31, 2010).

Virtual customer

The term 'virtual customer' describes the use of web-based tools for understanding **customer needs** by involving customers in the **innovation process** (Dahan and Hauser, 2002). These are significantly less expensive and much faster than conventional tools and can obtain a much broader range of feedback, even for targeted demographics (*ibid.*). Virtual methods can be valuable for identifying opportunities, improving the design and engineering of products, and testing ideas and concepts much earlier in the process (*ibid.*). Seven virtual customer methods are described in Table 2.

Further reading
Dahan, E. and Hauser, J. R. (2002) 'The Virtual Customer', *Journal of Product Innovation Management*, 19, 332–53.

V

Table 2 Virtual customer methods

Virtual Method	Description
Information pump (IP)	Participants formulate questions about product concepts and guess how others will react to their questions. The initial applications of the IP method were in the visual aesthetics of concept cars.
Fast polyhedral adaptive conjoint estimation (FP)	Screening a large number of product features inexpensively to identify and measure the importance of the most promising features for further development. Involves paired comparisons of attribute bundles where respondents are asked to select relative preference between two stimuli.
Web-based conjoint analysis (WCA)	Presenting product features, product use and marketing elements in streaming multimedia representations to gather customer input, and sorting and selecting bundles of desired attributes in products.
User design (UD)	A product is configured using visual drag and drop to find an ideal solution. Respondents trade off features against price or performance.
Virtual concept testing (VCT)	Testing concepts without actually building the product.
Securities trading of concepts (STOC)	Identifying winning concepts by allowing customers to interact with one another in a stock-market-like game. Each product concept is represented by a 'security' and is bought and sold by respondents interacting with one another.
Listening in	A four-step process that starts by finding and exploring new combinations of unmet customer needs by using a virtual adviser that selects questions adaptively to provide recommendations. An opportunity trigger then tracks recommendations and identifies opportunities based on new combinations of customer needs. A virtual engineer explores the opportunity with directed quantitative and qualitative questions and presents the customer with a design palette, giving them the opportunity to redesign a mobile phone, for example, according to their own specifications. A market-sizing mechanism then estimates the share of customers who desire the new combinations of needs.

Virtual customer environments (VCE)

Firms create and use virtual customer environments (VCE) to involve their customers in product design, testing and support activities through online discussion forums on virtual design and prototyping (Nambisan and Baron, 2009). Customer benefits from participating in virtual customer environments include (*ibid.*):

- Cognitive benefits of strengthening their understanding of the product and its environment.
- Social integrative benefits of strengthening consumers' ties with relevant others.

- Personal integrative benefits of strengthening the credibility, status and confidence of the individual.
- Hedonic or affective benefits that strengthen consumers' aesthetic or pleasurable experiences.

Italian motorcycle company Ducati calls its VCE the 'Tech Café' and has used its customers' deep technical knowledge to generate several ideas, including mechanical and technical designs (Nambisan and Nambisan, 2008). Microsoft leveraged its 'expert' customers in a VCE to maintain product support services and get product improvement suggestions for PowerPoint 97 and PowerPoint 2000 (*ibid*.).

Further reading
Nambisan, Satish and Baron, Robert A. (2009) 'Virtual Customer Environments: Testing a Model of Voluntary Participation in Value Co-creation Activities', *Journal of Product Innovation Management*, 26, 388–406.
Nambisan, Satish and Nambisan, Priya (2008) 'How to Profit From a Better Virtual Customer Environment',
http://sloanreview.mit.edu/the-magazine/articles/2008/spring/49313/how-to-profit-from-a-better-virtual-customer-environment/ (accessed on December 31, 2010).

Vision

Vision is 'one of the most overused and least understood terms in organizations' (Collins and Porras, 1996). Describing vision as a 'yin' and 'yang' system, Collins and Porras (1996) identify two of its components as core ideology and envisioned future. The basic dynamic of **visionary companies** is to preserve the core ideology and stimulate progress (*ibid*.). Collins and Porras (1996) describe the core ideology as the 'yin'. It is unchanging and complements the envisioned future. The core ideology of an organisation consists of two distinct parts:

- Core values – 'a system of guiding principles and tenets'.
- Core purpose – 'the organization's most fundamental reason for existence. It is the glue that holds the organization together through time.'

Collins and Porras (1996) describe the envisioned future as the 'yang' that consists of two parts: a 10–30-year 'audacious goal plus vivid descriptions of what it will be like to achieve' that goal.

Further reading
Collins, James C. and Porras, Jerry (1996) 'Building Your Company's Vision', *Harvard Business Review*, 74(5), 65–77.

Visionary companies

Visionary companies are premier institutions and leading innovators in their industries (Collins and Porras, 1994). According to Collins and Porras (1995), visionary companies can be built by a shift in thinking and realising that what is required to build a visionary company can be learned. The basic dynamic of visionary companies is to preserve their core ideology while achieving progress (Collins and Porras, 1994). Visionary companies such as 3M, Hewlett-Packard and Procter & Gamble (P&G) are widely admired by their peers and have a long track record of making a significant impact (*ibid*.). Visionary organisations often use bold missions. For example, NASA's 1960s moon mission was bold, it was easy to grasp, compelling in its own right and it could be easily understood by everyone (Collins and Porras, 1996).

See also: **vision**

Further reading
Collins, J. and Porras, Jerry (1994) *Built to Last: Successful Habits of Visionary Companies*, New York, Harper Business.
Collins, James C. and Porras, Jerry I. (1995) 'Building a Visionary Company', *California Management Review*, 37(2), 80–100
Collins, James C. and Porras, Jerry (1996) 'Building Your Company's Vision', *Harvard Business Review*, 74(5), 65–77.

Visualisation

Often described as *thinking in pictures*, visualisation refers to the generation, interpretation and manipulation of information through visual images and spatial representation (Yuan *et al.*, 2006). The use of advanced computing technologies and software for interpreting image data is now common practice (*ibid*.).

As a form of information processing, visualisation has been widely applied in **new product development** and marketing research, in areas such as advertising effectiveness, attitude development, consumption experience and satisfaction, and **creativity** in product design. *3D visualisation* is also in widespread use, in medical, scientific as well as engineering fields. In the offshore exploration of oilfields, BP AMOCO uses a visualisation technology called the Highly Immersive Visualisation Environment (HIVE) (Barrow, 2001). The HIVE technology includes three-dimensional stereo visualisation with stereo goggles and the team can explore a physical system from any direction and up close(*ibid*.). The visualisation experience has given the company a better appreciation of the physical challenges of developing an offshore oil field and has promoted teamwork, **innovation** and creativity (*ibid*.).

See also: **concept visualisation**

Further reading
Barrow, David C. (2001) 'Sharing Know-How at BP AMOCO', *Research Technology Management*, 44(3), 18–25.
Yuan, Feiniu, Liao, Guanfxuan, Fan, Weicheng, Lang, Wenhui and Liu, Zhengmin (2006) 'An Interactive 3D Visualization System Based on PC Using Intel SIMD, 3D Texturing and Thinning Techniques', *International Journal of Pattern Recognition & Artificial Intelligence*, 20(3), 393–416.

Voice of the customer (VoC)

Voice of the customer (VoC) is defined as a complete set of customer wants and needs, expressed in the customer's own words (Katz, 2004). The wants and needs are organised in the way the customers think about, use or interact with the product, and prioritised by importance and performance to the customer. VoC can be accessed through personal interviewing or **focus groups**. Personal interviewing of customers can provide rich and detailed information, but might be time consuming and costly. Abbie Griffin and John Hauser (1993) suggest that about 30 individual interviews, each lasting about three-quarters of an hour, produce close to 100 per cent of all customer needs, and 20 interviews produce about 90 per cent of those needs.

While it is commonly agreed that companies must listen to customers, there is disagreement about how to listen to the voice of the customer (Leonard, 2002). Customers are able to say what they want in a product or service within a familiar product category. For instance, Harley-Davidson's loyal customers can talk about what they want in a motor cycle because they have extensive experience with that product category (Leonard, 2002). However, customers are not adept at making recommendations for new products they have not seen or experienced because of two psychological blocks. The first block is a fixation about the way the product or service is normally used. This blocks the ability to imagine alternative functions. Leonard (2002) calls this functional fixedness. The second block occurs when people have conflicting needs. Parents told researchers at Kimberly-Clark that they did not want their toddlers to wear nappies any more; they also did not want their children to wet the bed (*ibid.*). Leonard (2002) found that successful companies go well beyond listening by understanding the difference between what customers say and what they want, and then acting on those unspoken wants.

See also: **empathic design**

Further reading
Brandt, D. Randall (2008) 'Getting More from the Voice of the Customer', *Marketing Management*, 17(6), 36–42.
Griffin, Abbie and Hauser, John (1993) 'The Voice of the Customer', *Marketing Science*, 12(1), 1–27.
Katz, Gerald M. (2004) 'The Voice of the Customer', in P. Belliveau, A. Griffin and S. M. Sommermeyer (eds), *PDMA ToolBook for New Product Development*, New York, John Wiley and Sons.
Leonard, Dorothy (2002) 'The Limitations of Listening', *Harvard Business Review*, 80(1), 93.

V

Appendix 1: The Thota-Munir innovation integration system

	Product/service development	Creativity	Entrepreneurship	Design	Business	Policy	People/culture	Engineering/technology	Tools and methods	Generally used terms definition	Innovation systems and standards	Theories	Innovation management
Absorptive capacity					√					√			
Accidental innovation	√	√	√							√		√	
Acquisition strategy					√								√
Adaptive design	√			√				√	√				
Adaptive learning					√		√					√	√
Adoption rate	√							√		√			√
Agile development	√							√		√			√
AK Paradigm			√		√							√	
Aligning innovation with market life cycle	√				√							√	√
Alliances			√		√					√			√
Alpha testing	√							√	√				√
Ambidextrous organisation					√		√						√
Analytic hierarchy process (AHP)	√							√	√	√			
Application innovation	√	√						√				√	
Architectural innovation		√						√				√	
Architecture	√			√				√		√			
ARIZ		√		√				√	√				√
A-T-A-R model	√				√				√				√
Autonomous innovation		√			√							√	√
Autonomous team					√		√						√
Axiomatic design (AD)	√	√		√				√	√	√			
Balanced portfolio	√				√							√	√
Balanced scorecard (BSC)					√				√				√
Base of the pyramid (BoP) consumer	√				√		√					√	

→

	Product/service development	Creativity	Entrepreneurship	Design	Business	Policy	People/culture	Engineering/technology	Tools and methods	Generally used terms definition	Innovation systems and standards	Theories	Innovation management
Base of the pyramid (BOP) innovation	√		√		√							^√	
Benchmarking	√				√				√	√			
Best practice	√				√					√			
Beta testing	√							√	√				√
Black-box design	√			√				√	√	√			
Blue ocean strategy		√	√		√				√			√	
Blue sky ideas	√	√	√										√
Bowling alley	√				√							√	√
Brainstorming	√	√							√				√
Brand-centred product development	√				√				√				√
Brand community					√		√						√
Brand innovation					√							√	√
Breakthrough growth	√		√		√								√
Breakthrough innovation	√	√	√					√				√	
Business					√					√			
Business angels			√		√					√			
Business case	√		√		√					√			√
Business charter	√		√		√					√			
Business intelligence (BI)					√				√	√			
Business model			√		√					√			√
Business model innovation	√	√			√							√	
Business plan	√				√				√	√			
Business process innovation (BPI)		√			√					√		√	√
Business process reengineering (BPR)					√			√	√	√			
Buyer utility map	√				√				√				√
Cannibalisation	√				√					√			
Capability maturity model (CMM)	√ f							√	√				√
Cash flow			√		√					√			

→

	Product/service development	Creativity	Entrepreneurship	Design	Business	Policy	People/culture	Engineering/technology	Tools and methods	Generally used terms definition	Innovation systems and standards	Theories	Innovation management
Catalytic innovation		√	√				√					√	
Champions	√						√			√			√
Change management				√			√			√			√
Chasm	√											√	√
Cloud computing								√		√			√
Cluster analysis	√								√	√			
Clusters										√	√		
Co-creation	√							√		√		√	√
Co-design	√			√				√		√			√
Cognitive modelling	√			√			√		√	√			
Co-ideation	√			√			√		√	√			√
Collaborative innovation	√		√		√		√					√	√
Collaborative new product and process development (NPPD)	√							√					√
Collective innovation	√			√			√					√	√
Co-location	√						√		√				√
Commercialisation	√		√							√			√
Commitment	√						√			√			√
Communities of practice (CoP)		√					√			√			
Community innovation survey (CIS)						√	√		√	√			
Comparative performance assessment study CPAS	√				√		√		√				√
Competitive advantage										√			√
Competitive strategy			√		√					√			√
Complex product systems (CoPS)	√							√			√		√
Computer-aided design (CAD)	√			√				√	√	√			√
Computer-enhanced creativity	√	√						√	√	√			
Concept	√									√			√
Concept definition	√							√		√			√
Concept testing	√						√	√					√

→

	Product/service development	Creativity	Entrepreneurship	Design	Business	Policy	People/culture	Engineering/technology	Tools and methods	Generally used terms definition	Innovation systems and standards	Theories	Innovation management
Concept visualisation	√							√	√				√
Concurrent engineering (CE)	√						√	√	√				√
Conjoint analysis	√							√	√				√
Connect and develop (P&G model for innovation)	√	√			√		√		√				√
Consumer innovativeness					√		√				√		
Consumer needs	√				√					√			√
Contracting for innovation					√					√			√
Contradiction matrix				√				√	√	√			
Convergent thinking		√					√		√	√			
Co-opetition					√								√
Core competence					√								√
Core purpose					√								√
Core strategic vision (CSV)					√				√				√
Core values					√		√			√			√
Corporate model of innovation					√		√				√	√	√
Corporate social networks					√		√				√		√
Corporate venturing (CV)			√		√								√
Cost innovation					√							√	√
Crawford Fellow	√						√						
Creative destruction		√	√		√							√	
Creative leadership		√	√		√		√						√
Creative problem-solving model	√	√	√	√				√	√			√	
Creative thinking methods		√	√	√								√	
Creativity		√	√		√		√			√			
Creativity in R&D	√	√	√				√			√		√	
Creativity measurement		√							√				√
Creativity techniques	√	√		√	√		√		√				√
Cross-functional interface management	√						√	√					√
Crowdsourcing	√	√					√		√			√	

→

	Product/service development	Creativity	Entrepreneurship	Design	Business	Policy	People/culture	Engineering/technology	Tools and methods	Generally used terms definition	Innovation systems and standards	Theories	Innovation management
Customer-centric enterprise	√				√		√			√		√	
Customer-driven innovation	√				√							√	√
Customer emotional clusters							√		√				√
Customer experience (CE)	√		√						√				
Customer feedback management	√				√				√	√			
Customer needs	√		√							√			√
Customer scenario mapping	√								√				√
Customer targeting					√		√		√				√
Customer value	√		√	√	√		√	√		√			
Cyclical economy						√	√				√		
Cyclical innovation model (CIM)	√				√	√			√			√	
Data mining	√							√	√	√			
Data visualisation	√							√	√	√			
Delphi method	√		√						√				√
Demand	√				√					√			√
Democratisation of innovation					√		√					√	√
Design	√			√				√		√			
Design anthropology				√			√					√	
Design–build–test cycle	√			√				√	√				√
Design-driven innovation	√	√		√			√					√	
Design for all (inclusive design)	√			√				√	√		√		
Design for assembly (DFA)	√			√				√	√		√		
Design for excellence (DFX)	√			√				√	√		√		
Design for manufacture (DFM)	√			√				√	√		√		
Design for manufacturing and assembly	√			√				√	√		√		
Design for reuse	√			√				√	√		√		
Design for X (DFX) tools	√			√				√	√				√
Design thinking				√				√	√		√		
Development	√			√						√			→

	Product/service development	Creativity	Entrepreneurship	Design	Business	Policy	People/culture	Engineering/technology	Tools and methods	Generally used terms definition	Innovation systems and standards	Theories	Innovation management
Development funnel	√							√	√			√	
Differentiation	√				√			√		√			
Differentiation strategy	√		√		√					√			√
Diffusion of innovations	√				√			√		√		√	
Discontinuities			√		√			√		√		√	
Discontinuous change			√		√			√		√		√	
Discovery driven growth (DDG)					√				√	√		√	
Discovery-driven planning					√				√	√			√
Disruptive innovation	√	√	√	√	√			√				√	
Divergent thinking		√							√	√			
Dominant design	√			√				√		√		√	
Due diligence (technology and patents)					√	√				√			√
Dynamic capabilities					√	√						√	√
Dynamics of innovation	√											√	√
Early supplier involvement (ESI)	√				√								√
Earned value management (EVM)	√				√				√				√
Ecodesign	√			√				√				√	
Eco-innovation	√									√	√	√	
Ecological technology innovation (ETI)	√							√			√		
Ecosystem	√				√					√	√		
E-innovation					√				√	√			
Empathic design	√		√						√	√			
Endogenous growth paradigm					√							√	
Enterprise Europe Network (EEN)								√			√		
Entrepreneur			√				√			√			
Entrepreneurial model of new ventures			√									√	

→

	Product/service development	Creativity	Entrepreneurship	Design	Business	Policy	People/culture	Engineering/technology	Tools and methods	Generally used terms definition	Innovation systems and standards	Theories	Innovation management
Entrepreneurial opportunity			√							√			√
Entrepreneurial orientation (EO)			√									√	√
Entrepreneurial process			√						√				√
Entrepreneurship			√							√			
Environmental innovation	√									√	√		
Ethnography research	√						√		√	√			√
European Innovation Scoreboard (EIS)						√			√		√		
Evolutionary innovation	√				√							√	
Expected commercial value (ECV)	√				√				√				√
Experience design	√			√			√		√	√			
Exploitation	√												
Exploration	√												√
Exploratory learning	√						√		√				√
Failure modes and effects analysis (FMEA)	√			√				√	√				√
Fast-follower strategy					√				√				
Field testing	√							√	√				√
Financial innovations					√				√				√
Firm growth					√							√	
First-mover advantage					√				√				√
Five generations of innovation models					√					√	√		
Flexibility		√		√						√			√
Focus group	√						√		√				√
Front end of innovation (fuzzy front end)	√			√					√			√	√
Game-changing innovations												√	√
Game theory					√							√	
Gamma testing	√								√				√
Gatekeepers	√			√						√			√

→

	Product/service development	Creativity	Entrepreneurship	Design	Business	Policy	People/culture	Engineering/technology	Tools and methods	Generally used terms definition	Innovation systems and standards	Theories	Innovation management
Generative learning					√		√			√		√	
GIM product design procedure	√								√		√		
Global product launch process	√								√				√
Global strategy					√				√			√	√
Globalisation of innovation					√							√	√
Green design	√			√	√					√			
Green engineering				√				√		√			
Green innovation	√			√				√		√			
Green process innovation	√									√			√
Green product innovation	√									√			√
Growth theory					√							√	
Harmony					√	√	√			√		√	
Horizons of growth (McKinsey framework)					√				√				√
Horizontal innovation	√				√				√				
Human-centred design	√		√						√		√		
Idea generation and enrichment	√	√							√	√			√
Idea management	√								√	√	√		√
IMP³rove®						√			√		√		
Incremental innovation	√									√			
Independent innovation							√			√		√	
Indigenous innovation							√			√		√	
Information technology (IT) innovation									√	√			
Innovation	√				√		√	√	√	√	√		
Innovation capitalist		√								√			
Innovation culture	√				√		√						√
Innovation diamond	√				√				√				√
Innovation economics					√	√						√	
Innovation ecosystem	√				√						√		→

	Product/service development	Creativity	Entrepreneurship	Design	Business	Policy	People/culture	Engineering/technology	Tools and methods	Generally used terms definition	Innovation systems and standards	Theories	Innovation management
Innovation engine	√								√				√
Innovation governance					√		√				√	√	
Innovation harmony						√	√			√		√	
Innovation killers					√		√						√
Innovation management	√				√			√	√				√
Innovation mindset					√		√						√
Innovation-oriented country						√	√				√		
Innovation outsourcing					√				√		√		√
Innovation platforms								√			√		
Innovation process	√								√	√			√
Innovation scoreboard						√					√		
Innovation strategy					√				√				√
Innovation structures									√			√	√
Innovation surveys							√		√				√
Innovation value chain					√				√	√			√
Innovator		√								√			
Innovator's dilemma					√							√	√
Integrated product and process development (IPPD)	√								√				√
Integrated project planning	√								√				√
Intellectual capital							√			√			
Intellectual property (IP)	√							√		√			√
Intelligent enterprise (IE)					√							√	
Interaction analysis				√				√	√	√			
Interactive value creation	√						√		√				√
Intrapreneur			√							√			
Invention								√		√			
Knowledge-driven innovation					√					√			√
Lateral thinking		√							√	√			
Launch (product launch)	√									√			√
Launch decision	√												√
Launch plan	√								√				√

	Product/service development	Creativity	Entrepreneurship	Design	Business	Policy	People/culture	Engineering/technology	Tools and methods	Generally used terms definition	Innovation systems and standards	Theories	Innovation management
Launch strategy	√				√		√		√				√
Lead users	√							√	√			√	√
Lean launch	√				√				√				√
Lean product development	√							√		√			
Linear model of innovation											√		
Living labs	√						√		√				√
Localisation	√				√		√		√				
Low-end disruption							√						√
Management role to support NPD							√	√		√			
Market definition								√		√			
Market orientation							√	√	√				
Market research	√						√		√				
Market testing	√			√					√				√
Marketing innovation					√				√				√
Marketing launch					√				√				√
Mass Customisation	√			√									√
Mergers and acquisitions (M&A)					√				√				√
Middle manager as innovator							√					√	√
Models of innovation					√				√				√
Modular product design	√			√					√	√			
National innovation systems (NIS)						√	√				√		
National innovative capacity						√	√				√		
National system of innovation (NSI)							√	√			√	√	
Network externalities							√			√			
New concept development model	√		√					√	√				√
New market disruption		√			√			√		√			
New product development	√		√										√

→

	Product/service development	Creativity	Entrepreneurship	Design	Business	Policy	People/culture	Engineering/technology	Tools and methods	Generally used terms definition	Innovation systems and standards	Theories	Innovation management
New product development process	√								√				√
New product development (NPD) team	√						√						√
New product strategy	√				√				√				√
New technology ventures			√		√			√		√			
New ventures			√		√			√		√			
Next practice										√	√		
Open innovation							√	√	√		√		√
Open-source innovation model	√												√
Opportunity analysis	√		√						√				√
Opportunity identification	√		√										√
Organisational innovation	√				√		√		√				
Organisational knowledge-creation theory					√		√					√	√
Organisational learning					√		√		√			√	
Osborn–Parnes creative problem-solving Model	√	√					√						√
Outcome-driven innovation					√				√				√
PACE (product and cycle time excellence)	√								√				√
Paradigm							√		√				
Parallel thinking		√							√	√			
Participatory design	√			√			√		√				√
People and teams	√						√		√				
Personas	√								√				√
Platform	√							√	√				√
Platform leadership	√								√				√
Portfolio management	√						√	√	√				√
Postponement	√								√				√
Prediction markets	√				√				√	√			
Process innovation	√								√				√

→

	Product/service development	Creativity	Entrepreneurship	Design	Business	Policy	People/culture	Engineering/technology	Tools and methods	Generally used terms definition	Innovation systems and standards	Theories	Innovation management
PRO INNO Europe						√				√			
Product and process innovation	√									√			√
Product and technology adoption curves	√							√	√				√
Product architecture	√			√						√			
Product autonomy	√									√			
Product champion	√									√			√
Product data management (PDM)	√							√		√			
Product definition	√								√	√			√
Product development	√				√	√	√	√		√			
Product development process	√				√		√		√	√			
Product development strategies	√				√				√				√
Product development team	√						√						√
Product family	√									√			
Product innovation	√				√			√					√
Product innovation charter (PIC)	√				√				√				√
Product life cycle (PLC)	√				√					√			√
Product life cycle management	√				√					√			√
Product management	√									√			√
Product platform strategy	√				√				√				√
Product strategy	√				√				√				√
Product visualisation	√							√	√				√
Productivity index	√				√								√
Product-service system (PSS)	√				√							√	
Product-servicisation	√				√							√	
Product-variety paradigm	√				√						√		
Profit					√						√	√	

	Product/service development	Creativity	Entrepreneurship	Design	Business	Policy	People/culture	Engineering/technology	Tools and methods	Generally used terms definition	Innovation systems and standards	Theories	Innovation management
Profiting from innovation					√								√
Programme management					√								√
Prototyping	√			√				√	√				√
Pseudo innovations					√							√	√
QualiQuant research	√								√	√			
Quality	√												√
Radical innovation	√							√	√				√
Rapid prototyping	√			√				√	√	√			
Real options analysis									√				√
Regional innovation systems						√					√		
Requirements engineering (RE)	√							√	√	√			
Research and development (R&D)	√	√						√					√
Research and development (R&D) Management	√	√						√					√
Research journals on innovation	√	√	√	√	√		√	√				√	√
Resource-based view of the firm					√						√	√	
Return on innovation investment (ROI2)					√				√				√
Reverse engineering	√			√				√	√	√			
Revolutionary innovation	√				√				√				√
Risk analysis	√				√				√	√			
Risk-quantification techniques					√				√				√
Roadmapping	√							√	√				√
Robust design	√			√					√				√
Sales forecast	√									√			√
Scenario analysis	√								√	√			
Scenario boards	√				√				√	√			
Scenarios	√		√						√	√			

➜

	Product/service development	Creativity	Entrepreneurship	Design	Business	Policy	People/culture	Engineering/technology	Tools and methods	Generally used terms definition	Innovation systems and standards	Theories	Innovation management
Schumpeterian framework					√							√	
Schumpeterian growth theory					√							√	
Scrum	√								√				√
Second-mover advantage					√					√			√
Service	√									√			
Service innovation	√									√			
Service life cycle	√				√								√
Service life-cycle flexibility	√				√								√
Service management maturity	√				√						√		
Service-productisation	√									√			√
Services innovation management model	√										√		
Services innovation theory	√									√	√		
Shared risk		√			√				√	√			
Simulation	√		√					√	√	√			
Six Sigma for Product Development (SSPD)	√								√	√			
Smart products	√							√		√			
Social capital					√	√					√		
Social entrepreneurship			√			√				√			
Social innovation					√	√				√			
Spiral development	√								√	√			
Stage-Gate™ new product development process													
Strategic alliances					√				√				√
Strategic innovation					√								√
Strategic management					√								√
Strategy	√				√								√
Strategy canvas					√					√			√
Strategy map	√				√					√			√
Sustainability					√	√		√		√			
Sustainable development	√				√	√		√		√	√		√
Sustaining innovation					√					√			√

→

	Product/service development	Creativity	Entrepreneurship	Design	Business	Policy	People/culture	Engineering/technology	Tools and methods	Generally used terms definition	Innovation systems and standards	Theories	Innovation management
SWOT Analysis					√					√	√		
Systemic innovation					√			√		√		√	√
Systems engineering (SE)								√	√	√			
Technological innovation								√		√			√
Technological regime					√								√
Technology								√	√				
Technology discontinuities								√				√	√
Technology fusion								√				√	√
Technology life cycle					√			√					√
Technology platforms								√		√			
Technology-push innovation					√			√				√	
Technology-Stage Gate™ (TSG)	√							√	√				√
Technology strategy					√			√					√
Technology transfer					√	√		√			√		
Three horizons model					√				√			√	
Time to market (TTM)	√	√								√			√
Types of development teams, structures and roles	√						√						√
Universal design	√			√				√				√	
Usability	√			√				√	√				
User-centred design	√			√					√		√		
User-centric innovation	√						√			√	√		
User-driven innovation	√		√		√			√				√	√
User-initiated innovation (UII)	√				√			√		√			
User innovations					√					√			√
User testing	√			√				√	√				√
Utterback and Abernathy model of product and process innovation	√											√	√
Value chain					√					√		√	
Value innovation	√				√					√		√	
Value proposition					√					√			√

→

	Product/service development	Creativity	Entrepreneurship	Design	Business	Policy	People/culture	Engineering/technology	Tools and methods	Generally used terms definition	Innovation systems and standards	Theories	Innovation management
Venture café			√				√			√			
Virtual customer	√						√	√	√	√			
Virtual customer environments (VCE)	√				√		√			√			
Vision			√		√					√			√
Visionary companies			√		√		√			√			√
Visualisation	√	√		√				√	√	√			√
Voice of the customer (VoC)	√								√				√

Appendix 2: Innovation in practice

Key Concept	Company Examples
Accidental innovation	3M
Acquisition strategy	AT&T, BASF, Cisco, EMC Corporation, IBM, Kapsch CarrierCom, Microsoft, Nortel's Carrier Networks Division
Agile development	Kitware, TransCanada Pipelines
Alliances	Microsoft, Nine, Optus, Volvo
Alpha testing	CTC Communications Group, Motorola's Healthcare Communications Solutions Group
Application innovation	Apple, Nokia
Autonomous innovation	Commercial Aircraft Corporation of China, Flickr, MySpace, Twitter, YouTube
Balanced scorecard	National Insurance
Base of the pyramid innovation	Hindustan Lever, Shakti Programme, Shakti Lady
Brand-centred product development	Apple, Harley-Davidson
Brand community	ilounge.com, Nike Talk
Brand innovation	Apple. Clorox, eBay, Harley-Davidson, Starbucks
Business model innovation	Disney Publishing Worldwide, FedEx, McDonald's, Simon & Schuster, Vook, Yahoo!
Business model	Dell
Co-creation	LEGO, Missha, Red Hat, Threadless
Collaborative innovation	Boeing
Collective innovation	Apache, Linux, MySQL
Competitive advantage	Apple, Wal-Mart
Connect and develop	Procter & Gamble
Core competence	Compaq, Qualcomm
Core purpose	3M, Hewlett-Packard, Sony, Walt Disney
Core strategic vision	Cisco
Core values	Merck, Sony, Walt Disney
Corporate social networks	Dow Chemical, Mercedes-Benz
Cost innovation	Dawning
Cost leaders	Black and Decker, DuPont, Emerson Electric, Texas Instruments
Crowdsourcing	Threadless
Customer-centric enterprise	Adidas, Land's End, Threadless
Customer experience	Verizon
Customer scenario mapping	Citigroup

→

Key Concept	*Company Examples*
Customer targeting	Barclays Stockbrokers, Citibank, National Trust, Omniture
Cyclical economy	Chinese government, Johnson & Johnson, P&G, US government
Data mining	Amazon, American Express, Wal-Mart
Design anthropology	IBM, Intel, Whirlpool
Design for all	Apple, Ford
Design-driven innovation	Alessi, Artemide
Differentiation strategy	Apple, Digital Equipment, Genentech, Hewlett-Packard, Lotus, Polaroid, Sharp, Stratus
Discovery-driven growth	Amazon, DuPont, Hewlett-Packard
Disruptive innovation	Canon, P&G, Toyota, Starbucks
Early supplier involvement	Tetra Pak
Eco-innovation	IBM, Nokia, Samsung, World Business Council for Sustainable Development
E-innovation	Amazon, Christie's, eBay, PayPal
Entrepreneurial model of innovation	Stemcell
Experience design	American Girl Stores, Hallmark Cards, LEGO, Microsoft, Starwood's W hotels
Fast-follower strategy	Litton, Matsushita, RCA, Sony, Toyota
Financial innovations	Kiwi Bank
First-mover advantage	Amazon, Austrian Airlines, Neumann, Wal-Mart
Game-changing innovations	Amazon, Apple, eBay, Intuit
Game theory	Amazon, Bell Atlantic, eBay, Google
Gatekeepers	Alessi
Green engineering	Nucor Steel
Green innovation	Greenbox, Toyota, Zipcar
Green process innovation	AstraZeneca, Bristol-Myers Squibb, Pfizer
Green product innovation	Philips
Information technology (IT) innovation	Equifax, Harley-Davidson, Harrah's, R. R. Donnelley, Wal-Mart
Innovation diamond	P&G
Innovation ecosystem	Nokia, Nortel, Philips, Sony, Thomson
Innovation governance	AstraZeneca
Innovation outsourcing	BenQ, Motorola
Innovation platform	TAL Apparel, UPS
Innovation strategy	3M, Adobe, Apple, Boeing, Canon, Cisco, DuPont, Gillette, Honda, IKEA, Intel, Microsoft, Nissan, Nokia, P&G, Pixar, SAP, Southwest Airlines, Syngenta, Tetra Pak, Toyota
Innovation structures	Dow Chemical, GE, Shell Oil
Innovation value chain	Monsanto, P&G, Sara Lee, Shell, Siemens

→

Key Concept	Company Examples
Lead users	3M
Lean product development	Dell Computer, Toyota
Living labs	Nokia
Low-end disruption	Kodak
Mass customisation	Adidas Salomon
Mergers and acquisitions	Cisco
New product development process	IDEO, PRTM
New product strategy	Neiman Marcus, Southwest Airlines
Open innovation	Deutsche Telekom
Open-source innovation model	Red Hat
Organisational learning	Apple, Hewlett-Packard, IBM
Osborn–Parnes creative problem-solving model	Frito-Lay
Participatory design	Yahoo Kids! (Korea)
Personas	Best Buy, Chrysler, Discover, Ford, Microsoft, Staples, Unilever, Whirlpool
Platform	Apple, Google, Intel, Microsoft
Platform leadership	Adobe, Facebook, Google, Intel, Microsoft
Portfolio management	Hewlett-Packard
Product architecture	Geely Group
Product autonomy	Electrolux
Product family	Airbus
Profit	General Electric, Hallmark, Intuit
Profiting from innovation	Merck
Programme management	Intel, Tektronix
Radical innovation	Kodak
Requirements engineering	ABB, Danaher Motion
Robust design	Honda
Scrum	Google, Honda, Toyota
Second-mover advantage	Chery, Hyundai, JVC, Microsoft
Simulation	Becton Dickinson and Company
Six Sigma for Product Development	Bristol-Myers Squibb, General Electric, GlaxoSmithKline, Motorola
Smart products	Electrolux, Mercedes-Benz
Social entrepreneurship	Grameen Bank
Social innovation	Aravind Eye Care, Narayana Hrudayalaya
Strategic alliance	GlaxoSmithKline, Oratech
Strategic innovation	Cisco, Gillette
Sustainability	3M, Dow Chemical, DuPont, GE
Technology platform	Apple, Intel, Linux, Microsoft
Usability	Apple
Visionary companies	3M, Hewlett-Packard, Procter and Gamble
Voice of the customer	Harley-Davidson, Kimberly-Clark

Index

Note: Page numbers in **bold** denote main concept entries.

320 Index